U.S.NRC
United States Nuclear Regulatory Commission

Protecting People and the Environment

NUREG-1556
Vol. 9, Rev. 2

I0488141

Consolidated Guidance About Materials Licenses

Program-Specific Guidance About Medical Use Licenses

Final Report

Manuscript Completed: January 2008
Date Published: January 2008

Prepared by
D.B. Howe, M. Beardsley, S. Bakhsh

Office of Federal and State Materials and
Environmental Management Programs

ABSTRACT

As part of its redesign of the materials licensing process, the United States Nuclear Regulatory Commission (NRC) consolidated and updated numerous guidance documents into a single comprehensive repository, as described in NUREG-1539, "Methodology and Findings of the NRC's Materials Licensing Process Redesign," dated April 1996, and draft NUREG-1541, "Process and Design for Consolidating and Updating Materials Licensing Guidance," dated April 1996. NUREG-1556, Volume 9, Revision 2, "Consolidated Guidance about Materials Licenses: Program-Specific Guidance about Medical Use Licenses," is the third version of the ninth program-specific guidance document developed for the new process; it is intended for use by applicants, licensees, and NRC staff and will also be available to Agreement States.

This document contains information that is intended to assist those preparing applications for licenses for the medical use of byproduct material. In particular, it describes the types of information needed to complete NRC Form 313, "Application for Materials License," and the NRC Form 313A series of forms: NRC Form 313A (RSO), "Radiation Safety Officer Training and Experience and Preceptor Attestation [10 CFR 35.50]"; NRC Form 313A (AMP), "Authorized Medical Physicist Training and Experience and Preceptor Attestation [10 CFR 35.51]"; NRC Form 313A (ANP), "Authorized Nuclear Pharmacist Training and Experience and Preceptor Attestation [10 CFR 35.55]"; NRC Form 313A (AUD), "Authorized User Training and Experience and Preceptor Attestation (for uses defined under 10 CFR 35.100, 35.200, and 35.500) [10 CFR 35.190, 35.290, and 35.590]"; NRC Form 313A (AUT), "Authorized User Training and Experience and Preceptor Attestation (for uses defined under 10 CFR 35.300) [10 CFR 35.390, 35.392, 35.394, and 35.396]"; and NRC Form 313A (AUS), "Authorized User Training and Experience and Preceptor Attestation (for uses defined under 10 CFR 35.400 and 35.600) [10 CFR 35.490, 35.491, and 35.690]."

The document provides an overview of the types of licenses issued by the NRC, the commitments and responsibilities that must be undertaken by a licensee, applicable regulations, the process for filing a license application, and the contents of applications for different types of medical uses of byproduct material. In particular, this document provides a description, on an item-by-item basis, of the information to be provided by an applicant on NRC Form 313. Because of the wide variety in the types of medical uses of byproduct material, indicators have been placed in the document to alert applicants for particular types of medical uses to material that pertains to those types of uses.

The document also contains appendices that include (1) copies of necessary forms; (2) a sample license application and sample licenses for different types of medical uses of byproduct materials; (3) examples of the types of supporting documents, such as implementing procedures, that may need to be prepared by applicants; and (4) information required by regulation for requesting authorization for preparation of Positron Emission Tomography (PET) radioactive drugs for noncommercial distribution to other members of a consortium. The NRC is placing added emphasis on conducting its regulatory activities in a risk-informed and performance-based manner. This approach is intended to be less prescriptive and to allow implementation by licensees that may be specific to their needs while meeting the regulatory requirements. By supplying examples, the NRC seeks to provide information to meet the needs of applicants for licensure, without being prescriptive. Guidance in this document represents one means acceptable to NRC staff of complying with NRC regulations and is not intended to be the only means of satisfying requirements for a license.

The original Volume 9 of NUREG-1556 provided guidance for licensure under revised Title 10, Part 35, "Medical Use of Byproduct Material." It combined and superseded guidance found in the documents listed below:

- Regulatory Guide (RG) 10.8, Revision 2, "Guide for the Preparation of Applications for Medical Use Programs";

- Appendix X to RG 10.8, Revision 2, "Guidance on Complying With New Part 20 Requirements";

- Draft RG DG-0009, "Supplement to Regulatory Guide 10.8, Revision 2, Guide for the Preparation of Applications for Medical Use Programs";

- Draft RG FC 414-4, "Guide for the Preparation of Applications for Licenses for Medical Teletherapy Programs";

- RG 8.23, "Radiation Safety Surveys at Medical Institutions, Revision 1";

- RG 8.33, "Quality Management Program";

- RG 8.39, "Release of Patients Administered Radioactive Materials";

- Policy and Guidance Directive (P&GD) 03-02, "Licensing Lixiscope and BMA";

- Policy and Guidance Directive (P&GD) 03-08, "Standard Review Plan for Teletherapy";

- Policy and Guidance Directive (P&GD) 3-17, "Review of Training and Experience Documentation Submitted by Proposed Physician User Applicants";

- Policy and Guidance Directive (P&GD) FC 87-2, "Standard Review Plan for License Applications for the Medical Use of Byproduct Material";

- Policy and Guidance Directive (P&GD) FC 86-4, Revision 1, "Information Required for Licensing Remote Afterloading Devices";

- Addendum to Revision 1 to P&GD FC 86-4, "Information Required for Licensing Remote Afterloading Devices-Increased Source Possession Limits";

- Policy and Guidance Directive (P&GD) FC 92-01 "Information Required for Licensing Mobile Nuclear Medicine Services"; and

- Policy and Guidance Directive (P&GD) 3-15, "Standard Review Plan for Review of Quality Management Programs."

Revision 1 of NUREG-1556, Volume 9, revised Volume 9 to reflect the March 30, 2005, Final Rule, Medical Use or Byproduct Material – Recognition of Specialty Boards (70 FR 16336), that revised the training and experience requirements for recognition of specialty boards. Revision 2 of NUREG-1556, Volume 9, revises Volume 9 to provide additional guidance to reflect regulatory changes made by the Naturally Occurring and Accelerator-Produced Material (NARM) Rule, "Requirements for Expanded Definition of Byproduct Material" (72 FR 55864), replaces NRC Form 313A with six new NRC Form 313A forms, makes additional changes to enhance clarification of the training and experience requirements, and removes all references to, and information contained in, 10 CFR Part 35, Subpart J, which expired on October 25, 2005.

Paperwork Reduction Act Statement

This NUREG contains information collection requirements that are subject to the Paperwork Reduction Act of 1995 (44 U.S.C. 3501 et seq.). These information collections were approved by the Office of Management and Budget (OMB), approval numbers 3150-0044; 3150-0014; 3150-0035; 3150-0017; 3150-0016; 3150-0001; 3150-0015; 3150-0010; 3150-0009; 3150-0008; 3150-0120; and 3150-0028.

Public Protection Notification

The NRC may not conduct or sponsor, and a person is not required to respond to, a request for information or an information collection requirement unless the requesting document displays a currently valid OMB control number.

CONTENTS

CONTENTS

APPENDICES

APPENDICES A-H   FORMS AND SAMPLES

APPENDICES I-W MODEL PROCEDURES FOR INFORMATION PURPOSES ONLY

CONTENTS

APPENDICES X-Z RECORDKEEPING AND REPORTING REQUIREMENTS AND DOT RULES FOR SHIPPING

FIGURES

TABLES

FOREWORD

This report, NUREG-1556, Volume 9, Revision 2, "Consolidated Guidance About Materials Licenses: Program-Specific Guidance About Medical Use Licenses," is one of twenty-one volumes in NRC's NUREG-1556 series addressing its materials licensing process. This report is intended for use by applicants, licensees, NRC license reviewers, and other NRC license personnel addressing the medical use of byproduct material. Below is a list of volumes currently included in the NUREG-1556 series:

Vol. No.	Volume Title	Status
1, Rev. 1	Program-Specific Guidance About Portable Gauge Licenses	Final Report
2	Program-Specific Guidance About Industrial Radiography Licenses	Final Report
3, Rev. 1	Applications for Sealed Source and Device Evaluation and Registration	Final Report
4	Program-Specific Guidance About Fixed Gauge Licenses	Final Report
5	Program-Specific Guidance About Self-Shielded Irradiators Licenses	Final Report
6	Program-Specific Guidance About 10 CFR Part 36 Irradiators Licenses	Final Report
7	Program-Specific Guidance About Academic, Research and Development, and Other Licenses of Limited Scope	Final Report
8	Program-Specific Guidance About Exempt Distribution Licenses	Final Report
9, Rev.2	Program-Specific Guidance About Medical Use Licenses	Final Report
10	Program-Specific Guidance About Master Materials Licenses	Final Report
11	Program-Specific Guidance About Licenses of Broad Scope	Final Report
12	Program-Specific Guidance About Possession Licenses for Manufacturing and Distribution	Final Report
13, Rev.1	Program-Specific Guidance About Commercial Radiopharmacy Licenses	Final Report
14	Program-Specific Guidance About Well Logging, Tracer, and Field Flood Study Licenses	Final Report
15	Program-Specific Guidance About Changes of Control and About Bankruptcy Involving Byproduct, Source, or Special Nuclear Material Licenses	Final Report
16	Program-Specific Guidance About Licenses Authorizing Distribution to General Licensees	Final Report
17	Program-Specific Guidance About Special Nuclear Material of Less Than Critical Mass Licenses	Final Report
18	Program-Specific Guidance About Service Provider Licenses	Final Report

Vol. No.	Volume Title	Status
19	Guidance For Agreement State Licensees About NRC Form 241 "Report of Proposed Activities in Non-Agreement States, Areas of Exclusive Federal Jurisdiction, or Offshore Waters" and Guidance for NRC Licensees Proposing to Work in Agreement State Jurisdiction (Reciprocity)	Final Report
20	Guidance About Administrative Licensing Procedures	Final Report
21	Program-Specific Guidance About Possession Licenses for Production of Radioactive Material Using an Accelerator	Final Report

Questions and Answers on the implementation of Part 35 of Title 10 of the Code of Federal Regulations (CFR) are posted on the NRC's Web site on the Medical Uses Licensee Toolkit http://www.nrc.gov/materials/miau/med-use-toolkit.html, serving as another source of guidance about implementation of revised 10 CFR Part 35.

After the October 2002 publication of NUREG-1556, Volume 9, the NRC amended 10 CFR Part 35, "Medical Use of Byproduct Material" (March 30, 2005; 70 FR 16335). The licensing guidance contained in NUREG-1556, Volume 9, Revision 1, included updated guidance on requirements for training and experience appearing in the amended rule. The guidance also reflected the extension of the effective date of Subpart J to October 24, 2005 (69 FR 55736).

Following the May 2005 publication of NUREG-1556, Volume 9, Revision 1, the NRC developed six new 313A Forms to record the training and experience of six different groups of individuals seeking recognition as authorized users, radiation safety officers, authorized nuclear pharmacists, and authorized medical physicists. On March 27, 2006, the NRC published a final rule to correct several minor errors in the CFR, update the address for Region III, and remove all references to Subpart J in 10 CFR Parts 32 and 35. Revision 2 of NUREG-1556, Volume 9, includes the new NRC Form 313A series of forms, provides guidance on how to fill them out, and removes references to 10 CFR Part 35, Subpart J.

On November 30, 2007, the NRC amended its regulations to include jurisdiction over certain radium sources, accelerator-produced radioactive materials, and certain naturally occurring radioactive material, as required by the Energy Policy Act of 2005 (EPAct), which was signed into law on August 8, 2005. The EPAct expanded the Atomic Energy Act of 1954 definition of byproduct material to include:

- any discrete source of radium-226 (Ra-226),

- any material made radioactive by use of a particle accelerator, and

- any discrete source of naturally occurring radioactive material, other than source material, that the Commission, in consultation with other Federal officials described in the EPAct, determines would pose a similar threat to the public health and safety or the common defense and security as a discrete source of radium-226

that are extracted or converted after extraction for use for a commercial, medical, or research activity.

In so doing, these materials were placed under the NRC's regulatory authority. Also as authorized by the EPAct, the NRC issued a waiver on August 31, 2005, to allow continued use and possession of naturally-occurring and accelerator-produced radioactive materials (NARM) while the NRC developed a regulatory framework for regulation of the new byproduct material. The NRC will terminate the waiver in phases, beginning November 30, 2007, and ending on August 7, 2009. On November 30, 2007, the NRC terminated the waiver for Federal Government agencies, Federally recognized Indian tribes, Delaware, the District of Columbia, Puerto Rico, the U.S. Virgin Islands, Indiana, Wyoming, and Montana. Waiver terminations for Connecticut, New Jersey, Pennsylvania, Vermont, Virginia, West Virginia, Michigan, Missouri, Alaska, Hawaii, Idaho, Pacific Trust Territories, and South Dakota will be executed for groups of States and U.S. Territories in phases between November 30, 2007 and August 7, 2009.

Upon waiver termination, all persons who possess the new byproduct materials in these States, U.S. Territories, or areas of exclusive Federal jurisdiction must be in compliance with NRC regulations. Being in compliance with the NRC regulations means that such persons are responsible for the proper handling, transfer, and disposal of these new byproduct materials as specified in the NRC's regulations. Some radioactive materials that fall under the newly expanded definition of byproduct materials may already be authorized on an existing NRC license, since the term "byproduct materials" will include the new NARM material. For those radioactive materials and uses of the new byproduct material that are not already on an NRC license, the person will either be required to: (1) apply for license amendments for the new byproduct material within 6 months from the date the waiver is terminated, or (2) submit a new license application for the new byproduct material within 12 months from the date the waiver is terminated. The person may continue to use the materials until NRC takes final licensing action, provided the amendment or new license request was made during the required time periods.

Revision 2 of NUREG-1556, Volume 9, includes updated guidance on requirements for licensing the accelerator-produced radioactive materials and discrete sources of radium-226 now included in the expanded definition of byproduct material.

In addition to combining and updating the guidance for applicants and licensees previously found in numerous Regulatory Guides, Policy and Guidance Directives, draft Regulatory Guides, Standard Review Plans, and Information Notices, this guidance incorporates input from stakeholders received in public workshops and written comments.

This report follows the risk-informed, performance-based approach adopted for revisions to 10 CFR Part 35. It reduces the amount of information submitted by an applicant seeking to possess and use certain quantities of byproduct material for medical use. In a number of instances, the regulations found in 10 CFR Part 35 and reflected in this report do not require the submission of detailed procedures. Instead, applicants are requested to confirm that they have developed and will implement and maintain procedures required by Part 35, but they are not required to submit those procedures as part of their license application. This report contains appendices containing suggested procedures that applicants may consider. The risk-informed, performance-based approach to the regulation of NRC-licensed materials is also being emphasized in the inspection and enforcement arena.

This document addresses those topics that an applicant must provide in preparing a license application on NRC Form 313. The report also includes descriptions of certain key elements of a medical use program that do not require a response on Form 313. This material is presented for clarification only.

Revision 2 of NUREG-1556, Volume 9, is not a substitute for NRC regulations. The approaches and methods described in this report are provided for information only. Guidance in this document represents one means acceptable to the staff of complying with NRC regulations and is not intended to be the only means of satisfying the requirements for licensing.

The NRC's "Procedures for Recognizing Certification Processes of Specialty Boards" may be found on NRC's Web site on the Medical Uses Licensee Toolkit http://www.nrc.gov/materials/miau/med-use-toolkit.html.

Complementary guidance on Inspection Procedures for inspections of medical use licensees is contained in the following documents available at NRC's Web site on the Medical Uses Licensee Toolkit http://www.nrc.gov/materials/miau/med-use-toolkit.html.

Inspection Procedures in the 87100 series:

- "Nuclear Medicine Programs — Written Directive Not Required,"

- "Nuclear Medicine Programs — Written Directive Required,"

- "Brachytherapy Programs,"

- "Medical Gamma Stereotactic Radiosurgery and Teletherapy Programs," and

- "Medical Broad-Scope Programs."

Charles L. Miller

Charles L. Miller, Director
Office of Federal and State Materials and
 Environmental Management Programs

ACKNOWLEDGMENTS

The guidance development team thanks the individuals listed below for assisting in the development and review of the report. All participants provided valuable insights, observations, and recommendations.

The Participants for Revision 2

Howe, Donna-Beth
Bakhsh, Sarah R.
Beardsley, Michelle
Taylor, Torre M.

The team also thanks Justine Cowan, Loleta Dixon, Agi Seaton, and Roxanne Summers of Computer Sciences Corporation for their assistance in the preparation of this document.

The Participants for Previous Versions

Bhalla, Neelam	Gabriel, Sandra L.	Null, Kevin G.
Bolling, Lloyd A.	Haney, Catherine	Psyk, Linda M.
Broseus, Roger	Henderson, Pamela J.	Rothschild, Marjorie U.
Brown, Carrie	Henson, Jay L.	Roe, Mary Louise
Brown, Frederick D.	Hill, Thomas E.	Simmons, Toye L.
Chidakel, Susan S.	Holahan, Patricia K.	Siegel, M.D., Barry
Cook, Jackie D.	Holden, Cornelius F., Jr.	Torres, Betty Ann
Cool, Donald A.	Howard, Marcia	Treby, Stuart A.
Decker, Thomas R.	Howe, Allen G.	Turner, Anita L.
DelMedico, Joseph R.	Howe, Donna-Beth	Walter, David
Euchner, Jennifer	Jones, Samuel Z.	Whitten, Jack E.
Flack, Diane S.	Lohaus, Paul H.	Young, Thomas F.
Frant, Susan M.	Merchant, Sally L.	Zelac, Ronald

We acknowledge both the assistance of Francis X. "Chip" Cameron for leading two facilitated round-table discussions and the participation of stakeholders in public meetings held at NRC headquarters on April 25 and 30, 2002.

The following individuals are recognized for their contribution to supporting documents that formed a basis for the original report:

Ayres, Robert	Minnick, Sheri A.
Bhalla, Neelam	Paperiello, Carl A.
Brown, Carrie	Schlueter, Janet R.
Brown, Keith D.	Smith, James A.
Frazier, Cassandra F.	Taylor, Torre M.
Fuller, Mike L.	Trottier, Cheryl A.
Merchant, Sally L.	

ABBREVIATIONS

AAPM	American Association of Physicists in Medicine
ACMUI	Advisory Committee on the Medical Use of Isotopes
ACR	American College of Radiology
ALARA	as low as is reasonably achievable
ALI	annual limit on intake
AMP	Authorized Medical Physicist
ANP	Authorized Nuclear Pharmacist
ANSI	American National Standards Institute
AU	Authorized User
bkg	background
BPR	Business Process Redesign
Bq	bequerel
CFR	Code of Federal Regulations
Ci	curie
cc	centimeter cubed
cm^2	square centimeter
Co-57	cobalt-57
Co-60	cobalt-60
cpm	counts per minute
Cs-137	cesium-137
DAC	derived air concentration
DOT	United States Department of Transportation
dpm	disintegrations per minute
EPAct	Energy Policy Act of 2005
F-18	fluorine-18
FDA	United States Food and Drug Administration
GM	Geiger-Mueller
GPO	Government Printing Office
GSR	gamma stereotactic radiosurgery
HDR	high dose-rate

ABBREVIATIONS

I-125	iodine-125
I-131	iodine-131
IN	Information Notice
IP	Inspection Procedure
Ir-192	iridium-192
LDR	low dose-rate
mCi	millicurie
ml	milliliter
Mo-99	molybdenum-99
mR	milliroentgen
mrem	millirem
mSv	millisievert
N–13	nitrogen-13
NaI(Tl)	sodium iodide (thallium doped)
NARM	Naturally Occurring and Accelerator-Produced Material
NCRP	National Council on Radiation Protection and Measurements
NIST	National Institute of Standards and Technology
NRC	United States Nuclear Regulatory Commission
NVLAP	National Voluntary Laboratory Accreditation Program
O-15	oxygen-15
OCFO	Office of the Chief Financial Officer
OCR	optical character reader
OMB	Office of Management and Budget
OSL	optically stimulated luminescence dosimeters
PET	Positron Emission Tomography
P-32	phosphorus-32
Pd-103	palladium-103
PDR	pulsed dose-rate
P&GD	Policy and Guidance Directive
QA	quality assurance
Ra-226	radium-226
Ru-82	rubidium-82

RG	Regulatory Guide
RIS	Regulatory Issue Summary
RSC	Radiation Safety Committee
RSO	Radiation Safety Officer
SDE	shallow-dose equivalent
SI	International System of Units (abbreviated SI from the French Le Système Internationale d'Unités)
Sr-82	strontium-82
Sr-85	strontium-85
Sr-90	strontium-90
SSDR	Sealed Source and Device Registry
std	standard
Sv	Sievert
TAR	Technical Assistance Request
Tc-99m	technetium-99m
TEDE	total effective dose equivalent
TI	Transport Index
TLD	thermoluminescent dosimeters
U-235	uranium-235
WD	written directive
Xe-133	xenon-133
Y-90	yttrium-90
μCi	microcurie
%	percent

1 OVERVIEW

1.1 PURPOSE OF REPORT

This report is intended to provide guidance on three topics to individuals who are preparing an application for a license for the medical use of byproduct material as well as to NRC staff who review applications:

Part 35	Applicability
100	✓
200	✓
300	✓
400	✓
500	✓
600	✓
1000	✓

(1) Preparation of a license application using NRC Form 313 "Application for Materials License," including supplemental forms:

- NRC Form 313A (RSO), "Radiation Safety Officer Medical Use Training and Experience Preceptor Attestation [10 CFR 35.50]";

- NRC Form 313A (AMP), "Authorized Medical Physicist Training and Experience and Preceptor Attestation [10 CFR 35.51]";

- NRC Form 313A (ANP), "Authorized Nuclear Pharmacist Training and Experience and Preceptor Attestation [10 CFR 35.55]";

- NRC Form 313A (AUD), "Authorized User Training and Experience and Preceptor Attestation (for uses defined under 10 CFR 35.100, 35.200, and 35.500) [10 CFR 35.190, 35.290, and 35.590]";

- NRC Form 313A (AUT), "Authorized User Training and Experience and Preceptor Attestation (for uses defined under 10 CFR 35.300) [10 CFR 35.390, 35.392, 35.394, and 35.396]"; and

- NRC Form 313A (AUS), "Authorized User Training and Experience and Preceptor Attestation (for uses defined under 10 CFR 35.400 and 35.600) [10 CFR 35.490, 35.491, and 35.690]."

(2) NRC criteria for evaluating a medical use license application. This report provides guidance for the following types of medical uses of byproduct material:

- Use of unsealed byproduct material for uptake, dilution, and excretion studies for which a written directive is not required under 10 CFR 35.40 (see Subpart D, 10 CFR 35.100-190);

- Use of unsealed byproduct material for imaging and localization studies for which a written directive is not required under 10 CFR 35.40 (see Subpart D, 10 CFR 35.200-290);

- Use of unsealed byproduct material for which a written directive is required under 10 CFR 35.40 (see Subpart E, 10 CFR 35.300-396);

- Use of sources for manual brachytherapy (see Subpart F, 10 CFR 35.400-491);

- Use of sealed sources for diagnosis (see Subpart G, 10 CFR 35.500-590);

- Use of a sealed source in a photon-emitting remote afterloader unit, teletherapy unit, or gamma stereotactic radiosurgery unit (see Subpart H, 10 CFR 35.600-690); and

- Other medical uses of byproduct material or radiation from byproduct material not specifically covered by 10 CFR Part 35, Subparts 35.100 through 35.600 (see 10 CFR 35.1000, Subpart K).

(3) The NRC criteria for evaluating an application for authorization of a medical facility to prepare PET radioactive drugs under 10 CFR 30.32(j) for noncommercial transfer to medical use licensees within its consortium.

To assist license applicants, this guide includes text boxes at the beginning of each section to indicate the type of use to which the guidance pertains (identified by the pertinent section of 10 CFR Part 35). These boxes are intended to guide the applicant through the sections of the guidance that are relevant to the applicant's particular type of use of byproduct material. A check indicates that applicants for that type of use should review the guidance section. Some of the checks have asterisks next to them. These asterisks indicate that there are conditions or limitations in that particular section of the guidance relating to the applicants who are subject to the checked section of the rule. Table 1.1 summarizes the material in the text boxes. The Table also includes Appendix AA because it includes information the applicant needs when requesting authorization under 10 CFR 30.32(j). Because this authorization is not an authorization for medical use, none of the medical uses were marked.

Table 1.1	Sections of NUREG-1556, Volume 9, Revision 2, that Applicants for a Particular Type of Use Should Review							
NUREG-1556 - Volume 9, Rev. 2 Section:		**Type of Use**						
		100	200	300	400	500	600	1000
8.1	License Action Type	●	●	●	●	●	●	●
8.2	Applicant's Name and Mailing Address	●	●	●	●	●	●	●
8.3	Address(es) Where Licensed Material Will Be Used or Possessed	●	●	●	●	●	●	●
8.4	Person to Be Contacted about This Application	●	●	●	●	●	●	●
8.5	Radioactive Material	●	●	●	●	●	●	●
8.6	Sealed Sources and Devices (including Ra-226 Sealed Sources and Devices)				●	●	●	●
8.7	Discrete Source of Ra-226 (other than Sealed Sources)	●	●	●				●
8.8	Recordkeeping for Decommissioning and Financial Assurance	●	●	●	●	●	●	●
8.9	Purpose(s) for which Licensed Material Will Be Used	●	●	●	●	●	●	●
8.10	Individual(s) Responsible for Radiation Safety Program and their Training and Experience	●	●	●	●	●	●	●
8.11	Radiation Safety Officer (RSO)	●	●	●	●	●	●	●

Table 1.1   Sections of NUREG-1556, Volume 9, Revision 2, that Applicants for a Particular Type of Use Should Review

NUREG-1556 - Volume 9, Rev. 2 Section:		Type of Use						
		100	200	300	400	500	600	1000
8.12	Authorized User (AU)	●	●	●	●	●	●	●
8.13	Authorized Nuclear Pharmacist (ANP)	●	●	●				●
8.14	Authorized Medical Physicist (AMP)				●		●	●
8.15	Facilities and Equipment	●	●	●	●	●	●	●
8.16	Facility Diagram	●	●	●	●	●	●	●
8.17	Radiation Monitoring Instruments	●	●	●	●	●	●	●
8.18	Dose Calibrator and Other Equipment Used to Measure Dosages of Unsealed Byproduct Material	●	●	●				●
8.19	Therapy Unit - Calibration and Use				●		●	●
8.20	Other Equipment and Facilities	●	●	●	●	●	●	●
8.21	Radiation Protection Program	●	●	●	●	●	●	●
8.22	Safety Procedures and Instructions						●	●
8.23	Occupational Dose	●	●	●	●	●	●	●
8.24	Area Surveys	●	●	●	●	●	●	●
8.25	Safe Use of Unsealed Licensed Material	●	●	●				●
8.26	Spill/Contamination Procedures	●	●	●	●	●	●	●
8.27	Installation, Maintenance, Adjustment, Repair, and Inspection of Therapy Devices Containing Sealed Sources						●	●
8.28	Minimization of Contamination	●	●	●	●	●	●	●
8.29	Waste Management	●	●	●	●	●	●	●
8.30	Fees	●	●	●	●	●	●	●
8.31	Certification	●	●	●	●	●	●	●
AA	Authorization under 10 CFR 30.32(j) to Prepare PET Radioactive Drugs for Noncommercial Transfer							
PROGRAM-RELATED GUIDANCE - NO RESPONSE FROM APPLICANTS ON NRC FORM 313								
8.32	Safety Instruction for Individuals Working In or Frequenting Restricted Areas	●	●	●	●	●	●	●
8.33	Public Dose	●	●	●	●	●	●	●
8.34	Opening Packages	●	●	●	●	●	●	●
8.35	Procedures for Administrations When a Written Directive Is Required			●	●		●	●

Table 1.1	Sections of NUREG-1556, Volume 9, Revision 2, that Applicants for a Particular Type of Use Should Review							
NUREG-1556 - Volume 9, Rev. 2 Section:		Type of Use						
		100	200	300	400	500	600	1000
8.36	Release of Patients or Human Research Subjects			●	●			●
8.37	Mobile Medical Service	●	●	●	●	●	●	●
8.38	Audit Program	●	●	●	●	●	●	●
8.39	Operating and Emergency Procedures	●	●	●	●	●	●	●
8.40	Material Receipt and Accountability	●	●	●	●	●	●	●
8.41	Ordering and Receiving	●	●	●	●	●	●	●
8.42	Sealed Source Inventory	●	●	●	●	●	●	●
8.43	Records of Dosages and Use of Brachytherapy Source	●	●	●	●			●
8.44	Recordkeeping	●	●	●	●	●	●	●
8.45	Reporting	●	●	●	●	●	●	●
8.46	Leak Tests	●	●	●	●	●	●	●
8.47	Safety Procedures for Treatments When Patients Are Hospitalized			●	●		●	●
8.48	Transportation	●	●	●	●	●	●	●

Applicants also should be aware that 10 CFR Part 35 contains general information, administrative requirements, and technical requirements that are pertinent to some or all of the types of use listed above (see 10 CFR 35.1 through 35.92).

This report is intended to consolidate, into one document, guidance that relates to satisfying regulations other than 10 CFR Part 35 that apply to medical use licensees, including the following:

- Provisions of 10 CFR Part 20 that relate to radiation safety;

- Provisions of 10 CFR Part 30 that relate to licensing (e.g., §30.33); and

- Provisions of 10 CFR 30.32(j) and 30.34(j) for preparation for noncommercial transfer of PET radioactive drugs to medical use licensees within a consortium.

This report does not address certain aspects of licensing and radiation safety for the medical use of byproduct materials. In particular, applicants and licensees should consider the following:

- NUREG-1556, Volume 11, "Consolidated Guidance about Materials Licenses: Program-Specific Guidance About Licenses of Broad Scope," dated April 1999, provides additional licensing guidance on medical use programs of broad scope. Section 1.2.1 below provides a general discussion on specific licenses of broad scope.

- 10 CFR Part 19, "Notices, Instructions and Reports to Workers: Inspection and Investigations."

- 10 CFR Part 20, "Standards for Protection Against Radiation," and other regulatory requirements potentially applicable to medical use licensees listed in Section 4 below.

- 10 CFR Part 21, "Reporting of Defects and Noncompliance."

- This report does not address the commercial aspects of manufacturing, distribution, and service of sources containing byproduct material in devices. Volumes 12, 13, and 18 of NUREG-1556, provide additional licensing guidance.

- This report does not address the accelerator production of radionuclides by the medical use licensee for either commercial or noncommercial distribution of radionuclides. Volume 21 of NUREG-1556, "Consolidated Guidance About Materials Licenses: Program-Specific Guidance About Possession Licenses for Production of Radioactive Materials Using an Accelerator," provides licensing guidance to applicants requesting a license to produce radioactive materials using an accelerator.

- This report does not describe the licensing, possession, or use of pacemakers, which are licensed under 10 CFR Part 70, "Domestic Licensing of Special Nuclear Material." However, a sample pacemaker license is included in Appendix F.

As a guidance document intended to assist a wide variety of applicants, this report contains a considerable amount of information about how licensees may choose to implement their programs to meet NRC regulatory requirements. The information in this document is not intended to impose any conditions beyond those required by the regulations in 10 CFR. This report provides specific guidance on what information should be submitted in an application to satisfy NRC requirements. Except for procedures required by Subpart H of 10 CFR Part 35, written procedures do not need to be submitted as part of the license application.

Guidance and model procedures provided in this NUREG that are not required to be submitted are for illustrative purposes to guide licensees in developing their programs. Use of the word "should" implies "may" and is not intended to mean "must" or "shall"; the procedures provided in this guidance are intended to serve only as examples.

Sections 1 through 7 of this document provide background information. Section 8 describes, item by item, the information that should be provided in Items 1 through 11 of NRC Form 313, in completing a license application. The format within this document for each item of technical information is as follows:

- **Regulations** – references the regulations applicable to the item;

- **Criteria** – outlines the criteria used to judge the adequacy of the applicant's response;

- **Discussion** – provides additional information on the topic sufficient to meet the needs of most readers; and

- **Response from Applicant** – provides suggested response(s) or indicates that no response is needed on that topic during the initial licensing process.

Some sections of the guidance include references to other documents that may be useful to the applicant. Appendix CC provides a complete list of documents used to prepare or referenced in the guidance. While specific availability information is included for some reference documents, the documents also may be accessed in the NRC Public Document Room, which is located at NRC Headquarters in Rockville, Maryland, or the NRC Electronic Reading Room at http://www.nrc.gov. See the Notice of Availability on the inside front cover of this report for more information.

When NRC Form 313 does not have sufficient space to provide full responses to Items 5-11, provide the information on separate attachments, label the attachments to indicate which item is being addressed, and submit the attachments with the completed NRC Form 313.

Appendix AA contains background information and item-by-item information that should be provided in Items 1 through 11 of NRC Form 313 for applicants requesting authorization under 10 CFR 30.32(j) for the production of PET radioactive drugs for noncommercial transfer to other medical use licensees within a consortium.

Other appendices to this report provide the following supplementary information:

- Appendices A and B provide sample application forms;

- Appendix C provides license application checklists for responding to Items 5-11 on NRC Form 313;

- Appendix D describes how to fill out the NRC Form 313A series of forms;

- Appendix E includes a sample application;

- Appendix F provides sample licenses;

- Appendices G and H provide information regarding required submissions;

- Appendices I through W provide model procedures;

- Appendices X through Z provide reference materials;

- Appendix BB, published as a separate document, provides a summary of public comments on drafts and NRC responses;

- Appendix CC provides a list of references; and

- Appendix DD provides a summary of public comments and NRC responses on draft NUREG-1556, Volume 9, Revision 2.

In this document, "dose" or "radiation dose" means absorbed dose, dose equivalent, effective dose equivalent, committed dose equivalent, committed effective dose equivalent, or total effective dose equivalent (TEDE). These quantities are defined in 10 CFR Part 20 and are expressed in units of rem and its SI equivalent, the Sievert (Sv) (1 rem = 0.01 Sv). (The quantities, absorbed dose and exposure, and their associated units, the rad and the roentgen, are not used in 10 CFR Part 20 to specify dose limits.) The byproduct materials commonly used in medicine emit beta and photon radiation, for which the quality factor is 1; a useful rule of thumb is an exposure of 1 roentgen is equivalent to an absorbed dose of 1 rad and dose equivalent of

1 rem. With the addition of accelerator-produced materials to the definition of byproduct material by the EPAct, licensees may see the development of alpha-emitting radioisotopes for medical uses that are under NRC jurisdiction. The quality factor used in 10 CFR Part 20 for alpha particles is 10.

This NUREG updates the information and guidance provided in Revision 2 of RG 10.8, "Guide for the Preparation of Applications for Medical Use Programs," revises the format in which the information is presented to assist with the preparation of a medical use license, and includes new guidance for the new byproduct material now under NRC jurisdiction in accordance with the expanded definition of byproduct material. Revision 2 of RG 10.8 was issued in August 1987 to provide guidance for the revised 10 CFR Part 35, which became effective April 1, 1987. Since then, 10 CFR Part 35 has been amended a number of times. Technology-specific information has been revised and expanded to include technologies that are now more commonly used; for example, computerized remote afterloading brachytherapy and gamma stereotactic radiosurgery (GSR). It has also been updated to include accelerator-produced radioactive materials and discrete sources of radium-226 (Ra-226) as a result of the expanded definition of byproduct material resulting from the EPAct.

Specific guidance for applicants requesting authorization to produce radioactive material using an accelerator is included in NUREG 1556, Volume 21, "Consolidated Guidance About Materials Licenses: Program-Specific Guidance About Possession Licenses for Production of Radioactive Materials Using an Accelerator," and is not within the scope of this guidance for medical use licensees. Note that this guidance (Volume 9) should be used for the activities that take place after the radiochemical is produced, which would include the radiochemistry or compounding of the radiochemical into a radiopharmaceutical by an authorized nuclear pharmacist (ANP) or qualified authorized user (AU) for the applicant's medical use.

1.2   TYPES OF LICENSES

Specific Medical Use License

The NRC defines "medical use" as "the intentional internal or external administration of byproduct material, or the radiation from byproduct material, to patients or human research subjects under the supervision of an authorized user" (10 CFR 35.2). An "authorized user" is defined as "a physician, dentist, or podiatrist" who meets the training and experience requirements specified in the board certification pathway in the applicable sections of 10 CFR Part 35 or who is identified as an AU (1) on an NRC or Agreement State license, (2) on a permit issued by an NRC master materials licensee or an NRC master materials broad-scope permittee that is authorized to permit the medical use of byproduct material, or (3) on a permit issued by an NRC or Agreement State broad-scope licensee authorized to permit the medical use of byproduct material (10 CFR 35.2). Section 10 CFR 35.57(b) also recognizes as an AU a physician, dentist, or podiatrist using only accelerator-produced radioactive materials, discrete sources of Ra-226, or both, for medical use under the provisions of the NRC waiver of August 31, 2005, for those same materials and uses.

The NRC issues two types of specific licenses for the medical use of byproduct material in medical practices and facilities:

- the specific license of limited scope (see Section 1.2.1), and
- the specific license of broad scope (see Section 1.2.2).

Medical use includes research involving human subjects, which may occur under either limited-scope or broad-scope specific licenses (see Section 1.2.3).

The NRC usually issues a single byproduct materials license to cover an entire radionuclide program. (Note, however, that nuclear-powered pacemakers are licensed separately under 10 CFR Part 70.) A license including teletherapy may also contain the authorization for source material (i.e., depleted uranium) used as shielding in many teletherapy units, and a license may include authorization for possession of sealed sources to be used to calibrate dose calibration devices.

The NRC may issue separate licenses to individual licensees for different medical uses. However, the NRC does not usually issue separate licenses to different departments in a medical facility or to individuals employed by a medical facility or with whom the medical facility has contracted. Only the facility's management may sign the license application.

General Laboratory License

The NRC also issues a general license pursuant to 10 CFR 31.11, under which a physician, veterinarian in the practice of veterinary medicine, clinical laboratory, or hospital may use byproduct material for certain *in vitro* clinical or laboratory testing. Such testing does not involve internal or external administration of byproduct material, or the radiation therefrom, to human beings or animals (see Section 1.2.4).

Positron Emission Tomography (PET) Radionuclide Production or Radioactive Drug Distribution Licenses and Authorizations

- A medical use licensee that possesses and uses an accelerator to produce radionuclides used in PET studies needs a separate license under 10 CFR Part 30 for the PET radionuclide production activities. Volume 21 of NUREG-1556 provides licensing guidance for this type of activity.

  A medical use licensee, using its PET radionuclide production facility in the preparation of PET radiopharmaceuticals for its own use, needs two licenses (i.e., the Part 30 production license and the Part 35 medical use license). The PET radioactive drugs are produced under the provisions of 10 CFR 35.100(b), 35.200(b), or 35.300(b), as appropriate.

- A medical use facility that is a member of a consortium that jointly owns, or shares in the operation and maintenance costs of, the PET radionuclide production facility, and receives PET radionuclides from that production facility to produce only PET pharmaceuticals for the consortium members' medical uses, needs an additional authorization under 10 CFR 30.32(j) for the noncommercial distribution of the PET radioactive drugs to its consortium members. See Appendix AA for additional information on this authorization.

- A medical use licensee with a PET radionuclide production facility that commercially distributes PET radionuclides to other licensees needs a 10 CFR Part 30 production license and an additional 10 CFR Part 32 commercial distribution license or authorization. Volume 12 of NUREG-1556 provides additional guidance on commercial distribution.

- A medical use facility that commercially distributes PET radioactive drugs to another medical use licensee needs a commercial medical distribution license either as a manufacturer or commercial nuclear pharmacy. Volumes 12 and 13 of NUREG-1556 provide additional guidance for this type of license application.

"Consortium" as used here and in 10 CFR Part 30 is defined as an association of medical use licensees and a PET radionuclide production facility in the same geographical area that jointly own or share in the operation and maintenance cost of the PET radionuclide production facility that produces PET radionuclides for use in producing radioactive drugs within the consortium for noncommercial distribution among its associated members for medical use. The PET radionuclide production facility within the consortium must be located at an educational institution, a Federal facility, or a medical facility.

Overview

Applicants should study this report, related guidance, and all applicable regulations carefully before completing NRC Form 313 and the NRC Form 313A series of forms. The NRC expects licensees to provide information on specific aspects of the proposed Radiation Protection Program in attachments to NRC Form 313. When necessary, the NRC may ask the applicant for additional information in order to gain reasonable assurance that an adequate Radiation Protection Program has been established.

After a license is issued, the licensee must conduct its program in accordance with the following:

- Statements, representations, and procedures contained in the application and in correspondence with NRC, when incorporated into a license by reference;

- Terms and conditions of the license; and

- NRC regulations.

In 10 CFR 30.9, the NRC requires that the information in the application be complete and accurate in all material aspects. Information is considered material if it has the ability to change or affect an agency decision on issuing the license.

1.2.1 SPECIFIC LICENSE OF LIMITED SCOPE

The NRC issues specific medical licenses of limited scope to private or group medical practices and to medical institutions. A medical institution is an organization in which more than one medical discipline is practiced. In general, individual physicians or physician groups located within a licensed medical facility (e.g., hospital) may not apply for a separate license because 10 CFR 30.33(a)(2) refers to the applicant's facilities. Since a physicians' group does not normally have control over the facilities, the hospital remains responsible for activities conducted on its premises and must apply for the license. On specific licenses of limited scope, the authorized users are specifically listed in the license.

Byproduct material may be administered to patients on an inpatient (i.e., hospitalized) or outpatient basis. For patients to whom byproduct material is administered and who are not releasable under 10 CFR 35.75, inpatient facilities are required. In general, facilities for private and group practices do not include inpatient rooms and, therefore, procedures requiring hospitalization of the patient under 10 CFR 35.75 cannot be performed.

A specific license of limited scope may also be issued to an entity requesting authorization to perform mobile medical services (10 CFR 35.80, 10 CFR 35.647). A medical institution or a private or group practice may apply for authorization to use byproduct material in a mobile medical service.

1.2.2 SPECIFIC LICENSE OF BROAD SCOPE

Medical institutions that provide patient care and conduct research programs that use radionuclides for *in vitro*, animal, and medical procedures may request a specific license of broad scope in accordance with 10 CFR Part 33. No medical use of byproduct material, including research involving human subjects, may be conducted without an authorization in a license from the NRC or an Agreement State as provided in 10 CFR Part 35. The criteria for the various types of broad-scope licenses are found in 10 CFR 33.13 through 10 CFR 33.17. Generally, the NRC issues specific licenses of broad scope for medical use (i.e., licenses authorizing multiple quantities and types of byproduct material for medical use under Part 35 as well as other uses) to institutions that: (1) have experience successfully operating under a specific license of limited scope, and (2) are engaged in medical research and routine diagnostic and therapeutic uses of byproduct material. Volume 11 of NUREG-1556 offers additional guidance to applicants for a specific license of broad scope.

1.2.3 RESEARCH INVOLVING HUMAN SUBJECTS

In 10 CFR 35.2, the definition of "medical use" includes the administration of byproduct material or radiation therefrom to human research subjects. Furthermore, 10 CFR 35.6, "Provisions for the protection of human research subjects," addresses the protection of the rights of human subjects involved in research by medical use licensees. For these licensees, prior NRC approval is not necessary if the research is conducted, funded, supported, or regulated by another Federal Agency that has implemented the Federal Policy for the Protection of Human Subjects. Otherwise, the licensee must apply for a specific amendment and receive approval for the amendment before conducting such research. Whether or not a license amendment is required, licensees must obtain informed consent from human subjects and prior review and approval of the research activities by an Institutional Review Board in accordance with the meaning of those terms under the Federal Policy. In accordance with 10 CFR 35.6(a), research involving human subjects shall be conducted only with byproduct materials listed in the license for the uses authorized in the license.

1.2.4 GENERAL *IN VITRO* LICENSE

In 10 CFR 31.11, "General License for Use of Byproduct Material for Certain *In Vitro* Clinical or Laboratory Testing," NRC establishes a general license authorizing physicians, veterinarians, clinical laboratories, and hospitals to receive, acquire, possess, or use small quantities of certain byproduct material for *in vitro* clinical or laboratory tests not involving "medical use" (i.e., not involving administration to humans). Section 31.11 explains the requirements for using the materials listed. If the general license alone meets the applicant's needs, only NRC Form 483, "Registration Certificate – *In Vitro* Testing With Byproduct Material Under General License," need be filed. Medical-use licensees authorized pursuant to 10 CFR Part 35 do not need to file the form.

The NRC limits possession to a total of 200 microcuries (7.4 megabecquerels (MBq)) of photon-emitting materials listed in 10 CFR 31.11 at any one time, at any one location of storage or use. The use of materials listed in 10 CFR 31.11 within the inventory limits of that section is subject only to the requirements of that section and not to the requirements of 10 CFR Parts 19, 20, and 21, except as set forth in 10 CFR 31.11.

An applicant needing more than 200 microcuries (7.4 MBq) of these materials must apply for a specific license and may request the increased inventory limit as a separate line item on NRC Form 313. This type of applicant generally requests an increased limit of 3 millicuries (111 MBq). If requesting an increased inventory limit, the applicant will be subject to the requirements of 10 CFR Parts 19, 20, and 21, including the requirements for waste disposal.

1.3 OTHER REQUIREMENTS

1.3.1 THE "AS-LOW-AS-REASONABLY-ACHIEVABLE (ALARA)" CONCEPT

In 10 CFR 20.1101, "Radiation Protection Programs," it is stated that "each licensee shall develop, document, and implement a Radiation Protection Program commensurate with the scope and extent of licensed activities ..." and "the licensee shall use, to the extent practical, procedures and engineering controls based upon sound radiation protection principles to achieve occupational doses and doses to members of the public that are ... ALARA." This section also requires that licensees review the content of the Radiation Protection Program and its implementation at least annually. The Radiation Safety Officer (RSO) is responsible for the day-to-day operation of the Radiation Protection Program.

References: The following documents contain information, methods, and references useful to those who are establishing Radiation Protection Programs to maintain radiation exposures at ALARA levels in medical facilities:

• RG 8.10, "Operating Philosophy for Maintaining Occupational Radiation Exposures ALARA."

1-11   NUREG - 1556, Vol. 9, Rev. 2

- RG 8.18, "Information Relevant to Ensuring that Occupational Radiation Exposures at Medical Institutions Will Be ALARA."

- NUREG-0267, "Principles and Practices for Keeping Occupational Radiation Exposures at Medical Institutions ALARA."

- NUREG-1134, "Radiation Protection Training for Personnel Employed in Medical Facilities."

- Information directly related to radiation protection standards in 10 CFR Part 20 is contained in NUREG 1736, "Consolidated Guidance: 10 CFR Part 20 - Standards for Protection Against Radiation."

Applicants should consider the ALARA philosophy detailed in these reports when developing plans to work with licensed radioactive materials.

1.3.2 WRITTEN DIRECTIVE PROCEDURES

In 10 CFR 35.41, certain medical use licensees are required to develop, implement, and maintain written procedures to provide high confidence that before each administration requiring a written directive (WD), the patient's identity is verified and the administration is in accordance with the WD. This regulation also specifies what an applicant must, at a minimum, address in these procedures. Appendix S provides further information on developing these procedures.

1.3.3 TIMELY NOTIFICATION OF TRANSFER OF CONTROL

Under 10 CFR 30.34(b) and 10 CFR 35.14(b), licensees must provide full information and obtain NRC's *written consent* before transferring control of the license, or, as some licensees refer to the process, "transferring the license."

Control may be transferred as a result of mergers, buyouts, or majority stock transfers. Although it is not NRC's intent to interfere with the business decisions of licensees, it is necessary for licensees to obtain NRC's written consent before transferring control of the license. This is to ensure the following:

- Radioactive materials are possessed, used, or controlled only by persons who have valid NRC licenses;

- Materials are properly handled and secured;

- Persons using these materials are competent and committed to implementing appropriate radiological controls;

- A clear chain of custody is established to identify who is responsible for final disposal of the material; and

- Public health and safety are not compromised by the use of such materials.

As provided in 10 CFR 35.14(b), if only the licensee's name or mailing address changes, and the name change does not constitute a transfer of control of the license as described in

10 CFR 30.34(b), a licensee must file a written notification with NRC no later than 30 days after the date(s) of the change(s). Otherwise, prior NRC written consent must be given before the transfer.

Guidance on information to be supplied to the NRC when seeking approval for transfer of control of licensed material is available in Appendix G.

Reference: See the Notice of Availability on the inside front cover of this report to obtain copies of IN 97-30, "Control of Licensed Material during Reorganizations, Employee-Management Disagreements, and Financial Crises," dated June 3, 1997, and NUREG-1556, Volume 15, "Program-Specific Guidance About Changes of Control and About Bankruptcy Involving Byproduct, Source, or Special Nuclear Material Licenses," dated November 2000.

These documents can also be accessed at NRC's Web site, in the Electronic Reading Room at http://www.nrc.gov/reading-rm/doc-collections/gen-comm/info-notices/1997/in97030.html and http://www.nrc.gov/reading-rm/doc-collections/nuregs/staff/. Appendix G, excerpted from Appendix F of NUREG-1556, Volume 15, identifies the information to be provided about transferring control.

1.3.4 TIMELY NOTIFICATION OF BANKRUPTCY PROCEEDINGS

Immediately following the filing of a voluntary or involuntary petition for bankruptcy for or against a licensee, the licensee is required by 10 CFR 30.34(h) to notify the appropriate NRC Regional Administrator, in writing, identifying the bankruptcy court in which the petition was filed and the date of the filing.

Even though the licensee may have filed for bankruptcy, the licensee remains responsible for compliance with all regulatory requirements. The NRC needs to know when licensees are in bankruptcy proceedings in order to determine whether all licensed material is accounted for and adequately controlled and whether there are any public health and safety concerns (e.g., contaminated facility). The NRC shares the results of its determinations with other entities involved (e.g., trustees) so that health and safety issues can be resolved before bankruptcy actions are completed.

Reference: See the Notice of Availability on the inside front cover of this report to obtain copies of NUREG-1556, Volume 15, "Consolidated Guidance About Materials Licenses: Program-Specific Guidance About Changes of Control and About Bankruptcy Involving Byproduct, Source, or Special Nuclear Material Licenses," dated November 2000.

1.4 OFFICE OF MANAGEMENT AND BUDGET CLEARANCES

The information collection requirements in 10 CFR Parts 30 and 35 and NRC Form 313 and the NRC Form 313A series of forms have been approved under the Office of Management and Budget Clearance Numbers 3150-0017, 3150-0010, and 3150-0120, respectively.

2 AGREEMENT STATES

Part 35	Applicability
100	✓
200	✓
300	✓
400	✓
500	✓
600	✓
1000	✓

Certain States, called Agreement States (see Figure 2.1), have entered into agreements with NRC that give them the authority to license and inspect byproduct, source, or special nuclear materials used or possessed within their borders. Any applicant, other than a Federal agency or Federally recognized Indian tribe, who wishes to possess or use licensed material in one of these Agreement States should contact the responsible officials in that State for guidance on preparing an application. These applications should be filed with State officials, not with NRC.

Locations of NRC Offices and Agreement States

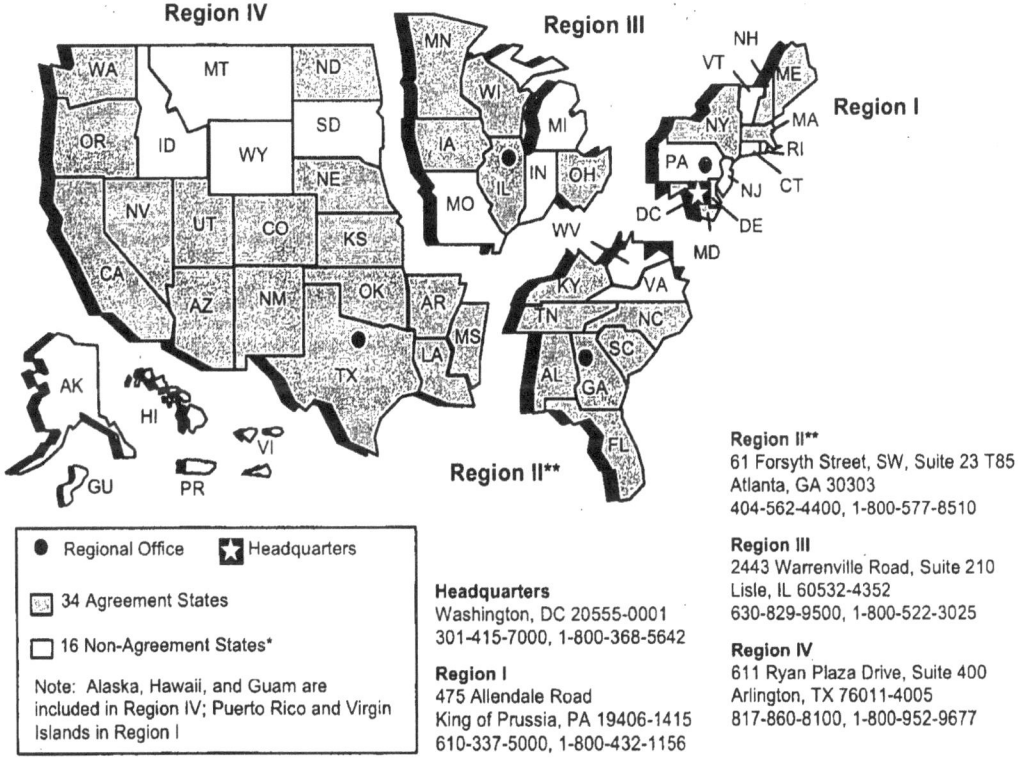

● Regional Office ⭐ Headquarters

[34] 34 Agreement States

□ 16 Non-Agreement States*

Note: Alaska, Hawaii, and Guam are included in Region IV; Puerto Rico and Virgin Islands in Region I

Headquarters
Washington, DC 20555-0001
301-415-7000, 1-800-368-5642

Region I
475 Allendale Road
King of Prussia, PA 19406-1415
610-337-5000, 1-800-432-1156

Region II*
61 Forsyth Street, SW, Suite 23 T85
Atlanta, GA 30303
404-562-4400, 1-800-577-8510

Region III
2443 Warrenville Road, Suite 210
Lisle, IL 60532-4352
630-829-9500, 1-800-522-3025

Region IV
611 Ryan Plaza Drive, Suite 400
Arlington, TX 76011-4005
817-860-8100, 1-800-952-9677

* The 16 Non-Agreement States include three States that have filed letters of intent: Pennsylvania, New Jersey, and Virginia.
** All applicants for materials licenses located in Region II's geographical area must send their applications to Region I.

1556-001q.ppt
053107

Figure 2.1   U.S. Map Location of NRC Offices and Agreement States.

Note: As of March 30, 2008, all Agreement States have to adopt the training and experience requirements in 10 CFR Part 35, Subparts B, D, E, F, G, and H. Before this date, some Agreement States may have additional training and experience criteria for certain medical uses such as the medical use of PET radiopharmaceuticals.

In the special situation of work at Federally controlled sites in Agreement States, it is necessary to know the jurisdictional status of the land in order to determine whether NRC or the Agreement State has regulatory authority. The NRC has regulatory authority over land determined to be "exclusive Federal jurisdiction," while the Agreement State has jurisdiction over nonexclusive Federal jurisdiction land. Applicants are responsible for finding out, in advance, the jurisdictional status of the specific areas where they plan to conduct licensed operations. The NRC recommends that applicants ask their local contact for the Federal agency controlling the site (e.g., contract officer, base environmental health officer, district office staff) to help determine the jurisdictional status of the land and to provide the information in writing, in order to comply with NRC or Agreement State regulatory requirements, as appropriate. Additional guidance on determining jurisdictional status is found in All Agreement States Letter, SP-96-022, dated February 16, 1996, which is available at http://nrc-stp.ornl.gov/asletters/other/sp96022.pdf.

Table 2.1 provides a quick way to check on which agency has regulatory authority.

Table 2.1   Who Regulates the Activity?

Applicant and Proposed Location of Work	Regulatory Agency
Federal agency or Federally recognized Indian tribe[1] regardless of location (except the Department of Energy and, under most circumstances, its prime contractors are exempt from licensing [10 CFR 30.12])	NRC
Non-Federal entity in non-Agreement State, District of Columbia, US territory, or possession, or in Offshore Federal Waters	NRC
Non-Federal entity in Agreement State at non-Federally controlled site	Agreement State
Non-Federal entity in Agreement State at Federally controlled site *not* subject to exclusive Federal jurisdiction	Agreement State
Non-Federal entity in Agreement State at Federally controlled site subject to exclusive Federal jurisdiction	NRC

[1] NRC exercises jurisdiction as the regulatory authority on land where a Federally recognized Indian tribe has tribal jurisdiction. Section 274b Agreements do not give States authority to regulate nuclear material in these areas. Companies owned or operated by native American Indians or non-Indians wishing to possess or use licensed material in these areas would contact the appropriate NRC Regional Office to request a license application.

Reference: A current list of Agreement States is available at the Office of Federal and State Materials and Environmental Management Programs' (FSME) public Web site, which is located at http://nrc-stp.ornl.gov. As an alternative, request the list from an NRC Regional Office.

3 MANAGEMENT RESPONSIBILITY

Part 35	Applicability
100	✓
200	✓
300	✓
400	✓
500	✓
600	✓
1000	✓

Regulations: 10 CFR 30.9, 10 CFR 35.12, 10 CFR 35.24.

The NRC endorses the philosophy that effective Radiation Protection Program management is vital to safe operations that comply with NRC regulatory requirements (see 10 CFR 35.24).

"Management" refers to the chief executive officer or other individual having the authority to *manage, direct, or administer the licensee's activities* or that person's delegate or delegates (see 10 CFR 35.2).

To ensure adequate management involvement in accordance with 10 CFR 35.12(a) and 35.24(a), a management representative (i.e., chief executive officer or delegate) must sign the submitted application acknowledging management's commitments to and responsibility for the following:

- Radiation protection, security and control of radioactive materials, and compliance with regulations;

- Completeness and accuracy of the radiation protection records and all information provided to NRC (10 CFR 30.9);

- Knowledge about the contents of the license application;

- Compliance with current NRC and United States Department of Transportation (DOT) regulations and the licensee's operating and emergency procedures;

- Provision of adequate financial and other resources (including space, equipment, personnel, time, and, if needed, contractors) to the Radiation Protection Program to ensure that patients, the public, and workers are protected from radiation hazards;

- Appointment of a qualified individual who has agreed in writing to work as the RSO;

- Approval of qualified individual(s) to serve as authorized medical physicists (AMPs), ANPs, and AUs for licensed activities.

For information on NRC inspection, investigation, enforcement, and other compliance programs, see the following:

- The NRC Enforcement Policy which is included on the NRC's Web site at http://www.nrc.gov/what-we-do/regulatory/enforcement/enforce-pol.html

- The NRC Inspection Manual, Chapter 2800, "Materials Inspection Program," and

- Inspection Procedures:

    83822 – "Radiation Protection,"

    84850 – "Radioactive Waste Management - Inspection of Waste Generator Requirements of 10 CFR Part 20 and 10 CFR Part 61,"

84900 – "Low-Level Radioactive Waste Storage,"

87130 – "Nuclear Medicine Programs — Written Directive Not Required,"

87131 – "Nuclear Medicine Programs — Written Directive Required,"

87132 – "Brachytherapy Programs,"

87133 – "Medical Gamma Stereotactic Radiosurgery and Teletherapy Programs," and

87134 – "Medical Broad-Scope Programs."

For availability of these documents, see the Notice of Availability on the inside front cover of this report. In addition, the Inspection Manual and procedures are available at http://www.nrc.gov/materials/miau/med-use-toolkit.html.

4 APPLICABLE REGULATIONS

Regulations applicable to medical use licensees are listed below.

Applicants should ensure the use of up-to-date versions of regulations, which are available at NRC's Web site at http://www.nrc.gov/reading-rm/doc-collections/cfr/ in the "Electronic Reading Room"; printed copies available from the U.S. Government Printing Office (GPO) are updated annually.

Part 35	Applicability
100	✓
200	✓
300	✓
400	✓
500	✓
600	✓
1000	✓

- 10 CFR Part 2, "Rules of Practice for Domestic Licensing Proceedings and Issuance of Orders"

- 10 CFR Part 19, "Notices, Instructions and Reports to Workers: Inspection and Investigations"

- 10 CFR Part 20, "Standards for Protection Against Radiation"

- 10 CFR Part 21, "Reporting of Defects and Noncompliance"

- 10 CFR Part 30, "Rules of General Applicability to Domestic Licensing of Byproduct Material"

- 10 CFR Part 31, "General Domestic Licenses for Byproduct Material"

- 10 CFR Part 32, "Specific Domestic Licenses to Manufacture or Transfer Certain Items Containing Byproduct Material"

- 10 CFR Part 33, "Specific Domestic Licenses of Broad Scope for Byproduct Material"

- 10 CFR Part 35, "Medical Use of Byproduct Material"

- 10 CFR Part 40, "Domestic Licensing of Source Material"

- 10 CFR Part 70, "Domestic Licensing of Special Nuclear Material" (for pacemaker devices)

- 10 CFR Part 71, "Packaging and Transportation of Radioactive Material"

  Part 71 requires that licensees or applicants who transport licensed material or who may offer such material to a carrier for transport must comply with the applicable DOT requirements in 49 CFR Parts 170 through 189.

- 10 CFR Part 150, "Exemptions and Continued Regulatory Authority in Agreement States and in Offshore Waters Under Section 274"

- 10 CFR Part 170, "Fees for Facilities, Materials, Import and Export Licenses, and Other Regulatory Services Under the Atomic Energy Act of 1954, as Amended"

- 10 CFR Part 171, "Annual Fees for Reactor Licenses and Fuel Cycle Licenses and Materials Licenses, Including Holders of Certificates of Compliance, Registrations, and Quality Assurance Program Approvals and Government Agencies Licensed by the NRC."

Availability: Copies of the above documents may be obtained by calling the GPO order desk in Washington, DC at (202) 512-1800, or online at http://www.bookstore.gpo.gov. A single copy of the above documents may be requested from NRC's Regional Offices (see Figure 2.1 for addresses and telephone numbers). In addition, 10 CFR Parts 1-199 can be found on NRC's Web site at http://www.nrc.gov/reading-rm/doc-collections/cfr/. Note that NRC and all other Federal agencies publish amendments to their regulations in the *Federal Register*.

5 HOW TO FILE

Part 35	Applicability
100	✓
200	✓
300	✓
400	✓
500	✓
600	✓
1000	✓

5.1 PREPARING AN APPLICATION

Applicants for an NRC materials license should do the following:

*   Use the most recent guidance in preparing an application, including Appendix AA of this document, if appropriate;

*   Complete NRC Form 313 (Appendix A), Items 1 through 4, 12, and 13, on the form itself;

*   Complete NRC Form 313, Items 5 through 11, on supplementary pages, or use Appendix C;

*   Complete the appropriate NRC Form 313A series of forms (Appendix B) to document training and experience, if electing to complete this optional form;

*   Provide sufficient detail for NRC to determine that equipment, facilities, training, experience, and the Radiation Safety Program are adequate to protect health and safety and minimize danger to life and property;

*   For each separate sheet, other than the NRC Form 313A series of forms or Appendix C, that is submitted with the application, identify and cross-reference it to the item number on the application or the topic to which it refers;

*   Submit all documents, typed, on 8-1/2 x 11-inch paper;

*   Avoid submitting proprietary information unless it is absolutely necessary;

*   If submitted, proprietary information and other sensitive information must be clearly identified (see Section 5.2 below);

*   Submit an original, signed application and one copy; and

*   Retain one copy of the license application for future reference.

Applications must be signed by the applicant's or licensee's management as required by 10 CFR 35.12(a); see Section 8.31, "Certification."

5.2 IDENTIFYING AND PROTECTING SENSITIVE INFORMATION

All licensing applications, except for portions containing sensitive information, will be made available for review in NRC's Public Document Rooms and electronically at the Public Electronic Reading Room. More information on the Public Electronic Reading Room is available at www.nrc.gov.

Several types of sensitive information need to be identified, marked, and protected against unauthorized disclosure to the public. Key examples are as follows:

- Proprietary Information/Trade Secrets: If it is necessary to submit proprietary information or trade secrets, follow the procedure in 10 CFR 2.390(b). Failure to follow this procedure could result in disclosure of the proprietary information to the public or substantial delays in processing the application.

- Private information: Personal information about employees or other individuals should not be submitted unless specifically requested by NRC. Examples of private information are: Social Security Number, home address, home telephone number, date of birth, and radiation dose information. If private information is submitted, it should be separated from the public portion of the application and clearly marked: "Privacy Act Information - Withhold Under 10 CFR 2.390."

- Security-Related Information: Following the events of September 11, 2001, the NRC changed its procedures to avoid release of information that terrorists could use to plan or execute an attack against facilities or citizens in the United States. As a result, certain types of information are no longer routinely released and are treated as sensitive unclassified information. For example, certain information about the quantities and locations of radioactive material at licensed facilities, and associated security measures, are no longer released to the public. Therefore, security-related sensitive information in an application should be marked as specified in Regulatory Issue Summary 2005-31, available at http://www.nrc.gov/reading-rm/doc-collections/gen-comm/reg-issues/2005/ri200531.pdf. Additional information on procedures and any updates are available at http://www.nrc.gov/reading-rm/sensitive-info.html.

5.3 PAPER FORMAT AND ELECTRONIC FORMAT

The NRC's new licensing process will be faster and more efficient, in part, through acceptance and processing of electronic applications at some future date. The NRC will continue to accept paper applications. However, these will be scanned through an optical character reader (OCR) to convert them to electronic format. To ensure a smooth transition to electronic applications, applicants should:

- Submit printed or typewritten – not handwritten – text on smooth, crisp paper that will feed easily into the scanner;

- Choose typeface designs that are sans serif, such as Arial, Helvetica, Futura, Univers; the text of this document is in a serif font called Times New Roman;

- Use 12-point or larger font;

- Avoid stylized characters such as script, italic, etc.;

- Use print that is clear and sharp;

- Ensure that there is high contrast between the ink and paper (black ink on white paper is best).

As the electronic licensing process develops, it is anticipated that NRC may provide mechanisms for filing applications via electronic media and through the Internet. Additional filing instructions will be provided as NRC implements these new mechanisms. When the electronic process becomes available, applicants may file electronically instead of on paper.

6 WHERE TO FILE

Part 35	Applicability
100	✓
200	✓
300	✓
400	✓
500	✓
600	✓
1000	✓

Applicants who wish to possess or use licensed material in any State or U.S. territory or possession subject to NRC jurisdiction must file an application with an NRC Regional Office for the locale in which the material will be possessed and/or used. Federally recognized Indian tribes must also file applications with the appropriate NRC Regional Office. Section 8.37 and Appendix V provide further information on filing procedures for applicants who wish to perform mobile medical services.

Figure 2.1 shows NRC's four Regional Offices and their respective areas for licensing purposes, and identifies Agreement States. Note that all materials applications are submitted to Regions I, III, or IV. All applicants for materials licenses located in Region II's geographical area should send their applications to Region I.

In general, applicants for possession or use of byproduct material in an Agreement State must file an application with the Agreement State, not NRC. However, if work will be conducted at Federally controlled sites in Agreement States, applicants must first determine the jurisdictional status of the land in order to determine whether NRC or the Agreement State has regulatory authority. Section 2, "Agreement States," has additional information.

7 LICENSE FEES

Application fees are required for new license applications and some other licensing actions. Each application for which a fee is specified must be accompanied by the appropriate fee. Refer to 10 CFR 170.31 to determine the amount of the fee. The NRC will not issue the licensing action before it receives the appropriate payment. Consult 10 CFR 170.11 for information on exemptions from fees. Once technical review has begun, no fees will be refunded. Application fees will be charged regardless of NRC's disposition of an application or the withdrawal of an application.

Part 35	Applicability
100	✓
200	✓
300	✓
400	✓
500	✓
600	✓
1000	✓

Most NRC licensees are also subject to annual fees; refer to 10 CFR 171.16. Consult 10 CFR 171.11 for information on exemptions from annual fees and 10 CFR 171.16(c) on reduced annual fees for licensees that qualify as "small entities."

Direct all questions about NRC fees or completion of Item 12 of NRC Form 313 (see Appendix A) to the Office of the Chief Financial Officer (OCFO) at NRC Headquarters in Rockville, Maryland, (301) 415-7554 (or toll free at (800) 368-5642, extension 7554). Information about fees may also be obtained by calling this NRC toll-free number or by sending an e-mail to fees@nrc.gov.

Enter the fee category and the amount of the fee enclosed with the application on NRC Form 313.

8 CONTENTS OF AN APPLICATION

This section explains, item by item, the information that medical use applicants must provide on NRC Form 313 (see Appendix A) and should provide on the appropriate NRC Form 313A series of forms if electing to use this optional form (see Appendices B and D). If an application contains security-related sensitive information (see Section 5.2), the cover letter should state that the "attached documents contain security-related sensitive information." If a cover letter is not used, NRC Form 313 should include this statement. The information needed to complete Items 5 through 11 on Form 313 describes the applicant's proposed medical use Radiation Safety Program. To assist the applicant in submitting complete information on these items, the applicable regulations are referenced in the discussion of each item. Appendix AA explains additional information the applicant must provide on NRC Form 313 when requesting authorization under 10 CFR 30.32(j) for preparing PET radioactive drugs for noncommercial distribution to medical use licensees within its consortium.

Table 1 in Appendix C is provided to help applicants determine which procedures must be developed, implemented, and maintained for the type of medical use requested. Several appendices in this report present sample procedures that applicants may use in developing their procedures. Suggested responses for each block on the NRC Form 313 appear under "Response from Applicant" in this guide.

If a particular part of a section does not apply, simply note "N/A" for "not applicable." If a particular section applies, but a procedure does not have to be developed, simply note "N" for "no response required." N/A, N, or short sentence responses to Items 5 through 11 should run consecutively on one or more sheets separate from responses provided on NRC Form 313. Lengthy responses should be appended as attachments.

As indicated on NRC Form 313 (see Appendix A), responses to Items 5 through 11 should be submitted on separate sheets of paper. Applicants may use the appropriate NRC Form 313A series of forms (see Appendix B) to document training and experience for new AUs, medical physicists, nuclear pharmacists, and RSOs. The NRC Form 313A series of forms may also be used by experienced individuals seeking additional authorizations. Applicants may use Appendix C to assist with completion of the application.

ITEMS FOR WHICH A RESPONSE FROM MEDICAL USE APPLICANT IS REQUIRED ON NRC FORM 313

(Also see Appendix AA for items requiring a response when applying for a 10 CFR 30.32(j) authorization)

8.1 ITEM 1: LICENSE ACTION TYPE

THIS IS AN APPLICATION FOR (Check appropriate item)

Type of Action	License No.
☐ A. New License	Not Applicable
☐ B. Amendment to License No.	XX-XXXXX-XX
☐ C. Renewal of License No.	XX-XXXXX-XX

Part 35	Applicability
100	✓
200	✓
300	✓
400	✓
500	✓
600	✓
1000	✓

Check A if the application is for a new license.

Check B for an amendment[1] to an existing license, and provide the license number.

Check C for a renewal of an existing license, and provide the license number.

8.2 ITEM 2: APPLICANT'S NAME AND MAILING ADDRESS

Part 35	Applicability
100	✓
200	✓
300	✓
400	✓
500	✓
600	✓
1000	✓

Regulations: 10 CFR 30.34(b), 10 CFR 30.34(h), 10 CFR 35.14(b).

List the legal name of the applicant's corporation or other legal entity with direct control over use of the radioactive material; a division or department within a legal entity may not be a licensee. An individual may be designated as the applicant only if the individual is acting in a private capacity and the use of the radioactive material is not connected with employment by a corporation or other legal entity. Provide the mailing address where correspondence should be sent. A post office box number is an acceptable mailing address. See Section 8.31, "Certification."

Note: NRC must be notified before control of the license is transferred or whenever bankruptcy proceedings are initiated. See Sections 1.3.3 and 1.3.4 for more details. The NRC's Information Notice (IN), IN 97-30, "Control of Licensed Material During Reorganizations, Employee-Management Disagreements, and Financial Crises," dated June 3, 1997, discusses the potential for the security and control of licensed material to be compromised during periods of organizational instability.

---

[1] See Section 9, "Amendments and Renewals to a License," in this document. Licensees may request an amendment to an existing license to add authorization for other uses of byproduct material.

CONTENTS OF AN APPLICATION

8.3 ITEM 3: ADDRESS(ES) WHERE LICENSED MATERIAL WILL BE USED OR POSSESSED

Part 35	Applicability
100	✓
200	✓
300	✓
400	✓
500	✓
600	✓
1000	✓

Regulations: 10 CFR 30.33(a)(2), 10 CFR 35.14(b)(2).

In order to ensure compliance with 10 CFR 30.33(a)(2) and as referenced in NRC Form 313, Item 3, specify the street address, city, and State or other descriptive address (e.g., on Highway 10, 5 miles east of the intersection of Highway 10 and State Route 234, Anytown, State) for each facility. The descriptive address should be sufficient to allow an NRC inspector to find the facility location. Sketches or street maps indicating the nearest intersection and the location of the proposed facility would be helpful but are not required. A post office box address is not acceptable. If byproduct material is to be used at more than one location under the license, the specific address (e.g., street and building) must be provided for each facility. If applying for a license for mobile medical services as authorized pursuant to 10 CFR 35.18(b), the applicant should refer to Section 8.37 and Appendix V of this report for specific licensing guidance. The NRC must be notified if the mailing address changes.

Notes:

- When responding to this section, follow the guidance in Section 5.2 to determine if the response includes security-related sensitive information and needs to be marked accordingly.

- If there is no cover letter, then Item 3 on NRC Form 313 should state "attached document contains security-related sensitive information" instead of the address if the address contains sensitive information, or in addition to the address, if the address is not sensitive but other information in the application is. Documents that give exact locations of use need to be marked "security-related information – withhold under 10 CFR 2.390."

> **Being granted an NRC license does not relieve a licensee from complying with other applicable Federal, State, or local regulations (e.g., local zoning requirements; a local ordinance requiring registration of a radiation-producing device).**

- The EPAct amended the definition of byproduct material in the Atomic Energy Act and gave NRC jurisdiction over accelerator-produced radioactive materials, discrete sources of Ra-226, and certain naturally occurring radioactive materials that are extracted or converted after extraction for use for a commercial, medical, or research activity. The definition encompasses those materials produced, extracted, or converted before, on, and after Section 651e of the EPAct was enacted. Therefore, NRC applicants and licensees who possess(ed) these materials must maintain permanent records on where the newly defined byproduct materials were used or stored prior to issuance of an NRC license, if they still possess the material once an NRC license is issued.

As discussed in Section 8.8, "Recordkeeping for Decommissioning and Financial Assurance," licensees must maintain permanent records on where the licensed material was used or stored

NUREG - 1556, Vol. 9, Rev. 2   8-6

while the license was in effect. These records are important for making future determinations about the release of these locations for unrestricted use (e.g., before the license is terminated). For medical use licensees, acceptable records include sketches and written descriptions of the specific locations where material is (or was) used or stored and any information relevant to spills (e.g., where contamination remains after cleanup procedures or when there is reasonable likelihood that contaminants may have spread), damaged devices, leaking radioactive sources, or contamination from Ra-226.

8.4 ITEM 4: PERSON TO BE CONTACTED ABOUT THIS APPLICATION

Part 35	Applicability
100	✓
200	✓
300	✓
400	✓
500	✓
600	✓
1000	✓

Identify the individual who can answer questions about the application and include his or her telephone number. This is typically the proposed RSO, unless the applicant has named a different person as the contact. The NRC will contact this individual if there are questions about the application.

Notify NRC of changes of contact name or telephone number so that NRC can contact the applicant or licensee in the future with questions, concerns, or information. This notice is for "information only" and does not require a license amendment or a fee.

The individual named in Item 4 may or may not be the same individual who signs the application as the "certifying officer" on behalf of the licensee with the authority to make commitments to NRC (see Item 13 on NRC Form 313).

The NRC recognizes that licensees may use a consultant or consultant group to help prepare the license application and provide support to the Radiation Protection Program. However, the NRC reminds licensees that regardless of the role of the consultant in radiation protection program management, the licensee remains responsible for all aspects of the licensed program, including the services performed by the consultant.

8.5 ITEM 5: RADIOACTIVE MATERIAL

Part 35	Applicability
100	✓
200	✓
300	✓
400	✓
500	✓
600	✓
1000	✓

Regulations: 10 CFR 30.32, 10 CFR 32.210, 10 CFR 35.65, 10 CFR 35.100, 10 CFR 35.200, 10 CFR 35.300, 10 CFR 35.400, 10 CFR 35.500, 10 CFR 35.600, 10 CFR 35.1000.

Criteria: Byproduct material for medical use in 10 CFR Part 35 is divided into seven types of use (10 CFR 35.100, 35.200, 35.300, 35.400, 35.500, 35.600, and 35.1000).

Discussion: The applicant should indicate the byproduct material requested. The amount and type of information necessary will vary according to the type of use and material requested.

Under Section 651c of the EPAct, the NRC now has regulatory authority over accelerator-produced byproduct material as well as discrete sources of Ra-226. Although sealed Ra-226 sources (e.g., Ra-226 needles) were once used for manual brachytherapy and are no longer believed to be used for medical uses, the medical use of discrete sources of Ra-226 is included in this guidance because its use for this purpose is not prohibited. The guidance also distinguishes between discrete sources of Ra-226 and sealed sources of Ra-226 because not all discrete sources are sealed sources.

The medical uses of the other new byproduct materials are essentially the same as the uses of the previously defined byproduct materials. However, some of the radionuclides now included in the expanded definition of byproduct material have significantly shorter half-lives and higher energy levels (e.g., PET radionuclides) that may result in delivery of the unsealed material by direct transfer tube from the accelerator production facility to the 35.100 and 35.200 medical use areas. This may result in higher potential radiation doses to workers and the public if additional handling and shielding precautions are not implemented, and licensees should consider this in evaluating their equipment, facilities, and programs.

35.100 and 35.200 Use: For 10 CFR 35.100 and 35.200 medical uses, the chemical/physical form may be "Any" unsealed byproduct material permitted by 10 CFR 35.100 or 35.200, as appropriate. For 10 CFR 35.100 and 35.200 medical uses, the total amount requested may be "As Needed."

The following format may be used:

Byproduct Material	Chemical/Physical Form	Maximum Amount
Any byproduct material permitted by 10 CFR 35.100	Any	As needed
Any byproduct material permitted by 10 CFR 35.200	Any	As needed[1]

35.300 Use: For 10 CFR 35.300 use, the chemical/physical form may be "Any" unsealed byproduct material permitted by 10 CFR 35.300. The total amount requested must be specified. The following format may be used:

Byproduct Material	Chemical/Physical Form	Maximum Amount
Any byproduct material permitted by 10 CFR 35.300	Any	300 millicuries

---

[1] Applicants that have their own cyclotrons and produce PET radionuclides that they use to produce PET radioactive drugs for their own use under the appropriate provisions of 10 CFR Part 35 may have different shielding or special equipment requirements than most medical use applicants who receive unit doses, multi-dosage vials, or generators from drug manufacturers or commercial nuclear pharmacies that are packaged in self-shielding radiation transport shields. Information needed for the different shielding or special equipment requirements can be found in Section 9.

35.400, 35.500, 35.600, and 35.1000 Use: For 10 CFR 35.400, 35.500, 35.600, and 35.1000 use, the radionuclide, the chemical/physical form (e.g., sealed source or device identified by manufacturer and model number), the total amount in becquerels (Bq), microcuries (µCi), millicuries (mCi), or curies (Ci), and the maximum number of sources or activity possessed at any one time must be specified. Sealed sources of Ra-226 may be used for 10 CFR 35.400, 35.500, and 35.1000 uses. Unsealed Ra-226 can only be used for medical use under 35.1000. Applicants should include all possible new sources they might use, in order to minimize the need for license amendments if they change model or vendor. The following format may be used:

Byproduct Material	Chemical/Physical Form	Maximum Amount
I-125 (specific radiation therapy system liquid brachytherapy source, 35.1000 use)	Liquid source (Manufacturer Name, Model #DEF)	2 curies total
Ra-226	Sealed source or device (Manufacturer Name, Model #HIJ)	Not to exceed 50 millicuries per source and 250 millicuries total
Cesium 137 (i.e., specific brachytherapy radionuclide, 35.400 use)	Sealed source or device (Manufacturer Name, Model #MNO)	2 curies total
Pd-103 (i.e., specific manual brachytherapy source, 35.400 use)	Sealed source or device (Manufacturer Name, Model #QRS)	Not to exceed 0.5 millicuries per source and 3 curies total
Gadolinium 153 (i.e., specific diagnostic sealed-source radionuclide, 35.500 use)	Sealed source or device (Manufacturer Name, Model #TUV)	Not to exceed 500 millicuries per source and 1 curie total
Cobalt 60 (i.e., specific teletherapy sealed-source radionuclide, 35.600 use)	Sealed source or device (Manufacturer Name, Model #XYZ)	Not to exceed 9,000 curies per source and 18,000 curies total
Iridium 192 (i.e., specific afterloader sealed-source radionuclide, 35.600 use)	Sealed source or device (Manufacturer Name, Model #XYZ)	Not to exceed 10 curies per source and 20 curies total
Cobalt 60 (i.e., specific gamma stereotactic radiosurgery sealed-source radionuclide, 35.600 use)	Sealed source or device (Manufacturer Name, Model #XYZ)	Not to exceed 36 curies per source and 6,600 curies total

For sealed sources used in devices, an applicant may wish to request a possession limit adequate to allow for the possession of a spare source, to accommodate the total quantity of material in the licensee's possession during replacement of the source in the device. The maximum activity for a single source or source loading may not exceed the activity specified by the manufacturer for the specific device and source combination as stated in the Sealed Source and Device Registry (SSDR) certificate. However, an applicant may request a maximum activity for the source in the shipping container that exceeds the maximum activity allowed in the device. To request this authorization, applicants should provide certification that the source transport container is approved for the requested activity. A source that is received with a higher activity than permitted in the device must be allowed to decay to or below the licensed activity limit prior to installation in the device.

Calibration, Transmission, and Reference Sources: For all calibration, transmission, and reference sources, including those with Ra-226, covered under 10 CFR 35.65, the specific sources do not need to be listed on the license as long as the licensee is authorized pursuant to 10 CFR 35.11 for the medical use of byproduct material.

Shielding Material/Depleted Uranium: Some high-activity radionuclide generators used to produce byproduct materials for 10 CFR 35.200 and 35.300 uses (e.g., Tc-99m generators) may include depleted uranium (i.e., uranium depleted in uranium-235 (U-235)) as shielding material. If a generator has depleted uranium shielding, an applicant should request authorization to possess depleted uranium as shielding material. Applicants receiving large therapy sources and devices also should determine if depleted uranium is used to shield the therapy sources and devices. This includes identifying depleted uranium used as shielding in linear accelerators because, even though NRC does not regulate the accelerator, it does regulate the depleted uranium in the accelerator. If applicable, the applicant should request authorization to possess depleted uranium (i.e., uranium depleted in U-235) in quantities sufficient to include shielding material in both the device(s) and source containers used for source exchange and shielding for other devices. The applicant should review the manufacturer's specifications for each device specified in the license request to determine: (1) if depleted uranium is used to shield the source(s) within the device; and (2) the total quantity of depleted uranium present in the device (in kilograms). The applicant should also consult the manufacturer's specifications or the source supplier to determine if depleted uranium is contained in shielding source containers used during source exchange, as well as the total quantity of depleted uranium in such containers (in kilograms). The following format may be used:

Byproduct Material	Chemical/Physical Form	Maximum Amount
Depleted Uranium	Metal	999 kilograms

Other Material: The applicant should make a separate entry for other required items (e.g., Ra-226 not previously described, more byproduct material for *in vitro* testing than is allowed under 10 CFR 31.11, survey meter calibration source, dosimetry system constancy check source, material for *in vitro*, animal, or human research studies). The following format may be used:

Byproduct Material	Chemical/Physical Form	Maximum Amount
Any byproduct material permitted by 10 CFR 31.11	Prepackaged kits	50 millicuries
Ra-226	unsealed	1 millicurie

Sources that are authorized by 10 CFR 35.65, "Authorization for calibration, transmission, and reference sources," should *not* be listed.

Applicants should number each line entry consecutively, following the 10 CFR Part 35 material.

Blood Irradiators: If the use of a device to irradiate blood is anticipated, the applicant should review NUREG-1556, Volume 5, "Consolidated Guidance About Materials Licenses: Program-Specific Guidance About Self-Shielded Irradiator Licenses."

Production of Radionuclides by Accelerators: If the applicant will use an accelerator to produce radionuclides, a separate license application will be needed for the production of the radionuclides. The applicant should review NUREG-1556, Volume 21, "Consolidated Guidance About Materials Licenses: Program-Specific Guidance about Possession Licenses for Production of Radioactive Materials Using an Accelerator."

Production of PET Radioactive Drugs for Noncommercial Distribution to Medical Use Licensees Within a Consortium: If the applicant will use PET radionuclides to produce PET radioactive drugs for its own medical use and noncommercial distribution to other members of its consortium, the applicant, to satisfy 10 CFR 30.33(a)(1), should identify the PET radionuclides, the proposed use of the material, and the maximum activity. The applicant should also review Appendix AA.

The following format may be used for unsealed PET radionuclides used to produce PET radioactive drugs for noncommercial transfer to other members within the consortium.

Byproduct Material	Chemical/Physical Form	Maximum Amount
PET Radionuclides for noncommercial distribution	Any	_____ curies

When applying for this authorization, the applicant should also consider applying for authorization to take back potentially contaminated transport shields from other consortium members. Each consortium member should dispose of unused dosages and used syringes and vials at its own facility.

When determining both individual radionuclide and total quantities, all materials to be possessed at any one time under the license should be included (i.e., materials received awaiting use (new teletherapy or brachytherapy sources for exchange), materials in use or possessed, material used for shielding, and materials classified as waste awaiting disposal or held for decay-in-storage).

Response from Applicant: The applicant should submit the information as described above. Certain information about quantities of radioactive materials is no longer released to the public and needs to be marked "security-related information – withhold under 10 CFR 2.390." Therefore, when responding to this section, follow the guidance in Section 5.2 to determine if the response includes security-related sensitive information and needs to be marked accordingly. Applicants requesting authorization for the medical use of a discrete source of Ra-226 (which includes a sealed source of Ra-226) or other NARM sources or devices containing NARM sources that do not have the information described above (e.g., manufacturer and model number from an SSDR certificate), or the information required in 10 CFR 30.32(g)(3), should consult the appropriate NRC Regional Office to discuss the contents of their application.

8.6 ITEM 5: SEALED SOURCES AND DEVICES (including Ra-226 sealed sources and devices)

Part 35	Applicability
100	
200	
300	
400	✓
500	✓
600	✓
1000	✓

Regulations: 10 CFR 30.32(g), 10 CFR 30.33(a)(2), 10 CFR 32.210.

Criteria: In accordance with 10 CFR 30.32(g), applicants must provide the manufacturer's name and model number for each requested sealed source and device (except for calibration, transmission, and reference sources authorized by 10 CFR 35.65, and certain NARM sources for which this information is not available). Licensees will be authorized to possess and use only those sealed sources and devices specifically approved or registered by NRC, an Agreement State or a non-Agreement State, or certain sources when information required in 10 CFR 30.32(g)(3) is provided.

Under the EPAct, the NRC was given regulatory authority over additional byproduct material including accelerator-produced radionuclides and discrete sources of Ra-226. See 10 CFR 30.4 for a complete definition of byproduct material.

Applicants and licensees should determine whether they possess, or will possess, sealed sources or devices containing this new byproduct material for uses under 10 CFR 35.400, 10 CFR 35.500, 10 CFR 35.600, or 10 CFR 35.1000, as well as check, calibration, transmission, and references sources that are not included in 10 CFR 35.65.

Applicants will need to request authorization for possession of these sealed source(s) or device(s). It should also be noted that NRC's regulatory authority includes the new byproduct material produced prior to August 8, 2005. As a result, neither the NRC, an Agreement State, nor a non-Agreement State, may have performed a safety evaluation of the sealed source or device and it may not have an Sealed Source and Device Registry (SSDR) certificate. Information that must be submitted for all sources is described in 10 CFR 30.32(g).

Discussion: The NRC or an Agreement State performs a safety evaluation of sealed sources and devices before authorizing a manufacturer to distribute the sources or devices to specific licensees. The safety evaluation is documented in an SSDR certificate. Some non-Agreement States may also have performed similar safety evaluations for sealed sources and devices containing NARM, and these safety evaluations may be documented in SSDR certificates.

Applicants must provide the manufacturer's name and model number for each requested sealed source and device so that NRC can verify whether they have been evaluated in an SSDR certificate or specifically approved on a license. Applicants should include all possible new sources they might use, in order to minimize the need for license amendments if they change model or vendor.

If such a review has not been conducted for the specific source/device model(s), licensees should request a copy of the latest version of NUREG-1556, Volume 3, Revision 1, "Consolidated Guidance about Materials Licenses: Applications for Sealed Source and Device Evaluation and

Registration," from an NRC Regional Office and submit the information requested therein to NRC for review.

If the sealed source or device that has not been reviewed contains NARM material and was produced before the effective date of the rule, November 30, 2007, the information required by 10 CFR 32.210 may not be available. If this is the case, the applicant must provide the information required in 10 CFR 30.32(g)(3).

An applicant may consult with the proposed supplier or manufacturer to ensure that requested sources and devices are compatible with each other and that they conform to the SSDR designations registered with NRC or an Agreement State. Licensees may not make any changes to the sealed source, device, or source-device combination that would alter the description or specifications from those indicated in the respective SSDR certificates without obtaining NRC's prior permission in a license amendment. Licensees providing information in accordance with the provisions of 10 CFR 30.32(g) may not make changes to the sealed sources, device, or source-device combination that would alter the description provided to NRC without obtaining NRC's prior permission in a license amendment. To ensure that sealed sources and devices are used in ways that comply with the SSDR certificates, applicants may want to review or discuss them with the manufacturer.

Response from Applicant: If the possession of a sealed source(s) or device(s) is requested, the applicant shall submit the information described above.

Reference: See the Notice of Availability on the inside front cover of this report to obtain a copy of NUREG-1556, Volume 3, Revision 1, "Consolidated Guidance About Materials Licenses: Applications for Sealed Source and Device Evaluation and Registration," and NUREG-1556, Volume 11, "Consolidated Guidance About Materials Licenses: Program-Specific Guidance About Licenses of Broad Scope."

Note: To obtain copies of the SSDR certificate, applicants should contact the manufacturer/distributor of the device or the appropriate NRC Regional Office (see Figure 2.1 for addresses and telephone numbers).

8.7 ITEM 5: DISCRETE SOURCE OF Ra-226 (OTHER THAN SEALED SOURCES)

Part 35	Applicability
100	✓
200	✓
300	✓
400	
500	
600	
1000	✓

Regulation: 10 CFR 30.33(a)(2)

Criteria: Licensees will be authorized to possess and use discrete sources of Ra-226 specifically authorized by the NRC or an Agreement State.

Response from Applicant: If possession of a discrete source of Ra-226 is requested, provide a complete description of the discrete source, including manufacturer, model number, activity, and intended use. Applicants who do not have this information for a discrete source of Ra-226 should consult with the appropriate NRC Regional Office to discuss the content of their application.

8.8 ITEM 5: RECORDKEEPING FOR DECOMMISSIONING AND FINANCIAL ASSURANCE

Part 35	Applicability
100	✓
200	✓
300	✓
400	✓
500	✓
600	✓
1000	✓

Regulations: 10 CFR 30.34(b), 10 CFR 30.35.

Criteria: All licensees are required to maintain records important to decommissioning. Licensees authorized to possess licensed material in excess of the limits specified in 10 CFR 30.35 must provide evidence of financial assurance for decommissioning.

Discussion: All licensees are required, under 10 CFR 30.35(g), to maintain records important to decommissioning in an identified location. These records must, in part, identify all areas where licensed material is (or was) used or stored and any information relevant to spills (e.g., where contamination remains after cleanup procedures or when there is a reasonable likelihood that contaminants may have spread), leaking sealed sources, and Ra-226 contamination. As an alternative to the potential need for site characterizations, some licensees prefer to maintain information on surveys and leak tests on an ongoing basis and as a low-cost means of providing evidence and assurance of an appropriate decommissioning status upon the termination of licensed activities and/or release of a site for nonlicensed use. Pursuant to 10 CFR 30.35(g), licensees must transfer the records important to decommissioning either to the new licensee before licensed activities are transferred or assigned in accordance with 10 CFR 30.34(b), and must transfer records to the appropriate NRC Regional Office before the license is terminated (see 30.51(b)).

The EPAct amended the definition of byproduct material in the Atomic Energy Act and gave NRC jurisdiction over accelerator-produced radioactive materials, discrete sources of Ra-226, and certain naturally occurring radioactive materials that are extracted or converted after extraction for use for a commercial, medical, or research activity. The expanded definition encompasses those materials produced, extracted, or converted before, on, and after Section 651(e) of the EPAct was enacted. Therefore, NRC applicants and licensees who possess(ed) these materials must maintain permanent records on where the newly defined byproduct materials were used or stored prior to issuance of an NRC license and any other information relevant to spills and leaking sealed sources that is important for decommissioning, if they still possess the material once an NRC license is issued.

Licensees using sealed sources authorized by 10 CFR Part 35 generally use licensed material in a manner that would preclude releases into the environment, would not cause the activation of adjacent materials, or would not contaminate work areas. The licensee's most recent leak test should demonstrate that there has been no leakage from the sealed sources while the sealed sources were in the licensee's possession. However, any leakage from Ra-226 sources or other sealed sources in excess of the regulatory limits would warrant further NRC review of decommissioning procedures on a case-by-case basis.

Licensees authorized to possess byproduct material in excess of the limits specified in 10 CFR 30.35 must also provide evidence of financial assurance for decommissioning. The requirements for financial assurance are specific to the types and quantities of byproduct material authorized on a license. Some medical use applicants and licensees may not need to take any action to comply with the financial assurance requirements because their total inventory of licensed material does not exceed the limits in 10 CFR 30.35 or because the half-life of the unsealed byproduct material used does not exceed 120 days. Applicants requesting licensed material with a half-life in excess of 120 days should determine whether financial assurance is necessary. In addition, applicants requesting more than one radionuclide must use the sum-of-the-ratios method to determine if financial assurance is needed.

Applications for authorization to possess and use unsealed byproduct material with a half-life exceeding 120 days must be accompanied by a decommissioning funding plan or certification of financial assurance when the trigger quantities given in 10 CFR 30.35(a) are exceeded. Acceptable methods of providing financial assurance include trust funds, escrow accounts, government funds, certificates of deposit, deposits of government securities, surety bonds, letters of credit, lines of credit, insurance policies, parent company guarantees, self guarantees, external sinking funds, statements of intent, special arrangements with government entities, and standby trust funds. Appendix A to Volume 3 of NUREG-1757, "Consolidated NMSS Decommissioning Guidance: Financial Assurance, Recordkeeping, and Timeliness," dated September 2003, contains acceptable wording for each mechanism authorized by the regulation to guarantee or secure funds.

The NRC will authorize sealed-source possession exceeding the limits given in 10 CFR 30.35(d) without requiring decommissioning financial assurance, for the purpose of a normal sealed-source exchange, for no more than 30 days.

Determining Need for Financial Assurance for Decommissioning

The half-lives of unsealed byproduct material used by medical licensees have traditionally been less than 120 days. Therefore, most medical use applicants need only consider Ra-226 and licensed material in sealed sources to evaluate the need for financial assurance. Use Table 8.1 to determine if financial assurance is required for the sealed sources listed. If requesting sealed sources other than those listed or any other unsealed byproduct material with a half-life greater than 120 days, refer to 10 CFR 30.35 and Appendix B to 10 CFR Part 30 for possession limits requiring financial assurance. The sum-of-the-fractions procedure is also depicted in Table 8.1 and must be used to determine the need for financial assurance for both sealed and unsealed byproduct material.

Step Number	Description	Cobalt-60	Cesium-137	Strontium-90
	Table 8.1 Worksheet for Determining Need for Financial Assurance for Sealed Sources			
1	Activity possessed, in curies*			
2	Activity requiring financial assurance, in curies	10,000	100,000	1,000
3	Divide data in Step 1 by data in Step 2 = FRACTION			
4	Add the fractions determined in Step 3			

*This table uses only conventional units. The conversion to the International System of units (SI) is:
1 curie = 37 gigabecquerel.

As 10 CFR 30.35 describes, if the sum of the fractions is greater than or equal to 1, the applicant will need to submit a decommissioning funding plan or financial assurance, as applicable.

Response from Applicant: No response is needed from most applicants. If financial assurance is required, applicants must submit evidence as described above and as provided for in NUREG-1757, Volume 3. If applicants have questions about financial assurance requirements associated with discrete sources of Ra-226, they should consult with the appropriate NRC Regional Office to discuss the contents of their application.

Reference: See the Notice of Availability on the inside front cover of this report to obtain copies of NUREG-1757, Volume 3, "Consolidated NMSS Decommissioning Guidance: Financial Assurance, Recordkeeping, and Timeliness," dated September 2003.

8.9   ITEM 6: PURPOSE(S) FOR WHICH LICENSED MATERIAL WILL BE USED

Part 35	Applicability
100	✓
200	✓
300	✓
400	✓
500	✓
600	✓
1000	✓

Regulations: 10 CFR 30.32(j), 10 CFR 30.33(a)(1), 10 CFR 35.100, 10 CFR 35.200, 10 CFR 35.300, 10 CFR 35.400, 10 CFR 35.500, 10 CFR 35.600, 10 CFR 35.1000.

Criteria: In 10 CFR Part 35, byproduct material for medical use is divided into seven types of use as follows:

10 CFR 35.100	Use of unsealed byproduct material for uptake, dilution, and excretion studies for which a written directive is not required
10 CFR 35.200	Use of unsealed byproduct material for imaging and localization studies for which a written directive is not required
10 CFR 35.300	Use of unsealed byproduct material for which a written directive is required
10 CFR 35.400	Use of sources for manual brachytherapy

10 CFR 35.500	Use of sealed sources for diagnosis
10 CFR 35.600	Use of a sealed source(s) in a device for therapy-teletherapy unit
	Use of a sealed source(s) in a device for therapy-remote afterloader unit
	Use of a sealed source(s) in a device for therapy-gamma stereotactic radiosurgery unit
10 CFR 35.1000	Other medical uses of byproduct material or radiation from byproduct material

Under 10 CFR 30.32(j), medical use licensees within a consortium are authorized to produce PET radioactive drugs for noncommercial distribution to medical use licensees within the consortium. Appendix AA provides additional information on this 10 CFR Part 30 use.

Discussion:

10 CFR 35.100, 35.200, and 35.300 Use: For 10 CFR 35.100, 35.200, and 35.300 use, the applicant should define the purpose of use by stating the applicable section of 10 CFR Part 35 (e.g., 10 CFR 35.100) and the description of the applicable modality (e.g., any uptake, dilution, and excretion procedure for which a written directive is not required).

The use of unsealed byproduct material in therapy (10 CFR 35.300) involves administering a byproduct material, either orally or by injection, to treat or palliate a particular disease. The most common form of use of unsealed byproduct material for therapy is the treatment of hyperthyroidism with iodine-131 (I-131) sodium iodide. Other therapeutic procedures include, for example, ablation of thyroid cancer metastasis, treatment of malignant effusions, treatment of polycythemia vera and leukemia, palliation of bone pain in cancer patients, and radiation synovectomy for rheumatoid arthritis patients. References to particular diagnostic or treatment modalities in this section are intended to be examples and are not intended to imply that licensees are limited to these uses.

If an applicant's request is limited to I-131 under 10 CFR 35.300, the license will be limited to that radionuclide.

35.400 Use: The applicant should define the purpose of use by stating that the applicable section of 10 CFR Part 35 is 10 CFR 35.400. If a source is to be used in a device, applicants may need to define the purpose of use by including the manufacturer's name and model number of the device. The licensee should relate the sealed sources, including sealed sources of Ra-226, listed in Item 5 to the devices described in this item.

In manual brachytherapy, several types of treatments are available. These may include, for example:

•   Interstitial Treatment of Cancer.

•   Eye Plaque Implants. This is considered interstitial, not topical, treatment.

- Intracavitary Treatment of Cancer. For purposes of NRC's sealed source and device evaluation on radiation safety issues, intraluminal use is considered analogous to intracavitary use.

- Topical (Surface) Applications.

35.500 Use: For 10 CFR 35.500 use, the applicant should define the purpose of use by stating that the applicable section of 10 CFR 35 is 10 CFR 35.500 and including the manufacturer's name(s) and model number(s) of devices containing sealed sources (where applicable). The licensee should correlate the sealed sources, including sealed sources of Ra-226, listed in Item 5 with the devices described in this item. Typically, a licensee should use the sealed sources according to the manufacturer's radiation safety and handling instructions and must use the sources as approved in the SSDR.

35.600 Use: For 10 CFR 35.600 use, the applicant should define the purpose of use by stating the applicable section of 10 CFR Part 35.600 (e.g., teletherapy, remote afterloading, GSR) and including the manufacturer's name(s) and model number(s) of the device(s) containing a sealed source(s) (e.g., for use in a [Manufacturer's Name and Unit Type, Model xxxx] radiation therapy unit for the treatment of humans). The applicant should correlate the sealed source(s) listed in Item 5 with the device described in this item. If applicable, the applicant should state that depleted uranium is used as shielding for the device and specify that authorization is being requested for an additional source to be stored in its shipping container, incident to source replacement.

35.1000 Use: Applicants must apply for authorization to use byproduct material, or radiation therefrom, in medical applications under 10 CFR 35.1000 when the type of use is not covered under 10 CFR 35.100-35.600. This includes the medical use of unsealed Ra-226 or of Ra-226 sealed sources for uses other than those described by 10 CFR 35.400 or 35.500.

When applying for use under the provisions of 10 CFR 35.1000, applicants should describe the purpose of use and submit the information required under Section 35.12(b) through (d), review regulatory requirements in other Subparts of 10 CFR Part 35, and use them as a guide on how to determine what should be included in an application that is required in Section 35.12. It is anticipated that many of the uses of byproduct material under the provisions of Section 35.1000 may involve research or product development; thus, applicants should ensure review and compliance with 10 CFR 35.6, "Provisions for the protection of human research subjects," and 10 CFR 35.7, "FDA, other Federal, and State requirements." Use of byproduct material in a source or device after approval by the U.S. Food and Drug Administration (FDA) (e.g., under an IDE (investigational device exemption) or an IND (investigational new drug exemption)), does not relieve individuals of the responsibility to obtain a license to use the byproduct material in medicine under the provisions of 10 CFR Part 35.

If the source for the type of use sought under 10 CFR 35.1000 is a sealed source, including sealed sources of Ra-226, Section 8.6 of this guide describes the information that must be provided at the time of application. Broad-scope licensees are exempted under 35.15(a) from requirements of 35.12(d) (which relates to the need to put into an application certain information about the radiation safety aspects of medical use under Section 35.1000). However, broad-scope licensees should ensure that the quantity needed for the proposed use is authorized on their license or apply for an increase if not. Applicants should refer to IN 99-024, "Broad-Scope Licensees' Responsibilities for Reviewing and Approving Unregistered Sealed Sources and Devices" for more information on sealed sources.

Applicants for uses under Section 35.1000 should consult with the appropriate NRC Regional Office to discuss the contents of their application.

Nonmedical Uses: Applicants may also describe nonmedical uses (e.g., survey meter calibrations with NIST-traceable brachytherapy sources) and reference the applicable radioactive material provided in response to Item 5. This would include the nonmedical use of discrete sources of Ra-226.

Authorization under 10 CFR 30.32(j) to produce PET radioactive drugs for noncommercial transfer to licensees in its consortium for medical use is another nonmedical use. Applicants intending to produce PET radioactive drugs under this provision should include this use under this section, list the applicable radioactive materials under Item 5, and review Appendix AA for additional information.

Radionuclide Production by an Accelerator: Production of radionuclides for both medical and nonmedical uses is beyond the scope of this guidance and a medical use license. See NUREG-1556, Volume 21, "Consolidated Guidance About Materials Licenses: Program-Specific Guidance about Possession Licenses for Production of Radioactive Materials Using an Accelerator."

Response from Applicant: The applicant must submit information regarding the purpose for which the licensed material will be used. The applicant should consider including the information described above, as applicable to the type of use(s) proposed.

When responding to this section, follow the guidance in Section 5.2 to determine if the response includes security-related sensitive information and needs to be marked accordingly.

8.10 ITEM 7: INDIVIDUAL(S) RESPONSIBLE FOR RADIATION SAFETY PROGRAMS AND THEIR TRAINING AND EXPERIENCE

Part 35	Applicability
100	✓
200	✓
300	✓
400	✓
500	✓
600	✓
1000	✓

Regulations: 10 CFR 30.33(a)(3), 10 CFR 30.34(j), 10 CFR 33.13, 10 CFR 35.24, 10 CFR 35.50, 10 CFR 35.51, 10 CFR 35.55, 10 CFR 35.57, 10 CFR 35.59, 10 CFR 35.190, 10 CFR 35.290, 10 CFR 35.390, 10 CFR 35.392, 10 CFR 35.394, 10 CFR 35.396, 10 CFR 35.490, 10 CFR 35.491, 10 CFR 35.590, 10 CFR 35.690.

Criteria: The RSO, AUs, AMPs, and ANPs must have adequate training and experience.

Discussion: "Authorized user (AU)" is not defined for nonmedical use, but for purposes of this discussion, the term AU will be used to also mean individuals who are authorized for such nonmedical uses. The requirements in 10 CFR 35.24 describe the authority and responsibilities for the Radiation Protection Program, including those of the licensee's management and the RSO appointed by licensee management. Other personnel who have a role in the Radiation Protection Program are AUs, AMPs, ANPs, and members of the Radiation Safety Committee (RSC) (if the licensee is required to establish an RSC). In 10 CFR 30.33(a)(3), the NRC requires that an applicant be qualified by training and experience to use licensed materials for the purposes requested in such a manner as to protect health and minimize danger to life or property. Subparts B, D, E, F, G, and H of 10 CFR Part 35 give specific criteria for acceptable training and experience for AUs for medical use, ANPs, the RSO, and AMPs; AUs for nonmedical uses must meet the criteria in 10 CFR 30.33(a)(3).

A résumé or a curriculum vitae is likely to be insufficient because such documents usually do not supply all the information needed to evaluate an individual's training and experience for NRC purposes. Applicants should ensure that they submit the specific training information required by NRC regulations in 10 CFR Part 35. The NRC Form 313A series of forms provides a convenient format for submitting the information required in 10 CFR Part 35, Subparts B, D, E, F, G, and H. For nonmedical use AUs, the information provided should focus on educational training and radiation safety training and experience specific to the radionuclides and uses requested.

Licensees are responsible for their Radiation Protection Programs; it is essential that strong management control and oversight exist to ensure that licensed activities are conducted properly. The licensee's management must appoint an RSO, who agrees in writing to be responsible for implementing the Radiation Protection Program, and must provide the RSO sufficient authority, organizational freedom, time, resources, and management prerogative to communicate with personnel and direct personnel regarding NRC regulations and license provisions, including: identifying radiation safety problems; initiating, recommending, or providing corrective actions; stopping unsafe operations; and verifying the implementation of corrective actions. Nevertheless, the licensee retains the ultimate responsibility for the conduct of licensed activities.

Licensees that are authorized for two or more different types of uses of byproduct material under Subparts E, F, and H, or two or more types of units under Subpart H are required under 10 CFR 35.24(f) to establish an RSC to oversee all uses of byproduct material permitted by the license. Membership in the committee must include an AU for each type of use permitted by the license, the RSO, a representative of the nursing service, and a representative of management who is neither an AU nor the RSO. The committee may include other members the licensee considers appropriate.

Licensees may contract for medical use services, including those involving patient services. However, the licensee should not assume that, by hiring a contractor to provide certain services, it has satisfied all regulatory requirements or that it has transferred responsibility for the licensed program to the contractor. Licensee management should ensure that adequate mechanisms for oversight are in place to determine that the Radiation Protection Program, including the training of contractor staff, is effectively implemented by the appropriate individuals.

Training for an experienced RSO, teletherapy or medical physicist, AU or nuclear pharmacist; recentness of training. Under 10 CFR 35.57(a)(1) and (a)(2), experienced individuals, who may be candidates to serve as RSO, AMP, or ANP, are not required to meet the requirements of Sections 35.50, 35.51, or 35.55, respectively (are "grandfathered"), under certain conditions (e.g., the individual is named on an NRC or Agreement State license). Under 10 CFR 35.57(b)(1) and (b)(2), AUs are also not required to meet the requirements in Subparts D-H of 10 CFR Part 35 under certain conditions (e.g., if they are named on an NRC or Agreement State license). The individuals must have been named on a license or permit before the applicable date in Section 35.57.

Subsequent to the EPAct, RSOs, medical physicists, nuclear pharmacists, physicians, podiatrists, and dentists that only used accelerator-produced radioactive material, discrete sources of Ra-226, or both, are also grandfathered, under NRC regulations in 10 CFR 35.57(a)(3) and (b)(3), for medical uses or the practice of nuclear pharmacy when using materials for the same uses performed before or under NRC's waiver issued August 31, 2005. The requirements in 10 CFR 35.59 (that the training and experience specified in 10 CFR 35, Subparts B, D, E, F, G, and H, must have been obtained within 7 years preceding the date of application or the individual must have related continuing education and experience) do not apply to those individuals "grandfathered" under the regulations implementing the EPAct. Also, 10 CFR 35.57 provides that nuclear pharmacists, medical physicists, physicians, dentists, and podiatrists that meet the criteria in 10 CFR 35.57(a)(3) and (b)(3) qualify as ANPs, AMPs, and AUs for those materials and uses performed before or under NRC's waiver of August 31, 2005.

Response from Applicant: Refer to the subsequent sections specific to the individuals described above.

8.11   ITEM 7: RADIATION SAFETY OFFICER (RSO)

Part 35	Applicability
100	✓
200	✓
300	✓
400	✓
500	✓
600	✓
1000	✓

Regulations: 10 CFR 30.33(a)(3), 10 CFR 35.2, 10 CFR 35.14, 10 CFR 35.24, 10 CFR 35.50, 10 CFR 35.57, 10 CFR 35.59, 10 CFR 35.2024.

Criteria: The RSOs must have adequate training and experience. The training and experience requirements for the RSO are described in 10 CFR 35.50 and allow for the following training pathways:

• Certification as provided in 10 CFR 35.50(a) by a specialty board whose certification process has been recognized by the NRC or an Agreement State, plus a written attestation signed by a preceptor RSO as provided in 35.50(d) and training as specified in 35.50(e); or

• Completion of classroom and laboratory training (200 hours) and 1 year of full-time radiation safety experience as described in 10 CFR 35.50(b)(1) plus a written attestation signed by a preceptor RSO as provided in 10 CFR 35.50(d) and training as specified in 35.50(e); or

- Certification as provided in 10 CFR 35.50(c)(1) as a medical physicist under 35.51(a), plus a written attestation signed by a preceptor RSO as provided in 10 CFR 35.50(d) and training as specified in 35.50(e); or

- Identification as provided in 10 CFR 35.50(c)(2) on the licensee's license as an AU, AMP, or ANP with experience in the radiation safety aspects of similar types of byproduct material use for which the individual has RSO responsibilities, with a written attestation signed by a preceptor RSO as provided in 10 CFR 35.50(d) and training as specified in 35.50(e).

The licensee must also establish, in writing, the authority, duties, and responsibilities of the RSO as required by 10 CFR 35.24(b).

Discussion: The RSO is responsible for day-to-day oversight of the Radiation Protection Program. In accordance with 10 CFR 35.24, the licensee must provide the RSO sufficient authority, organizational freedom, time, and resources to perform his or her duties. Additionally, the RSO must have a sufficient commitment from management to fulfill the duties and responsibilities specified in 10 CFR 35.24 to ensure that radioactive materials are used in a safe manner. The NRC requires the name of the RSO on the license, and an agreement in writing from the RSO, to ensure that licensee management has identified a responsible, qualified person and that the named individual knows of his or her designation and assumes the responsibilities of an RSO.

Usually, the RSO is a full-time employee of the licensed facility. The NRC has authorized individuals who are not employed by the licensee, such as a consultant, to fill the role of RSO or to provide support to the facility RSO. In order to fulfill the duties and responsibilities, the RSO should be on site periodically to conduct meaningful, person-to-person interactions with licensee staff, commensurate with the scope of licensed activities, to satisfy the requirements of 10 CFR 35.24. Appendix I contains a model RSO Delegation of Authority. Appendix B contains NRC Form NRC 313A (RSO), "Medical Use Training and Experience and Preceptor Attestation [35.50]," which can be used to document the RSO's training and experience.

RSO Responsibilities: Some of the typical duties and responsibilities of RSOs include ensuring the following:

- Unsafe activities involving licensed materials are stopped;
- Radiation exposures are ALARA;
- Material accountability and disposal;
- Interaction with NRC;
- Timely and accurate reporting and maintenance of appropriate records;
- Annual program audits;
- Proper use and routine maintenance;
- Personnel training; and
- Investigation of incidents involving byproduct material (e.g., medical events).

Appendix I contains a detailed list of typical duties and responsibilities of the RSO.

Applicants are reminded of recentness of training requirements described in 10 CFR 35.59. Specifically, RSO applicants must have successfully completed the applicable training and experience described in 10 CFR Part 35 within 7 years preceding the date of the application. Alternatively, RSO applicants must have had related continuing education and experience since completing the required training and experience. This time provision applies to board certification as well as to other pathways to meeting requirements for training and experience.

In implementing the EPAct, the NRC "grandfathered" RSOs that performed as RSOs for medical uses of only accelerator-produced radioactive material, discrete sources of Ra-226, or both. These individuals do not have to meet the requirements in either 10 CFR 35.59 or 10 CFR 35.50; however, the applicant must document that the individual meets the criteria in 10 CFR 35.57 (a)(3).

Response from Applicant: Provide the following:

- Name of the proposed RSO.

AND

For an individual previously identified as an RSO on an NRC or Agreement State license or permit:

- Previous license number (if issued by the NRC) or a copy of the license (if issued by an Agreement State) or a copy of a permit issued by an NRC master materials licensee on which the individual was named as the RSO.

For an individual qualifying under 10 CFR 35.57 (a)(3):

(*Note:* This is only for a new medical use license requesting use of only accelerator-produced radioactive material, discrete sources of Ra-226, or both, for the same uses authorized under NRC's waiver of August 31, 2005.)

- Documentation that this individual functioned as an RSO for only accelerator-produced radioactive materials, discrete sources of Ra-226, or both, before or during the effective period of NRC's waiver of August 7, 2005;

AND

- Documentation that the individual performed as the RSO for the same medical uses requested.

For an individual qualifying under 10 CFR 35.50(a):

- Copy of certification by a specialty board whose certification process has been recognized[2] by the NRC or an Agreement State under 10 CFR 35.50(a);

AND

---

[2] The names of board certifications that have been recognized by the NRC or an Agreement State are posted on NRC's Web site http://www.nrc.gov/materials/miau/med-use-toolkit.html.

- Description of the training and experience specified in 10 CFR 35.50(e) demonstrating that the proposed RSO is qualified by training in the radiation safety, regulatory issues, and emergency procedures as applicable to the types of use for which the applicant seeks approval of an individual to serve as RSO;

AND

- Written attestation, signed by a preceptor RSO, that the individual has successfully completed the training and experience specified for certification, as well as the required training and experience in radiation safety, regulatory issues, and emergency procedures for the types of use for which the licensee seeks approval and has achieved a level of radiation safety knowledge sufficient to function independently as an RSO.

AND

- If applicable, description of recent related continuing education and experience as required by 10 CFR 35.59.

For an individual qualifying under 10 CFR 35.50(b):

- Description of the training and experience specified in 10 CFR 35.50(b) demonstrating that the proposed RSO is qualified by training and experience as applicable to the types of use for which the applicant seeks approval of an individual to serve as RSO;

AND

- Description of the training and experience specified in 10 CFR 35.50(e) demonstrating that the proposed RSO is qualified by training in radiation safety, regulatory issues, and emergency procedures as applicable to the types of use for which the applicant seeks approval of an individual to serve as RSO;

AND

- Written attestation, signed by a preceptor RSO, that the individual has successfully completed the training and experience in 10 CFR 35.50(b), as well as the required training and experience in radiation safety, regulatory issues, and emergency procedures for the types of use for which the licensee seeks approval and has achieved a level of radiation safety knowledge sufficient to function independently as an RSO.

AND

- If applicable, description of recent related continuing education and experience as required by 10 CFR 35.59.

For an individual qualifying under 10 CFR 35.50(c)(1):

- Copy of the certification(s) as a medical physicist by a board whose certification process has been recognized[3] by the NRC or an Agreement State under 10 CFR 35.51(a) and description of the experience specified in 35.50(c)(1) demonstrating that the proposed RSO

---

[3] The names of board certifications that have been recognized by the NRC or an Agreement State are posted on the NRC's Web site http://www.nrc.gov/materials/miau/med-use-toolkit.html.

is qualified by experience applicable to the types of use for which the applicant seeks approval of an individual to serve as RSO;

AND

- Description of the training and experience specified in 10 CFR 35.50(e) demonstrating that the proposed RSO is qualified by training in radiation safety, regulatory issues, and emergency procedures as applicable to the types of use for which the applicant seeks approval of an individual to serve as RSO;

AND

- Written attestation, signed by a preceptor RSO, that the individual has satisfactorily completed the requirements in 35.50(c)(1), as well as the required training and experience in radiation safety, regulatory issues, and emergency procedures for the types of use for which the licensee seeks approval and has achieved a level of radiation safety knowledge sufficient to function independently as an RSO.

AND

- If applicable, description of recent related continuing education and experience as required by 10 CFR 35.59.

For an individual qualifying under 10 CFR 35.50(c)(2):

- Copy of the licensee's license indicating that the individual is an AU, AMP, or ANP identified on the licensee's license and has experience with the radiation safety aspects of similar types of use of byproduct material for which the applicant seeks approval of an individual to serve as RSO;

AND

- Description of the training and experience specified in 10 CFR 35.50(e) demonstrating that the proposed RSO is qualified by training in radiation safety, regulatory issues, and emergency procedures applicable to the types of use for which the applicant seeks approval of an individual to serve as RSO;

AND

- Written attestation, signed by a preceptor RSO, that the individual satisfactorily completed the requirements in 35.50(c)(2), as well as the required training and experience in radiation safety, regulatory issues, and emergency procedures for the types of use for which the licensee seeks approval and has achieved a level of radiation safety knowledge sufficient to function independently as an RSO;

AND

- If applicable, description of recent related continuing education and experience as required by 10 CFR 35.59.

Notes:

- NRC Form 313A (RSO), "Radiation Safety Officer Training and Experience and Preceptor Attestation [10 CFR 35.50]," may be used to document training and experience for those individuals qualifying under 10 CFR 35.50.

- The licensee must notify the NRC within 30 days if, under 10 CFR 35.14, an RSO permanently discontinues his or her duties under the license or has a name change; licensees must also request an amendment to change an RSO under 10 CFR 35.13.

- An AU for medical uses, AMP, or ANP may be designated as the RSO on the license if the individual has experience with the radiation safety aspects of similar types of byproduct material use for which he or she has RSO responsibilities (see 10 CFR 35.50(c)(2)) and, as required by 10 CFR 35.24(g), has sufficient time, authority, organizational freedom, resources, and management prerogative to perform the duties.

- Descriptions of training and experience will be reviewed using the criteria listed above. The NRC will review the documentation to determine if the applicable criteria in 10 CFR Part 35, Subpart B, are met. If the training and experience do not appear to meet the criteria in Subpart B, the NRC may request additional information from the applicant or may request the assistance of the Advisory Committee on the Medical Uses of Isotopes (ACMUI) in evaluating such training and experience.

- The training and experience for the RSO of a medical use broad-scope license will be reviewed using the above criteria as well as criteria in 10 CFR Part 33.

8.12   ITEM 7: AUTHORIZED USERS (AUs)

Part 35	Applicability
100	✓
200	✓
300	✓
400	✓
500	✓
600	✓
1000	✓

Regulations: 10 CFR 30.33(a)(3), 10 CFR 35.2, 10 CFR 35.11, 10 CFR 35.14, 10 CFR 35.27, 10 CFR 35.57, 10 CFR 35.59, 10 CFR 35.190, 10 CFR 35.290, 10 CFR 35.390, 10 CFR 35.392, 10 CFR 35.394, 10 CFR 35.396, 10 CFR 35.490, 10 CFR 35.491, 10 CFR 35.590, 10 CFR 35.690.

Criteria: Training and experience requirements for AUs for medical uses are described in 10 CFR 35.190, 10 CFR 35.290, 10 CFR 35.390, 10 CFR 35.392, 10 CFR 35.394, 10 CFR 35.396, 10 CFR 35.490, 10 CFR 35.491, 10 CFR 35.590, or 10 CFR 35.690.

Discussion: Although NRC does not define "AU" for nonmedical uses, for purposes of this discussion the term AU will be used to also mean individuals authorized for such nonmedical uses.

AU for Medical Uses: The responsibilities of AUs involved in medical use include the following:

- Radiation safety commensurate with use of byproduct material;

- Administration of a radiation dose or dosage and how it is prescribed;

- Direction of individuals under the AU's supervision in the preparation of byproduct material for medical use and in the medical use of byproduct material;

- Preparation of written directives (WD), if required.

Applicants must meet recentness of training requirements described in 10 CFR 35.59. The AU applicants must have successfully completed the applicable training and experience criteria described in 10 CFR Part 35 within 7 years preceding the date of the application. Alternatively, applicants must have had related continuing education and experience since completing the required training and experience. This time provision applies to board certification as well as to other training pathways.

Section 35.57 of 10 CFR Part 35 provides that experienced AUs who are named on a license or permit are not required to comply with the training requirements in Subparts D through H to continue performing those medical uses for which they were authorized before the effective date of changes to the regulations in Section 35.57 (check the regulations to determine this date). For example, a physician who was authorized to use sodium iodine-131 for imaging and localization, involving greater than 30 microcuries (a quantity for which a written directive is required under 10 CFR 35.40), would continue to be authorized for this use.

In implementing the EPAct, the NRC "grandfathered" physicians, podiatrists, and dentists using only accelerator-produced radioactive materials, discrete sources of Ra-226, or both, for medical use, for the same uses performed before or under the NRC waiver of August 31, 2005. These individuals do not have to meet the requirements in 10 CFR 35.59, 35.190, 35.290, 35.390, 35.396, or 35.490. However, the applicant must document that the individual meets the criteria in 10 CFR 35.57(b)(3). This Section also states that physicians, dentists, and podiatrists who met certain criteria will qualify as AUs for those materials and uses performed before NRC's waiver was terminated for them.

Technologists, therapists, or other personnel may use byproduct material for medical use under an AU's supervision in accordance with 10 CFR 35.27, "Supervision," and in compliance with applicable FDA, other Federal, and State requirements (10 CFR 35.7). Examples include FDA requirements for the conduct of certain types of clinical research after the submission of applications for Investigational New Drugs (IND) and under the auspices of a Radioactive Drug Research Committee (21 CFR 361.1).

There is no NRC requirement that an AU must render an interpretation of a diagnostic image or results of a therapeutic procedure. The NRC recognizes that the AU may or may not be the physician who interprets such studies. Additionally, NRC regulations do not restrict who can read and interpret diagnostic scans or the results of therapeutic procedures involving the administration of byproduct material to individuals.

AU for Nonmedical Uses: For *in vitro* studies, animal research, calibration of survey instruments, and other uses that do not involve the intentional exposure of humans, the list of proposed AUs should include the individuals who will actually be responsible for the safe use of the byproduct material for the requested use. This includes the individuals responsible for the production of PET radioactive drugs for noncommercial transfer to other medical users within a consortium (see Appendix AA).

An applicant should note which user will be involved with a particular use by referring to Items 5 and 6 of the application and providing information about the user's training and experience.

Authorized nonmedical use or uses that do not involve the intentional exposure of humans (e.g., *in vitro* and animal research, calibration, dosimetry research) will be reviewed on a case-by-case basis.

Response from Applicant:

AU for Medical Uses: Provide the following:

- Name of the proposed AU and uses requested;

AND

- Medical, podiatry, or dental license number and issuing entity;

AND

For an individual previously identified as an AU on an NRC or Agreement State license or permit:

- Previous license number (if issued by the NRC) or a copy of the license (if issued by an Agreement State) or a copy of a permit issued by an NRC master materials licensee, a permit issued by an NRC or Agreement State broad-scope licensee, or a permit issued by an NRC Master Materials License broad-scope permittee on which the physician, dentist, or podiatrist was specifically named as an AU for the uses requested;

AND

- For an AU requesting a medical use not currently authorized on a license or permit, a description of the additional training and experience is needed to demonstrate the AU is also qualified for the new medical uses requested (e.g., training and experience needed to meet the requirements in 10 CFR 35.290(b), 35.396, 35.390(b)(1)(ii)(G) or 35.690(c)). A preceptor attestation may also be required. (For example, a preceptor attestation is needed to meet the requirements of 10 CFR 35.396 and 35.690.)

For an individual qualifying under 10 CFR 35.57(b)(3):

- Documentation that the physician, dentist, or podiatrist used only accelerator-produced radioactive materials, discrete sources of Ra-226, or both, for medical uses before or during the effective period of NRC's waiver of August 31, 2005;

AND

- Documentation that the physician, dentist, or podiatrist used these materials for the same medical uses requested;

AND

- For an AU requesting a medical use for which he or she is not currently authorized on a license or permit, a description of the additional training and experience to demonstrate the AU is also qualified for the new medical uses requested. A preceptor attestation may also be required. (For example, training, experience, and attestations are needed to meet the requirements in 10 CFR 35.290(b), 35.396, 35.390(b)(1)(ii)(G) or 35.690(c).)

For an individual qualifying under 10 CFR Part 35, Subparts D, E, F, G, and/or H, who is board-certified:

- A copy of the certification(s) by a specialty board(s) whose certification process has been recognized[4] by the NRC under 10 CFR Part 35, Subpart D, E, F, G, or H, as applicable to the use requested;

AND

- For a physician with a board certification recognized under 10 CFR 35.390, a description of the supervised work experience administering dosages of radioactive drugs required in 10 CFR 35.390(b)(1)(ii)(G) demonstrating that the proposed AU is qualified for the types of administrations for which authorization is sought;

AND

- For a physician with a board certification recognized under 10 CFR 35.390 for medical uses described in 10 CFR 35.200, a description of the supervised work experience eluting generator systems required in 10 CFR 35.290(c)(1)(ii)(G) demonstrating that the proposed AU is also qualified for imaging and localization medical uses;

AND

- For a physician with a board certification recognized under 10 CFR 35.490 or 10 CFR 35.690 for medical uses described in 10 CFR 35.396, a description of the training and supervised work experience and a copy of the attestation required in 10 CFR 35.396(d) to demonstrate qualifications for administering parenteral administrations of unsealed byproduct material requiring a written directive;

AND

- For an individual seeking authorization under 10 CFR Part 35, Subpart H, a description of the training specified in 10 CFR 35.690 (c) demonstrating that the proposed AU is qualified for the type(s) of use for which authorization is sought;

AND

- A written attestation, signed by a preceptor physician AU, that the training and experience specified for certification have been satisfactorily completed and that a level of competency sufficient to function independently as an AU for the medical uses authorized has been achieved. For individuals seeking authorization under 10 CFR 35.390, 10 CFR 35.396, and 10 CFR 35.690, the attestation must also include successful completion of the clinical case work in 10 CFR 35.390(b)(1)(ii)(G), or training and experience required by 10 CFR 35.396(d), or training for 10 CFR 35.600 types of use, as appropriate;

AND

---

[4] The names of board certifications that have been recognized by the NRC or an Agreement State are posted on the NRC's Web site http://www.nrc.gov/materials/miau/med-use-toolkit.html.

- If applicable, a description of recent related continuing education and experience as required by 10 CFR 35.59.

For an individual qualifying under 10 CFR Part 35, Subparts D, E, F, G, and/or H, who is not board-certified:

- A description of the training and experience identified in 10 CFR Part 35, Subparts D, E, F, G, and H, demonstrating that the proposed AU is qualified by training and experience for the use(s) requested;

AND

- For an individual seeking authorization under 10 CFR Part 35, Subpart H, a description of the training specified in 10 CFR 35.690(c), demonstrating that the proposed AU is qualified for the type(s) of use for which authorization is sought;

AND

- A written attestation, signed by a preceptor physician AU, that the above training and experience have been satisfactorily completed and that a level of competency sufficient to function independently as an AU for the medical uses authorized has been achieved;

AND

- If applicable, a description of recent related continuing education and experience as required by 10 CFR 35.59.

Notes:

- NRC Form 313A (AUD), "Authorized User Training and Experience and Preceptor Attestation (for uses defined under 35.100, 35.200, and 35.500) [10 CFR 35.190, 35.290, and 35.590]"; or NRC Form 313A (AUT), "Authorized User Training and Experience and Preceptor Attestation (for uses defined under 35.300) [10 CFR 35.390, 35.392, 35.394, and 35.396]"; or NRC Form 313A (AUS), "Authorized User Training and Experience and Preceptor Attestation (for uses defined under 35.400 and 35.600) [10 CFR 35.490, 35.491, and 35.690]" may be used as appropriate to document training and experience for those individuals qualifying under 10 CFR Part 35, Subparts D, E, F, G, and/or H.

- Licensees must notify the NRC within 30 days if an AU permanently discontinues his or her duties under the license or has a name change under 10 CFR 35.14.

- Descriptions of training and experience will be reviewed using the criteria listed above. The NRC will review the documentation to determine if the applicable criteria in 10 CFR Part 35 are met. If the training and experience do not appear to meet the 10 CFR Part 35 criteria, the NRC may request additional information from the applicant or may request the assistance of the ACMUI in evaluating such training and experience.

Note to reviewers: Licenses will reflect any limitations on use for listed AUs (e.g., whether administrations in excess of 33 mCi of iodine-131 are allowed and specific uses under 10 CFR 35.600).

AU for Nonmedical Uses: Provide the following:

- Name of the proposed nonmedical use AU,

- Description of types, quantities, and proposed nonmedical uses for which the individual is responsible, and

- Description of individual's educational and radiation safety training and experience with the types of materials and uses requested. This may include:

  — A copy of the NRC or Agreement State License listing the individual as an AU for the same types, quantities, and uses requested.

  — A permit issued by a Master Materials License licensee or broad-scope licensee or broad-scope permittee identifying the individual as an AU for the types, quantities, and uses requested.

Note: Authorized nonmedical use or uses that do not involve the intentional exposure of humans (e.g., *in vitro* and animal research, calibration, dosimetry research) will be reviewed on a case-by-case basis.

8.13 ITEM 7: AUTHORIZED NUCLEAR PHARMACIST (ANP)

Part 35	Applicability
100	✓
200	✓
300	✓
400	
500	
600	
1000	✓

Regulations: 10 CFR 30.33(a)(3), 10 CFR 32.72(b)(2), 10 CFR 35.2, 10 CFR 35.11, 10 CFR 35.14, 10 CFR 35.27, 10 CFR 35.55, 10 CFR 35.57, 10 CFR 35.59.

Criteria: Training and experience requirements for ANPs are described in 10 CFR 35.55.

Discussion: At many licensed medical facilities, an ANP is directly involved with the preparation of radiopharmaceuticals under the provisions of 10 CFR 35.100(b), 35.200(b), or 35.300(b). This may include the production of PET radioactive drugs under the provisions of 10 CFR 30.32(j).

Technologists, or other personnel, may prepare byproduct material for medical use under an ANP's supervision in accordance with 10 CFR 35.27, "Supervision," and in compliance with applicable FDA, other Federal, and State requirements (10 CFR 35.7). (Preparation of byproduct material for medical use may also be performed under the supervision of a physician who is an AU.)

Applicants are reminded that the recentness of training requirements described in 10 CFR 35.59 also apply to training and experience requirements in 10 CFR Part 35, Subpart B. Specifically, nuclear pharmacist applicants must have successfully completed the applicable training and experience criteria described in 10 CFR Part 35 within 7 years preceding the date of the application. Alternatively, nuclear pharmacist applicants must have had related continuing education and experience since initially completing the required training and experience. This time provision applies to board certification as well as to other training pathways for meeting requirements for training and experience.

In implementing the EPAct, the NRC "grandfathered" nuclear pharmacists using only accelerator-produced radioactive materials, discrete sources of Ra-226, or both, in the practice of nuclear pharmacy for the uses performed before or under the NRC waiver of August 31, 2005. These individuals do not have to meet the requirements of 10 CFR 35.59 or 10 CFR 35.55. The applicant must, however, document that the individual meets the criteria in 10 CFR 35.57(a)(3). Section 35.57 also provides that nuclear pharmacists who met certain criteria will qualify as ANPs for those materials and uses performed before or under NRC's waiver of August 31, 2005.

Response from Applicant: Provide the following:

- Name of the proposed ANP;

AND

- Pharmacist's license number and issuing entity;

AND

For an individual previously identified as an ANP on an NRC or Agreement State license or permit or by a commercial nuclear pharmacy that has been authorized to identify ANPs:

- Previous license number (if issued by the NRC) or a copy of the license (if issued by an Agreement State) or a copy of a permit issued by an NRC master materials licensee, a permit issued by an NRC or Agreement State broad-scope licensee, or a permit issued by an NRC Master Materials License broad-scope permittee on which the individual was named an ANP or a copy of an authorization as an ANP from a commercial nuclear pharmacy that has been authorized to identify ANPs.

OR

For an individual qualifying under 10 CFR 35.57(a)(3):

- Documentation that the nuclear pharmacist used only accelerator-produced radioactive material, discrete sources of Ra-226, or both, in the practice of pharmacy before or during the effective period of NRC's waiver of August 31, 2005;

AND

- Documentation that the nuclear pharmacist used these materials for the same uses as requested.

OR

For an individual qualifying under 10 CFR 35.55(a):

- Copy of the certification of the specialty board whose certification process has been recognized[5] under 10 CFR 35.55(a);

AND

---

[5] The names of board certifications that have been recognized by the NRC or an Agreement State are posted on the NRC's Web site http://www.nrc.gov/materials/miau/med-use-toolkit.html.

- Written attestation, signed by a preceptor ANP, that training and experience required for certification have been satisfactorily completed and that a level of competency sufficient to function independently as an ANP has been achieved.

AND

- If applicable, description of recent related continuing education and experience as required by 10 CFR 35.59.

OR

For an individual qualifying under 10 CFR 35.55(b):

- Description of the training and experience specified in 10 CFR 35.55(b) demonstrating that the proposed ANP is qualified by training and experience;

AND

- Written attestation, signed by a preceptor ANP, that the above training and experience have been satisfactorily completed and that a level of competency sufficient to function independently as an ANP has been achieved;

AND

- If applicable, description of recent related continuing education and experience as required by 10 CFR 35.59.

Notes:

- NRC Form 313A (ANP), "Authorized Nuclear Pharmacist Training and Experience and Preceptor Attestation [10 CFR 35.55]" may be used to document training and experience for those individuals qualifying under 10 CFR 35.55.

- Under 10 CFR 35.14, licensees must notify the NRC within 30 days if an ANP permanently discontinues his or her duties under the license or has a name change.

- Descriptions of training and experience will be reviewed using the criteria listed above. The NRC will review the documentation to determine if the applicable criteria in 10 CFR Part 35, Subpart B, are met. If the training and experience do not appear to meet the criteria in Subpart B, the NRC may request additional information from the applicant or may request the assistance of the ACMUI in evaluating such training and experience.

8.14 ITEM 7: AUTHORIZED MEDICAL PHYSICIST (AMP)

Part 35	Applicability
100	
200	
300	
400	✓
500	
600	✓
1000	✓

Regulations: 10 CFR 30.33(a)(3), 10 CFR 35.2, 10 CFR 35.14, 10 CFR 35.51, 10 CFR 35.57, 10 CFR 35.59, 10 CFR 35.433.

Criteria: Training and experience requirements for AMPs are described in 10 CFR 35.51.

Discussion: While the AMP may not administer the dose, at licensed medical facilities conducting radiation therapy treatments, an AMP is directly involved with the calculation and other tasks associated with the administration of the radiation dose. The American Association of Physicists in Medicine (AAPM) suggests that a medical physicist limit his or her involvement in radiation therapy to areas for which he or she has established competency.

Applicants are reminded of recentness of training requirements described in 10 CFR 35.59. Specifically, medical physicist applicants must have successfully completed the applicable training and experience criteria described in 10 CFR Part 35 within 7 years preceding the date of the application. Alternatively, medical physicist applicants must have had related continuing education and experience since completing the required training and experience. This time provision applies to board certification as well as to other training pathways for meeting requirements for training and experience.

In implementing the EPAct, the NRC "grandfathered" medical physicists using only accelerator-produced radioactive materials, discrete sources of Ra-226, or both, for medical uses performed before or under the NRC waiver of August 31, 2005. These individuals do not have to meet the requirements of 10 CFR 35.59 or 10 CFR 35.51. The applicant must, however, document that the individual meets the criteria in 10 CFR 35.57(a)(3). Section 35.57 also provides that medical physicists who met certain criteria will qualify as AMPs for those materials and uses performed before or under NRC's waiver of August 31, 2005. *Note:* Although there may be a number of medical physicists working with manual brachytherapy sources during the waiver, the NRC only requires AMPs for the medical use of strontium-90 eye applicators, teletherapy units, remote afterloader units, and gamma stereotactic radiosurgery units. Because none of these devices are known to contain only NARM material, the NRC expects few, if any, medical physicists to meet the criteria in 10 CFR 35.57 of an AMP.

Response from Applicant: Provide the following:

- Name of the proposed AMP.

AND

For an individual previously identified as an AMP on an NRC or Agreement State license or permit:

- Previous license number (if issued by the NRC) or a copy of the license (if issued by an Agreement State) or a copy of a permit issued by an NRC master materials licensee, a permit issued by an NRC or Agreement State broad-scope licensee, or a permit issued by an NRC Master Materials License broad-scope permittee on which the individual was specifically named an AMP for the uses requested.

OR

For an individual qualifying under 10 CFR 35.57(a)(3):

- Documentation that the medical physicist used only accelerator-produced radioactive material, discrete sources of Ra-226, or both, for medical uses before or during the effective period of NRC's waiver of August 31, 2005;

AND

- Documentation that the medical physicist used these materials for the same medical uses as requested.

OR

For an individual qualifying under 10 CFR 35.51(a):

- Copy of the certification(s) of the specialty board(s) whose certification process has been recognized[6] under 10 CFR 35.51(a);

AND

- Description of the training and experience specified in 10 CFR 35.51(c) demonstrating that the proposed AMP is qualified by training in the types of use for which he or she is requesting AMP status, including hands-on device operation, safety procedures, clinical use, and operation of a treatment planning system;

AND

- Written attestation, signed by a preceptor AMP, that the required training and experience required for certification, as well as the required training in 10 CFR 35.51(c) for the types of uses specified, have been satisfactorily completed and that a level of competency sufficient to function independently as an AMP has been achieved;

AND

- If applicable, a description of recent related continuing education and experience as required by 10 CFR 35.59.

OR

For an individual qualifying under 10 CFR 35.51 (b):

- Description of the training and experience demonstrating that the proposed AMP is qualified by training and experience identified in 10 CFR 35.51(b)(1) for the uses requested;

AND

- Description of the training and experience specified in 10 CFR 35.51(c) demonstrating that the proposed AMP is qualified by training in the types of use for which the licensee seeks approval of an individual as AMP, including hands-on device operation, safety procedures, clinical use, and operation of a treatment planning system;

AND

- Written attestation, signed by a preceptor AMP, that the training and experience required in 10 CFR 35.51(b)(1), as well as the training in 10 CFR 35.51(c) for the types of use specified, have been satisfactorily completed and that a level of competency sufficient to function independently as an AMP has been achieved;

AND

---

[6] The names of board certifications that have been recognized by the NRC or an Agreement State are posted on the NRC's Web site http://www.nrc.gov/materials/miau/med-use-toolkit.html.

- If applicable, a description of recent related continuing education and experience as required by 10 CFR 35.59.

Notes:

- NRC Form 313A (AMP), "Authorized Medical Physicist Training and Experience and Preceptor Attestation [10 CFR 35.51]," may be used to document training and experience for those individuals qualifying under 10 CFR 35.51.

- Under 10 CFR 35.14, licensees must notify NRC within 30 days if an AMP permanently discontinues his or her duties under the license or has a name change.

- Descriptions of training and experience will be reviewed using the criteria listed above. The NRC will review the documentation to determine if the applicable criteria in 10 CFR Part 35, Subpart B, are met. If the training and experience do not appear to meet the criteria in Subpart B, the NRC may request additional information from the applicant or may request the assistance of the ACMUI in evaluating such training and experience.

8.15   ITEM 9: FACILITIES AND EQUIPMENT

Regulations: 10 CFR 30.33(a)(2), 10 CFR 35.12(b)(1), 10 CFR 35.18(a).

Criteria: Facilities and equipment must be adequate to protect health and minimize danger to life or property.

Part 35	Applicability
100	✓
200	✓
300	✓
400	✓
500	✓
600	✓
1000	✓

Discussion: Requirements to provide information about the design and construction of facilities and safety equipment are contained in 10 CFR 30.33(a)(2), 35.12(b)(1), and 35.18(a). Applications will be approved if, among other things, "the applicant's proposed equipment and facilities are adequate to protect health and minimize danger to life or property." Facility and equipment requirements depend on the scope of the applicant's operations (e.g., planned use of the material, types of radioactive emissions, quantity and form of radioactive materials possessed, production of PET radioactive drugs under 30.32(j) authorization). Applicants should focus particularly on operations using large quantities of radioactive materials; preparation steps involving liquids, gases, and volatile radioactive materials; and the use of alpha-emitters, high-energy photon-emitters, and high-energy beta-emitters.

Response from Applicant: Refer to Sections 8.16 through 8.20 for guidance.

8.16   ITEM 9: FACILITY DIAGRAM

Regulations: 10 CFR 20.1003, 10 CFR 20.1101, 10 CFR 20.1201, 10 CFR 20.1301, 10 CFR 20.1302, 10 CFR 20.1601, 10 CFR 20.1602, 10 CFR 20.1901, 10 CFR 20.1902, 10 CFR 20.2102, 10 CFR 30.33(a)(2),

Part 35	Applicability
100	✓
200	✓
300	✓
400	✓
500	✓
600	✓
1000	✓

10 CFR 35.12, 10 CFR 35.14, 10 CFR 35.18(a)(3), 10 CFR 35.75, 10 CFR 35.315(a), 10 CFR 35.415, 10 CFR 35.615.

Criteria: In order to issue a license, the NRC must find that facilities and equipment must be adequate to protect health and minimize danger to life or property as required under 10 CFR 30.33(a) and/or 35.18(a).

Discussion: Applicants must describe the proposed facilities and equipment as required by 10 CFR 30.33(a)(2) and 10 CFR 35.12. The facility diagram should include the room or rooms and adjacent areas where byproduct material is prepared, used, administered, and stored, at a level of detail that is sufficient to demonstrate that the facilities and equipment are adequate to protect health and minimize danger to life or property.

Drawings and diagrams that provide the exact location of materials or depict specific locations of safety or security equipment should be marked as "security-related information – withhold under 10 CFR 2.390." (See Section 5.2.)

If the applicant receives PET radionuclides from either an offsite or onsite PET radionuclide production facility by direct transfer tube to a PET radioactive drug production area, the facility diagram should include the direct transfer tube as well as a diagram of the PET radioactive production area.

For types of use permitted by 10 CFR 35.100 and 35.200, applicants should provide room numbers for areas in which byproduct materials are used or prepared for use (i.e., "hot labs"). (See Figure 8.1 for a sample attachment to 9.1.) If the applicant has a radionuclide delivery line from a PET radionuclide/radioactive drug production area in the 10 CFR 35.100 or 35.200 medical use area, a description of the room, location, and delivery line should be provided. A discussion of the shielding associated with the delivery line, including shielding calculations, should also be provided.

Attachment 9.1
SECURITY-RELATED INFORMATION – WITHHOLD UNDER 10 CFR 2.390*

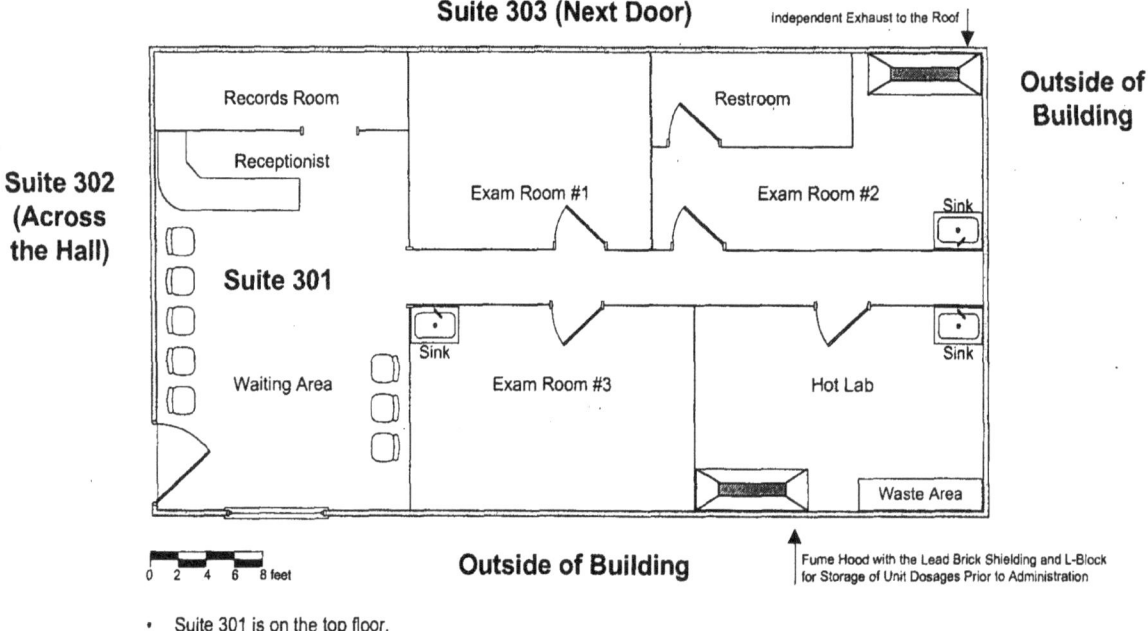

SECURITY-RELATED INFORMATION – WITHHOLD UNDER 10 CFR 2.390*

- Suite 301 is on the top floor.
- Suite 301 is located at a corner of the building.
- Suite 302 is occupied by an accounting firm.
- Suite 303 is occupied by a law firm.
- Directly below Suite 301 is an insurance company.

1556-095.ppt
10142002

*For the purposes of this NUREG, the facility diagram is marked appropriately for an application. This particular diagram does not contain real security-related information.

Figure 8.1   Facility Diagram for Nuclear Medicine Suite

Most applicants requesting the use of PET radioactive drugs will designate an area or room as a "quiet room" where patients wait after the PET radioactive drug is administered. This room should be included in the facility diagram. The location and design of the "quiet room" should be considered when implementing the ALARA requirements in 10 CFR 20.1101. The applicable public dose limits are discussed in Section 8.33 of this document.

When information regarding an area or room is provided, adjacent areas and rooms, including those above and below, should be described. For types of use permitted by 10 CFR 35.300 and 35.400, applicants should provide the above information and, in addition, they should provide the locations where sources are stored. Describe the rooms where patients will be housed if they cannot be released under 10 CFR 35.75. The discussion should include a description of shielding, if applicable. For types of use permitted by 10 CFR 35.500, the applicant should provide the room numbers of use.

For types of use permitted by 10 CFR 35.600, and production of PET radioactive drugs, the applicant should provide all of the information discussed above and the shielding calculations for the facility as described in the diagram. Applicants should also describe the equipment used in the PET radioactive drug production area (e.g., hot cells, remote manipulation devices in the hot cells, equipment and/or method used to physically transfer PET radionuclides during the chemical synthesis, "real-time" effluent (stack) monitoring equipment). When preparing applications for use under 10 CFR 35.1000, applicants should review the above to determine the type of information appropriate to evaluate the adequacy of the facilities.

All limited specific medical use licensees are required by 10 CFR 35.13 to obtain a license amendment before adding to or changing an area of use identified in the application or on the license. This includes additions and relocations of areas where PET radionuclides are produced or additions and locations of a radionuclide/radioactive drug delivery line from the PET radionuclide production area to a 10 CFR 35.100 or a 35.200 medical use area. However, other changes and additions to the 10 CFR 35.100 and 35.200 medical use areas do not require a license amendment and can be made, provided NRC is notified as required by 10 CFR 35.14 within 30 days following the changes. The broad-scope medical use licensee does not have to notify NRC of changes that do not require a license amendment.

Regulatory requirements, the principle of ALARA, good medical care, and access control should be considered when determining the location of the therapy patient's room or a therapy treatment room.

The applicant should demonstrate that the limits specified in 10 CFR 20.1301(a) will not be exceeded. If the calculations demonstrate that these limits cannot be met, indicate any further steps that will be taken to limit exposure to individual members of the public. The applicant may consider the following options:

- Adding shielding to the barrier in question, with corresponding modification of the facility description if necessary.

  Note: If applicants are proposing to use portable shielding to protect health and minimize danger to life or property, they should describe the alternative equipment and administrative procedures they propose to use for evaluation and approval by NRC. If applicants elect to use portable shielding, they should commit to having administrative procedures to control configuration management to maintain dose within regulatory limits.

- Requesting prior NRC authorization to operate up to an annual dose limit for an individual member of the public of 5 mSv (0.5 rem) and demonstrating that the requirements of 10 CFR 20.1301 will be met. The applicant must demonstrate the need for and the expected duration of operations that will result in an individual dose in excess of the limits specified in 10 CFR 20.1301(a). A program to assess and control dose within the 5 mSv (0.5 rem) annual limit and procedures to be followed to maintain the dose ALARA (10 CFR 20.1101) must be developed (see 10 CFR 20.1301(d)).

If radiopharmaceutical therapy and brachytherapy patient rooms are added after the initial license is issued, additional room diagrams should be submitted if the room design (including shielding) and the occupancy of adjacent areas are significantly different from the original diagrams provided. A written description should be submitted for simple changes.

For teletherapy units, it may be necessary to restrict use of the unit's primary beam if the treatment room's walls, ceiling, or floor will not adequately shield adjacent areas from direct or scattered radiation. Electrical, mechanical, or other physical means (rather than administrative controls) must be used to limit movement or rotation of the unit (e.g., electrical or mechanical stops). Some applicants have found it helpful to have a sample response for guidance. The following is an example of an acceptable response on the use of a rotational unit with an integral beam absorber (also called a beam catcher).

- "For the primary beam directed toward the integral beam absorber, electrical or mechanical stops are set so that the primary beam must be centered (within plus or minus 2 degrees) on the integral beam absorber and, in that configuration, the attenuated primary beam may be rotated 360 degrees pointing toward the floor, east wall, ceiling, and west wall."

- "For the primary beam directed away from the integral beam absorber, electrical or mechanical stops permit the unattenuated primary beam to be directed in a 95-degree arc from 5 degrees toward the west wall to vertically down toward the floor to 90 degrees toward the east wall."

Experience has shown that, given this type of example, many applicants can make changes to accommodate their own situations (e.g., use of a vertical unit, use of a rotational unit without an integral beam absorber).

Response from Applicant: All medical use applicants, including broad-scope medical use applicants, are required to provide facility diagrams. The applicant should follow the guidance in Section 5.2 to determine if the response includes security-related sensitive information and needs to be marked accordingly. Provide the following on the facility diagrams:

- Drawings should be to scale, and the scale used should be indicated;

- Location, room numbers, and principal use of each room or area where byproduct material is prepared, used or stored; location of direct transfer delivery tubes from a PET radionuclide/radioactive drug production facility, or production area of PET radioactive drugs under 10 CFR 30.32(j), as provided above under the heading "Discussion"; and areas where higher energy gamma-emitting radionuclides (e.g., PET radionuclides) are used, including a "quiet room";

- Location, room numbers, and principal use of each adjacent room (e.g., office, file, toilet, closet, hallway), including areas above, beside, and below therapy treatment rooms, indicating whether the room is a restricted or unrestricted area as defined in 10 CFR 20.1003; and

- Shielding calculations, including information about the type, thickness, and density of any necessary shielding to enable independent verification of shielding calculations, and a description of any portable shields used (e.g., shielding of proposed patient rooms used for implant therapy, including the dimensions of any portable shield, if one is used; source storage safe; shielding for PET radionuclide direct transfer tubes; PET radioactive drug production areas).

In addition to the above, for teletherapy and GSR facilities, applicants should provide the directions of primary beam usage for teletherapy units and, in the case of an isocentric unit, the plane of beam rotation.

References: National Council on Radiation Protection and Measurements (NCRP) Report 49, "Structural Shielding Design and Evaluation for Medical Use of X-Rays and Gamma Rays of Energies up to 10 MeV"; Report 102, "Medical X-Ray, Electron Beam and Gamma Ray Protection for Energies up to 50 MeV (Equipment Design, Performance and Use)"; and Report 40, "Protection Against Radiation from Brachytherapy Sources," may be helpful in responding to the items above. In addition, NUREG/CR-6276, "Quality Management in Remote Afterloading Brachytherapy," and NUREG/CR-6324, "Quality Assurance for Gamma Knives," may also be helpful in responding to the items above. However, it should be noted that references to 10 CFR Part 35 in the NUREGs may be outdated because the rule was amended after these documents were published.

8.17   ITEM 9: RADIATION MONITORING INSTRUMENTS

Regulations: 10 CFR 20.1101, 10 CFR 20.1501, 10 CFR 20.2102, 10 CFR 20.2103(a), 10 CFR 30.33(a)(2), 10 CFR 35.27, 10 CFR 35.61, 10 CFR 35.2061.

Part 35	Applicability
100	✓
200	✓
300	✓
400	✓
500	✓
600	✓
1000	✓

Criteria: All licensees shall possess calibrated radiation detection and measuring instruments that will be used for radiation protection, including survey and monitoring instruments and quantitative measuring instruments needed to monitor the adequacy of radioactive materials containment and contamination control.

Discussion: The Radiation Protection Program that licensees are required to develop, document, and implement in accordance with 10 CFR 20.1101 must include provisions for survey instrument calibration (10 CFR 20.1501). Licensees shall possess instruments used to measure radiation levels, radioactive contamination, and radioactivity, as applicable. Instruments used for quantitative radiation measurements must be calibrated for the radiation measured. The instruments should be available for use at all times when byproduct material is in use. The licensee should possess survey instruments sufficiently sensitive to measure the type and energy of radiation used, including survey instruments used to locate low-energy or low-activity seeds (e.g., I-125, Pd-103) if they become dislodged in the operating room or patient's room.

For the purposes of this document, radiation monitoring instruments are defined as any device used to measure the radiological conditions at a licensed facility. Some of the instruments that may be used to perform the above functions include:

- Portable or stationary count rate meters,
- Portable or stationary dose rate or exposure rate meters,
- Area Monitors,
- Single or multichannel analyzers,
- Liquid Scintillation Counters (LSC),
- Gamma counters,

- Proportional counters,

- Solid state detectors, and

- Hand and foot contamination monitors.

Usually, it is not necessary for a licensee to possess a survey meter solely for use during sealed-source diagnostic procedures, unless the diagnostic study involves localization of radioactive seeds, since it is not expected that a survey will be performed each time such a diagnostic study is performed. In these cases, it is acceptable for the meter to be available on short notice in the event of an accident or malfunction that could reduce the shielding of the sealed source(s). Surveys may be required to verify source integrity of the diagnostic sealed source and to ensure that dose rates in unrestricted areas and public and occupational doses are within regulatory limits. For localization studies using sealed sources, survey meters are needed to verify source integrity and assist in source accountability.

Survey meter calibrations must be performed by persons, including licensed personnel, who are qualified to perform calibrations. One method a licensee may use to determine if the service is qualified to perform these activities is to determine that it has an NRC (or an equivalent Agreement State) license. Alternatively, an applicant may choose to develop, implement, and maintain procedures to ensure instruments are calibrated, or propose an alternate method for calibration.

Appendix K provides guidance regarding appropriate instrumentation and model survey instrument calibration procedures to meet the requirements detailed in 10 CFR 35.61.

Response from Applicant: Provide the following:

- A statement that: "Radiation monitoring instruments will be calibrated by a person qualified to perform survey meter calibrations."

AND/OR

- A statement that: "We have developed and will implement and maintain written survey meter calibration procedures in accordance with the requirements in 10 CFR 20.1501 and that meet the requirements 10 CFR 35.61."

AND

- A description of the instrumentation (e.g., gamma counter, solid state detector, portable or stationary count rate meter, portable or stationary dose rate or exposure rate meter, single or multichannel analyzer, liquid scintillation counter, proportional counter) that will be used to perform required surveys.

AND

- A statement that: "We reserve the right to upgrade our survey instruments as necessary as long as they are adequate to measure the type and level of radiation for which they are used."

Note: If calibrations will not be performed by the licensee or by a person qualified to perform survey meter calibrations, the applicant should propose an alternate method of calibration for review by NRC.

References: See the Notice of Availability on the inside front cover of this report to obtain a copy of NUREG-1556, Volume 18, "Consolidated Guidance About Materials Licenses: Program-Specific Guidance About Service Provider Licenses," dated November 2000.

8.18 ITEM 9: DOSE CALIBRATOR AND OTHER EQUIPMENT USED TO MEASURE DOSAGES OF UNSEALED BYPRODUCT MATERIAL

Part 35	Applicability
100	✓*
200	✓*
300	✓*
400	
500	
600	
1000	✓*

*If applicant will measure patient dosages or use other than unit dosages.

Regulations: 10 CFR 30.3, 10 CFR 30.33, 10 CFR 30.34(j), 10 CFR 35.27, 10 CFR 35.41, 10 CFR 35.60, 10 CFR 35.63, 10 CFR 35.2060, 10 CFR 35.2063.

Criteria: In 10 CFR 35.60 and 10 CFR 35.63, the NRC describes requirements for the use, possession, calibration, and check of instruments (e.g., dose calibrators) used to measure patient dosages. Section 10 CFR 30.34(j) contains requirements for possession, calibration, and check of instruments used to measure dosages of PET radioactive drugs noncommercially transferred to other members of the consortium.

Discussion: If the licensee produces PET radioactive drugs for noncommercial distribution to other consortium members under 10 CFR 30.32(j), the licensee is required by 10 CFR 30.34(j) to possess and calibrate all instruments used for measuring dosages (see Appendix AA).

As described in 10 CFR 35.63, dosage measurement is required for licensees who prepare patient dosages.

- If the licensee uses only unit dosages made by a manufacturer or preparer licensed under 10 CFR 32.72, or a PET radioactive drug producer authorized under 10 CFR 30.32(j) (and does not split, combine, or otherwise modify unit dosages), the licensee is not required to possess an instrument to measure the dosage. Furthermore, licensees may rely on the provider's dose label for the measurement of the dosage and decay-correct the dosage to the time of administration.

- If the licensee performs direct measurements of dosages in accordance with 10 CFR 35.63 (e.g., prepares its own dosages, breaks up unit dosages for patient administration, or decides to measure unit dosages), the licensee is required to possess and calibrate all instruments used for measuring patient dosages.

Equipment used to measure dosages must be calibrated in accordance with nationally recognized standards (e.g., ANSI) or the manufacturer's instructions. The measurement equipment may be a well ion chamber, a liquid scintillation counter, etc., as long as the instrument can be calibrated appropriately for the type and energy of radiation emitted and is both accurate and reliable.

For other than unit dosages, the activity must be determined by direct measurement, by a combination of radioactivity measurement and mathematical calculation, or by a combination of volumetric measurement and mathematical calculation. However, there are inherent technical difficulties to overcome. For beta-emitting radionuclides, these difficulties include dependence on geometry, lack of an industry standard for materials used in the manufacture of vials and syringes, and lack of a NIST-traceable standard for some radionuclides used. For instance, when determining the dosage of P-32, assays with a dose calibrator may result in inaccuracies caused by inherent variations in geometry; therefore, a volumetric measurement and mathematical calculation may be more accurate. Licensees must assay patient dosages in the same type of vial and geometry as used to determine the correct dose calibrator settings. Using different vials or syringes may result in measurement errors due, for example, to the variation of bremsstrahlung created by interaction between beta particles and the differing dosage containers. Licensees are reminded that beta emitters should be shielded using a low-atomic-numbered material to minimize the production of bremsstrahlung. When a high-activity source is involved, consideration should be given to adding an outer shield made from material with a high atomic number to attenuate bremsstrahlung.

The inherent technical difficulties in measuring alpha-emitting radionuclides are even greater than those of measuring beta emissions. In the absence of an additional photon, gamma, or beta particle emission that can be measured and quantified in relation to the alpha particle emissions, most alpha measuring instruments (e.g., gas proportional counters and liquid scintillation counters) will require preparation and measurement of an aliquot of the unsealed byproduct material. Measurement of aliquots introduces additional uncertainties associated with removing precise and reproducible volumes from homogeneous samples. To avoid these difficulties, the best method is to use unit dosages and the manufacturer's or commercial nuclear pharmacy's dose label for measurement of the dosage and decay-correct the dosage to the time of administration. These difficulties can also be avoided when not using unit dosages by relying on the provider's dose label for measurement of the radioactivity and a combination of volumetric measurement and mathematical calculation.

Response from Applicant: If applicable.

For the administration of gamma- and beta-emitting unsealed byproduct materials, provide the following:

- A statement that: "Equipment used to measure dosages will be calibrated in accordance with nationally recognized standards or the manufacturer's instructions."

For the administration of the alpha-emitting unsealed byproduct material in other than unit dosages made by a manufacturer or preparer licensed under 10 CFR 32.72, provide the following:

- A statement that: "Dosages will be determined relying on the provider's dose label for measurement of the radioactivity and combination of volumetric measurement and mathematical calculation."

OR

- A description of the equipment used to measure the dosages. Identify the nationally recognized standard used to calibrate the instrument or provide a copy of the manufacturer's instructions to calibrate the instrument for the alpha-emitters measured, and provide a description of the procedures to be followed when measuring the dosage.

8.19 ITEM 9: THERAPY UNIT — CALIBRATION AND USE

Regulations: 10 CFR 30.33(a)(2), 10 CFR 35.27, 10 CFR 35.432, 10 CFR 35.630, 10 CFR 35.632, 10 CFR 35.633, 10 CFR 35.635, 10 CFR 35. 642, 10 CFR 35.643, 10 CFR 35.645, 10 CFR 35.2432, 10 CFR 35.2630, 10 CFR 35.2632, 10 CFR 35.2642, 10 CFR 35.2643, 10 CFR 35.2645.

Part 35	Applicability
100	
200	
300	
400	✓*
500	
600	✓*
1000	✓

* Special requirements re: brachytherapy and LDR afterloader sources and Sr-90 sources.

Criteria: The above regulations contain NRC requirements, including recordkeeping requirements, for verification and periodic spot-checks of source activity or output. To perform these measurements, the applicant must possess appropriately calibrated dosimetry equipment. For manual brachytherapy sources and low dose-rate (LDR) remote afterloader sources, licensees may use source activity or output determined by the manufacturer, provided that the manufacturer's measurements meet applicable requirements.

Discussion: Except for manual brachytherapy sources and LDR remote afterloader sources, where the source output or activity is determined by the manufacturer in accordance with 10 CFR Part 35, the applicant must possess a calibrated dosimetry system (e.g., Farmer chamber, electrometer, well-type ionization chamber) that will be used to perform calibration measurements of sealed sources to be used for patient therapy. Dosimetry systems and/or sealed sources used to calibrate the licensee's dosimetry systems must be traceable to NIST or to a laboratory accredited by AAPM, pursuant to 10 CFR 35.630. The licensee must maintain records of calibrations of dosimetry equipment for the duration of the license.

The licensee's AMP must perform full calibrations of sealed sources and devices used for therapy in accordance with published protocols currently accepted by nationally recognized bodies (e.g., AAPM, ACR, ANSI). (Note: Calibration by an AMP is not required for manual brachytherapy sources, except for calculating the activity of strontium-90 sources.) The licensee's AMP must calculate the activity of each strontium-90 source that is used to determine the treatment times for ophthalmic treatments. In addition, the licensee must perform spot-check measurements of sealed sources and devices used for therapy in accordance with written procedures established by the AMP (10 CFR 35.642, 10 CFR 35.643, and 10 CFR 35.645). Calibration procedures described by the AAPM or any published protocol approved by a nationally recognized body, as applicable, may be used.

The calibration procedures should address, in part, the method used to determine the exposure rate (or activity) under specific criteria (i.e., distances used for the measurement, whether the measurement is an "in air" measurement or done using a phantom configuration of the chamber with respect to the source(s) and device, scatter factors used to compute the exposure rate, etc.).

Full calibrations must be performed before first medical use[7], whenever spot-check measurements (if required) indicate that the output differs by more than 5% from the output obtained at the last full calibration corrected mathematically for decay, following replacement of the sources or reinstallation of the unit in a new location not previously described in the license, following any repairs of the unit that include removal of sealed sources or major repair of the components associated with the source exposure assembly, and at intervals as defined in 10 CFR 35.632, 10 CFR 35.633, and 10 CFR 35.635. Manual brachytherapy sources must be calibrated only initially, prior to use.

For sealed sources used in therapy, and in particular, for new types of use, licensees should select dosimetry equipment that will accurately measure the output or the activity of the source. Contact a licensing specialist at an NRC Regional Office for additional assistance.

Response from Applicant: Provide the following:

- The applicant must provide the procedures required by 10 CFR 35.642, 10 CFR 35.643, and 10 CFR 35.645, if applicable to the license application.

References:

- AAPM Task Group No. 21, "A Protocol for the Determination of Absorbed Dose from High-Energy Photon and Electron Beams";

- AAPM Task Group No. 40, "Comprehensive QA for Radiation Oncology," AAPM Report No. 54, "Stereotactic Radiosurgery";

- AAPM Task Group No. 56, "Code of Practice for Brachytherapy Physics."

Copies of these documents and many other documents from AAPM referenced in this guide may be obtained from Medical Physics Publishing (MPP), 4513 Vernon Boulevard, Madison, WI 53705-4964 or ordered electronically from http://www.medicalphysics.org.

8.20 ITEM 9: OTHER EQUIPMENT AND FACILITIES

Part 35	Applicability
100	✓
200	✓
300	✓
400	✓
500	✓
600	✓
1000	✓

Regulations: 10 CFR 20.1101, 10 CFR 20.1801, 10 CFR 30.33(a)(2), 10 CFR 30.34, 10 CFR 35.12, 10 CFR 35.315, 10 CFR 35.415, 10 CFR 35.457, 10 CFR 35.615, 10 CFR 35.647, 10 CFR 35.657.

Criteria: Facilities and equipment must be adequate to protect health and minimize danger to life or property.

---

[7] For brachytherapy sources, "first medical use" is defined as the first use following the effective date of the revised 10 CFR Part 35, October 24, 2002.

Discussion: The applicant should describe, in Item 9 of the application, other equipment and facilities available for safe use and storage of byproduct material listed in Item 5 of this application. This description should be identified as Attachment 9.4.

The applicant must describe additional facilities and equipment for PET radionuclide and radiopharmaceutical therapy programs to safely receive, use, store, and dispose of radioactive material. The applicant should focus on facilities to be used for radioactive drug therapy administration and patient accommodations (e.g., private room with private bath). The most widely used source of radiopharmaceutical therapy is I-131 sodium iodide. If the radionuclide is administered in volatile liquid form, it is important to place the patient dosage in a closed environment (e.g., a fume hood). Also note there are hazards associated with volatile iodine in pill form; applicants should consider this in establishing their radiological controls. When patients are treated with I-131 sodium iodide, sources of contamination include airborne I-131, urine, perspiration, saliva, and other secretions.

For **PET radionuclide use** and **PET radioactive drug production areas**, the applicant should focus on the need for (1) additional shielding, (2) hot cells containing remote handling devices, (3) other remote handling devices that may be needed when handling and storing the higher energy emissions of these materials, and (4) special delivery systems if the applicant prepares its own PET radionuclides or has them delivered by a direct transfer tube or system from a PET radionuclide producer. Applicants synthesizing PET radioactive drugs should also focus on volatility issues and releases.

For **teletherapy, GSR,** and **high dose-rate (HDR) facilities**, the licensee shall require any individual entering the treatment room to ensure, through the use of appropriate radiation monitors, that radiation levels have returned to ambient levels. One method of meeting the requirements of 10 CFR 35.615(c) is a beam-on radiation monitor permanently mounted in each therapy treatment room that is equipped with an emergency power supply separate from the power supply for the therapy unit. Such beam-on monitors can provide a visible indication (e.g., flashing light) of an exposed or partially exposed source. Applicants may propose an alternative to a permanently mounted monitor.

Section 10 CFR 35.615(d) requires that, except for LDR units, each licensee shall construct or equip each treatment room so as to permit continuous observation of the patient while the patient is in the treatment room. If a shielded viewing window will be used, the thickness, density, and type of material used should be specified. If a closed-circuit television system (or some other electronic system) will be used to view the patient, the backup system or procedure to be used in case the electronic system malfunctions should be specified, or the applicant must commit to suspending all treatments until the electronic system is repaired and functioning again. The communications system should allow the patient to communicate with the unit operator in the event of medical difficulties. An open microphone system can be used to allow communication without requiring a patient to move to activate controls.

The regulations require adequate equipment and controls to maintain exposures of radiation to workers ALARA and within regulatory limits. Section 10 CFR 35.615(b), in part, requires that each door leading into the treatment room be provided with an electrical interlock system to control the on-off mechanism of the therapy unit. The interlock system must cause the source(s) to be shielded if the door to the treatment room is opened when the source is exposed. The interlock system must also prevent the operator from initiating a treatment cycle unless the treatment room entrance door is closed. Further, the interlock must be wired so that the source(s) cannot be exposed after interlock interruption until the treatment room door is closed and the on-off control for the source(s) is reset at the console.

Due to the unique characteristics of **pulsed dose-rate (PDR) remote afterloaders** and the lack of constant surveillance of their operation, a more sophisticated alarm system is essential to ensure the patient is protected during treatment. In addition to the above, consider the following:

- The PDR device control console is *not* accessible to unauthorized personnel during treatment.

- A primary care provider checks the patient to ensure that the patient's device has not been moved, kinked, dislodged, or disconnected.

- A more sophisticated interlock/warning system is normally installed for PDR devices. This system should perform the following functions or possess the following characteristics:

  - The signal from the PDR device and the signal from the room radiation monitor should be connected in such a manner that an audible alarm sounds if the room monitor indicates the presence of radiation and the device indicates a "safe" or retracted position.

  - The alarm circuit should also be wired in such a manner that an audible alarm is generated for any device internal error condition that could indicate the unintended extension of the source. This would constitute a circuit that generates the audible alarm when either the "source retracted and radiation present" or the appropriate internal error condition(s) exists.

  - The "source safe and radiation present" signal should also be self-testing. If a "source not safe" input is received without a corresponding "radiation present" signal, the circuit should generate an interlock/warning circuit failure signal that will cause the source to retract. Reset this circuit manually before attempting to continue treatment.

  - The audible alarm should be sufficiently loud to be clearly heard by the facility's responsible device/patient monitoring staff at all times.

  - No provisions for bypassing this alarm circuit or for permanently silencing the alarm should be made to the circuit as long as the room radiation monitor is indicating the presence of radiation. If any circuitry is provided to mute the audible alarm, such circuitry should not mute the alarm for a period of more than 1 minute. Controls that disable this alarm circuit or provide for silencing the alarm for periods in excess of 1 minute should be prohibited.

If the alarm circuit is inoperative for any reason, licensees should prohibit further treatment of patients with the device until the circuit has been repaired and tested. If the alarm circuit fails during the course of a patient treatment, the treatment in progress may continue as long as continuous surveillance of the device is provided during each treatment cycle or fraction.

Applicants may submit information on alternatives to fixed shielding as part of their facility description. This information must demonstrate that the shielding will remain in place during the course of patient treatment.

For patient rooms where **LDR remote afterloader** use is planned, neither a viewing nor an intercom system is required. However, the applicant should describe how the patient and device will be monitored during treatment to ensure that the sources and catheter guide tube are not disturbed during treatment and to provide for prompt detection of any operational problems with the LDR device during treatment.

Response from Applicant: Follow the guidance in Section 5.2 to determine if the response to this section includes security-related sensitive information and needs to be marked accordingly.

For PET radionuclide use, PET radioactive drug production, and radiopharmaceutical therapy programs, describe the additional facilities and equipment for these uses.

For manual brachytherapy facilities, provide a description of the emergency response equipment.

For teletherapy, GSR, and remote afterloader facilities, provide a description of the following:

- Warning systems and restricted area controls (e.g., locks, signs, warning lights and alarms, interlock systems) for each therapy treatment room;

- Area radiation monitoring equipment;

- Viewing and intercom systems (except for LDR units);

- Steps that will be taken to ensure that no two units can be operated simultaneously, if other radiation-producing equipment (e.g., linear accelerator, X-ray machine) is in the treatment room;

- Methods to ensure that whenever the device is not in use or is unattended, the console keys will be inaccessible to unauthorized persons; and

- Emergency response equipment.

8.21 ITEM 10: RADIATION PROTECTION PROGRAM

Part 35	Applicability
100	✓
200	✓
300	✓
400	✓
500	✓
600	✓
1000	✓

Regulations: 10 CFR 20.1101, 10 CFR 20.2102, 10 CFR 30.33, 10 CFR 30.34(e), 10 CFR 35.24, 10 CFR 35.26, 10 CFR 35.610, 10 CFR 35.2024, 10 CFR 35.2026.

Criteria: The regulations in 10 CFR 20.1101 state that each licensee must develop, document, and implement a Radiation Protection Program commensurate with the scope of the licensed activity. The program must be sufficient to ensure compliance with the provisions of 10 CFR Part 20 regulations. The licensee is responsible for the conduct of all licensed activities and the acts and omissions of individuals handling licensed material.

Under 10 CFR 30.34(e), the NRC may incorporate into byproduct materials licenses, at the time of issuance or thereafter, additional requirements and conditions that it deems appropriate or necessary to protect health or to minimize danger to life and property. Licensee management's authorities and responsibilities for the Radiation Protection Program are described in 10 CFR 35.24, while 10 CFR 35.26 sets forth four circumstances in which the licensee may revise its Radiation Protection Program without NRC approval. For example, no NRC approval is required when the revision does not require a license amendment.

Discussion: Applicants/licensees must abide by all applicable regulations, develop, implement, and maintain procedures when required, and/or provide requested information about the proposed Radiation Protection Program during the licensing process. Tables C.1 and C.2 in Appendix C may be helpful in determining what information should be provided when requesting a license. If the licensee has authority for the production of PET radioactive drugs under 10 CFR 30.32(j), the radiation production program must include radiation safety issues associated with this nonmedical use.

Response from Applicant: Respond to subsequent sections of this document regarding Item 10 of the application.

8.22 ITEM 10: SAFETY PROCEDURES AND INSTRUCTIONS

Part 35	Applicability
100	
200	
300	
400	
500	
600	✓
1000	✓

Regulations: 10 CFR 30.34(j), 10 CFR 35.12(c)(2), 10 CFR 35.610, 10 CFR 35.642, 10 CFR 35.643, 10 CFR 35.645.

Criteria: When applying for authorization under 10 CFR 30.32(j) to produce PET radioactive drugs for noncommercial distribution to other medical use licensees in the consortium, the applicant must develop, document, and implement certain procedures. See Appendix AA for discussion and response from applicant.

Before using materials under 10 CFR 35.600, the applicant must develop, document, submit, and implement written safety procedures for emergency response. Section 10 CFR 35.610 requires, in part, that written procedures be developed, implemented, and maintained for responding to an abnormal situation involving a remote afterloader unit, a teletherapy unit, or a gamma stereotactic radiosurgery unit. The procedures needed to meet 10 CFR 35.610 must include:

- Instructions for responding to equipment failures and the names of the individuals responsible for implementing corrective actions,

- The process for restricting access to and posting of the treatment area to minimize the risk of inadvertent exposure, and

- The names and telephone numbers of AUs, AMPs, and the RSO to be contacted if the unit or console operates abnormally.

A copy of these procedures must be physically located at the therapy unit console. The instructions must inform the operator of procedures to be followed if the operator is unable to place the source(s) in the shielded position, or remove the patient from the radiation field with controls from outside the treatment room.

Discussion: The applicant must establish and follow written procedures for emergencies that may occur (e.g., a therapy source fails to retract or return to the shielded position, or a GSR couch fails to retract). A copy of the manufacturer's recommendations and instructions should be given to each individual performing therapy treatments or operating the therapy device. Practice drills, using nonradioactive (dummy) sources (when possible), must be practiced annually or more frequently, as needed. The drills should include dry runs of emergency procedures that cover stuck or dislodged sources and applicators (if applicable), and emergency procedures for removing the patient from the radiation field. Team practice may also be important for adequate emergency coordination for such maneuvers as removing a patient from a malfunctioning GSR unit and manual movement of the patient treatment table. These procedures, designed to minimize radiation exposure to patients, workers, and the general public, should address the following points, as applicable to the type of medical use:

- When the procedures are to be implemented, such as any circumstance in which the source becomes dislodged, cannot be retracted to a fully shielded position, or the patient cannot be removed from the beam of radiation.

- The actions specified for emergency source recovery or shielding that primarily consider minimizing exposure to the patient and health care personnel while maximizing the safety of the patient.

- The step-by-step actions for single or multiple failures that specify the individual(s) responsible for implementing the actions. The procedures should clearly specify which steps are to be taken under different scenarios. The procedure should specify situations in which surgical intervention may be necessary and the steps that should be taken in that event.

- Location of emergency source recovery equipment, specifying what equipment may be necessary for various scenarios. Emergency equipment should include shielded storage containers, remote handling tools, and if appropriate, supplies necessary to surgically remove applicators or sources from the patient and tools necessary for removal of the patient from the device.

- Radiation safety priorities, such as giving first consideration to minimizing exposure to the patient, usually by removing the patient from the room (rather than using tools to attempt to return the source to the off position). *Note:* If the first step of the emergency procedures for teletherapy units specifies pressing the emergency bar on the teletherapy unit console, the applicant is advised that this action may cause the source to return to the off position but may also cut power to the entire teletherapy unit or to the gantry or the couch.

- Instructing the staff to act quickly and calmly, and to avoid the primary beam of radiation.

- Specifying who is to be notified.

- Requirements to restrict (lock, as necessary) and post the treatment area with appropriate warning signs as soon as the patient and staff are out of the treatment room.

Response from Applicant: Provide procedures required by 10 CFR 35.610. See Appendix AA for responses required by 10 CFR 30.32(j).

8.23 ITEM 10: OCCUPATIONAL DOSE

Regulations: 10 CFR 20.1003, 10 CFR 20.1101, 10 CFR 20.1201, 10 CFR 20.1202, 10 CFR 20.1204, 10 CFR 20.1207, 10 CFR 20.1208, 10 CFR 20.1501, 10 CFR 20.1502, 10 CFR 20.2102, 10 CFR 20.2106.

Part 35	Applicability
100	✓
200	✓
300	✓
400	✓
500	✓
600	✓
1000	✓

Criteria: Applicants must do either of the following:

- Demonstrate that unmonitored individuals are not likely to receive, in 1 year, a radiation dose in excess of 10 % of the allowable limits as shown in Figure 8.2.

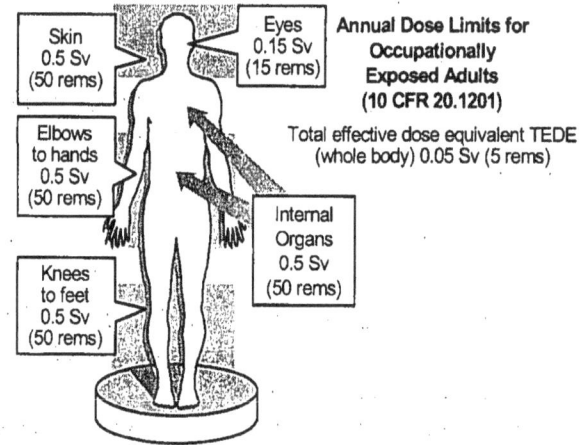

Figure 8.2   Annual Occupational Dose Limits for Adults

> TOTAL EFFECTIVE DOSE EQUIVALENT (TEDE) = DEEP DOSE FROM EXTERNAL EXPOSURE + DOSE FROM INTERNALLY DEPOSITED RADIONUCLIDES

OR

- Monitor external and/or internal occupational radiation exposure, if required by 10 CFR 20.1502.

Discussion: The NRC was given regulatory authority over accelerator-produced radioactive materials and discrete sources of Ra-226 by the EPAct. For individuals working with or near NRC-regulated materials, 10 CFR Part 20 has always included the radiation exposure from radiation sources NRC did not regulate (e.g., x-rays, radiation from NARM materials). With the new definition of byproduct material, workers that previously were not subject to the requirements in 10 CFR Part 20 because they did not use NRC-regulated materials will now be subject to these requirements if they work with or near accelerator-produced radioactive materials or Ra-226. Applicants should review the use of all NRC-regulated materials (including the new accelerator-produced and discrete Ra-226 byproduct materials) when determining, for NRC requirements, who is an occupationally exposed individual.

The Radiation Protection Program that licensees are required to develop, document, and implement in accordance with 10 CFR 20.1101 must include provisions for monitoring occupational dose. The licensee must evaluate the exposure of all occupational workers (e.g., nurses, technologists, and individuals producing PET radioactive drugs under a 10 CFR 30.32(j) authorization) to determine if monitoring is required to demonstrate compliance with Subpart F of 10 CFR Part 20. Licensees must consider the internal and external dose and the occupational workers' assigned duties when evaluating the need to monitor occupational radiation exposure. Review of dosimetry histories for workers previously engaged in similar duties may be helpful in assessing potential doses.

When evaluating an external dose from xenon gas, the licensee may take credit for the reduction of dose resulting from the use of xenon traps. Additionally, periodic checks of the trap effluent may be used to ensure proper operation of the xenon trap. Licensees may vent xenon gas directly to the atmosphere as long as the effluent concentration is within 10 CFR Part 20 limits.

When evaluating doses from aerosols, licensees may take credit for the reduction of dose resulting from the use of aerosol traps. Licensees may vent aerosols directly to the atmosphere as long as the effluent concentration is within 10 CFR Part 20 limits.

Appendix M provides a model procedure for monitoring external occupational exposure.

If external dose monitoring is necessary, the applicant should describe the type of personnel dosimetry, such as film badges, optically stimulated luminescence (OSL) dosimeters, and thermoluminescent dosimeters (TLDs), that personnel will use. If occupational workers handle licensed material, the licensee should evaluate the need to provide extremity monitors, which are required if workers are likely to receive a dose in excess of 0.05 Sv (5 rems) shallow-dose equivalent (SDE), in addition to whole-body badges. Additionally, applicants should ensure that their personnel dosimetry program contains provisions that personnel monitoring devices be worn in such a way that the part of the body likely to receive the greatest dose will be monitored.

Some licensees use self-reading dosimeters in lieu of processed dosimetry. This is acceptable if the regulatory requirements are met. See American National Standards Institute (ANSI) N322, "Inspection and Test Specifications for Direct and Indirect Reading Quartz Fiber Pocket Dosimeters," for more information. If pocket dosimeters are used to monitor personnel exposures, applicants should state the useful range of the dosimeters, along with the procedures and frequency for their calibration (10 CFR 20.1501(b)).

When personnel monitoring is needed, most licensees use either film badges or TLDs that are supplied by a processor holding current personnel dosimetry accreditation from the National Voluntary Laboratory Accreditation Program (NVLAP). Under 10 CFR 20.1501, licensees must verify that the processor is accredited by NVLAP for the type of radiation for which monitoring will be performed. Consult the NVLAP-accredited processor for its recommendations for exchange frequency and proper use. Also, it is recommended that ANPs, AUs, radiopharmacy technologists, and individuals producing PET radioactive drugs under 10 CFR 30.32(j) wear a pocket/audible dosimeter in addition to their personal dosimeter(s) when they are working with high-energy gamma-emitting radionuclides such as positron-emitting radionuclides.

It may be necessary to assess the intake of radioactivity for occupationally exposed individuals in accordance with 10 CFR 20.1204 and 20.1502. If internal dose assessment is necessary, the applicant shall measure the following:

- Concentrations of radioactive material in air in work areas, or

- Quantities of radionuclides in the body, or

- Quantities of radionuclides excreted from the body, or

- Combinations of these measurements.

The applicant should describe in its procedures the criteria used to determine the type of bioassay and the frequencies at which bioassays (both *in vivo* and *in vitro*) will be performed to evaluate intakes. The criteria also should describe how tables of investigational levels are derived, including the methodology used by the evaluated internal dose assessments (i.e., the empirical models used to interpret the raw bioassay data). The bioassay procedures should provide for baseline, routine, emergency, and follow-up bioassays. If a commercial bioassay service will be used, the applicant should ensure that the service is licensed by NRC (or an equivalent Agreement State) for that service or provide an alternative for NRC to review.

Acceptable criteria that applicants may use in developing their bioassay programs are outlined in RG 8.9, Revision 1, "Acceptable Concepts, Models, Equations, and Assumptions for a Bioassay Program," and NUREG/CR-4884, "Interpretation of Bioassay Measurements." *Note:* These documents predate the EPAct and may not address the criteria for accelerator-produced radionuclides or for discrete sources of Ra-226.

Regulatory Issue Summary (RIS) 2002-06, "Evaluating Occupational Dose for Individuals Exposed to NRC-Licensed Material and Medical X-Rays," provides guidance for evaluating occupational dose when some exposure is due to X-rays and dosimeters are used to measure exposure behind lead aprons and elsewhere.

Note: The definition of "shallow-dose equivalent" in 10 CFR 20.1003 was revised, effective June 4, 2002[8], to change the area for averaging dose to skin from 1 square centimeter to 10 square centimeters (see NRC Regulatory Issue Summary 2002-10, "Revision of the Skin Dose Limit in 10 CFR Part 20").

Response from Applicant: If personnel monitoring is required, provide the following:

- A statement that: "Either we will perform a prospective evaluation demonstrating that unmonitored individuals are not likely to receive, in 1 year, a radiation dose in excess of 10% of the allowable limits in 10 CFR Part 20, or we will provide dosimetry that meets the requirements listed under 'Criteria' in NUREG-1556, Volume 9, Revision 1, 'Consolidated

---

[8] 67 FR 16298

Guidance About Materials Licenses: Program-Specific Guidance About Medical Use Licenses.' "

OR

- A description of an alternative method for demonstrating compliance with the referenced regulations.

References:

- National Institute of Standards and Technology (NIST) Publication 810, "National Voluntary Laboratory Accreditation Program Directory," is published annually and is available for purchase from GPO and on the Internet at http://ts.nist.gov/ts/htdocs/Standards/scopes/programs.htm.

- ANSI N322, "Inspection and Test Specifications for Direct and Indirect Reading Quartz Fiber Pocket Dosimeters," may be obtained from the American National Standards Institute, 1430 Broadway, New York, NY 10018, or ordered electronically from http://www.ansi.org.

- NUREG/CR-4884, "Interpretation of Bioassay Measurements."

- RG 8.9, Revision 1, "Acceptable Concepts, Models, Equations, and Assumptions for a Bioassay Program."

- "Evaluating Occupational Dose for Individuals Exposed to NRC-Licensed Material and Medical X-Rays."

- NRC Regulatory Issue Summary 2002-06, "Evaluating Occupational Dose for Individuals Exposed to NRC-Licensed Material and Medical X-Rays."

- NRC Regulatory Issue Summary 2002-10, "Revision of the Skin Dose Limit in 10 CFR Part 20."

See the Notice of Availability on the inside front cover of this report to obtain copies of the NRC documents. Copies of Regulatory Issue Summaries are also available on the NRC's Web site in the Electronic Reading Room at http://www.nrc.gov/reading-rm/doc-collections/gen-comm/reg-issues/.

8.24   ITEM 10: AREA SURVEYS

Regulations: 10 CFR 20.1003, 10 CFR 20.1101, 10 CFR 20.1201, 10 CFR 20.1301, 10 CFR 20.1302, 10 CFR 20.1501, 10 CFR 20.1801, 10 CFR 20.1802, 10 CFR 20.2102, 10 CFR 20.2103, 10 CFR 20.2107, 10 CFR 35.70, 10 CFR 35.315, 10 CFR 35.404, 10 CFR 35.604, 10 CFR 35.2070.

Part 35	Applicability
100	✓
200	✓
300	✓
400	✓
500	✓
600	✓
1000	✓

Criteria: Licensees are required to make surveys of potential radiological hazards in their workplace. Licensed material now also includes accelerator-produced radionuclides and discrete sources of Ra-226. For example, licensees must perform surveys to:

- Ensure that licensed material will be used, transported, and stored in such a way that doses to members of the public do not exceed 1 mSv per year (100 millirem/year) and that the dose in any unrestricted area will not exceed 0.02 mSv (2 mrem) in any 1 hour from licensed operations;

- Ensure that licensed material will be used, transported, and stored in such a way that occupational doses to individuals will not exceed the limits specified in 10 CFR 20.1201;

- Control and maintain constant surveillance over licensed material that is not in storage and secure licensed material from unauthorized access or removal; and

- Ensure that licensed material will be used, transported, and stored in such a way that the air emissions do not exceed the constraint value in 10 CFR 20.1101.

Discussion: The Radiation Protection Program that licensees are required to develop, document, and implement in accordance with 10 CFR 20.1101 must include provisions for area surveys. Surveys are evaluations of radiological conditions and potential hazards. These evaluations may be measurements (e.g., radiation levels measured with survey instruments or results of wipe tests for contamination), calculations, or a combination of measurements and calculations. The selection and proper use of appropriate instruments is one of the most important factors in ensuring that surveys accurately assess radiological conditions.

There are many different kinds of surveys performed by licensees:

- Contamination:
    - Fixed, or
    - Removable.
- Air Effluent,
- Water Effluent,
- Leak Test,
- Bioassays,
- Air Sample,
- Restricted Areas,
- Unrestricted Areas, and
- Personnel (during use, transfer, or disposal of licensed material).

Surveys are required when it is reasonable under the circumstances to evaluate a radiological hazard and when necessary for the licensee to comply with the appropriate regulations. The most important types of surveys are as follows:

- Surveys for radioactive contamination that could be present on surfaces of floors, walls, laboratory furniture, and equipment;

- Measurements of radioactive material concentrations in air for areas where radiopharmaceuticals are handled or processed in unsealed form and where operations could expose workers to the inhalation of radioactive material (e.g., radioiodine) or where licensed material is or could be released to unrestricted areas;

- Bioassays to determine the kinds, quantities, or concentrations, and in some cases, the location of radioactive material in the human body. Radioiodine uptake in a worker's thyroid gland is commonly measured by external counting using a specialized thyroid detection probe;

- Surveys of external radiation exposure levels in both restricted and unrestricted areas;

- Surveys of radiopharmaceutical packages entering (e.g., from suppliers) and departing (e.g., returned radiopharmaceuticals to the supplier); and

- Surveys (by licensees authorized under 10 CFR 30.32(j) for noncommercial distribution of PET radioactive drugs) of PET packages being sent to other members of the consortium.

The frequency of routine surveys depends on the nature, quantity, and use of radioactive materials, as well as the specific protective facilities, equipment, and procedures that are designed to protect workers and the public from external and internal exposure. Also, the frequency of the survey depends on the type of survey. Appendix R contains model procedures that represent one acceptable method of establishing survey frequencies for medical use ambient radiation levels and contamination surveys. Appendix R contains some of the information an applicant requesting authorization under 10 CFR 30.32(j) must include in its procedures to meet survey requirements.

For example, medical use licensees are required to perform daily surveys in all areas used for the preparation and administration of radiopharmaceuticals for which a written directive is required (diagnostic activities exceeding 30 µCi of I-131 and all therapy treatments); when the licensee administers radiopharmaceuticals requiring a WD in a patient's room, the licensee is not required to perform a survey of the patient's room. Licensees should perform surveys after the patient's release. Licensees must perform surveys prior to the release of the room for unrestricted use. Licensees should be cognizant of the requirement to perform surveys to demonstrate that public dose limits are not exceeded. Licensees should survey areas near direct transport tubes used to transfer PET radionuclides or radiopharmaceuticals to administration areas from onsite or offsite PET radionuclide production facilities.

Because therapy sealed sources (including applicators, catheters, and therapy sources used for diagnostic purposes) may become dislodged during implantation or after surgery, and inadvertently lost or removed, the following surveys shall be performed:

- Immediately after implanting sources in a patient or a human research subject, the licensee shall make a survey to locate and account for all sources that have not been implanted.

- Immediately after removing the last temporary implant source from a patient or human research subject, the licensee shall make a survey of the patient or human research subject with a radiation detection survey instrument to confirm that all sources have been removed.

In addition, licensees should also consider the following:

- The patient's bed linens before removing them from the patient's room,

- The operating room and the patient's room after source implantation (e.g., radiation level and/or visual check),

- All trash exiting the patient's room, and

- Areas of public access in and around the patient's room.

Response from Applicant: Provide the following statement:

"We have developed and will implement and maintain written procedures for area surveys in accordance with 10 CFR 20.1101 that meet the requirements of 10 CFR 20.1501 and 10 CFR 35.70."

8.25   ITEM 10: SAFE USE OF UNSEALED LICENSED MATERIAL

Part 35	Applicability
100	✓
200	✓
300	✓
400	
500	
600	
1000	✓

Regulations: 10 CFR 20.1101, 10 CFR 20.1301, 10 CFR 20.1302, 10 CFR 20.2102, 10 CFR 20.2103, 10 CFR 30.33(a)(2), 10 CFR 30.34(e), 10 CFR 35.27, 10 CFR 35.69, 10 CFR 35.70, 10 CFR 35.310.

Criteria: Before using licensed material, the licensee must develop and implement a Radiation Protection Program that includes safe use of unsealed licensed material. Unsealed licensed material now also includes unsealed quantities of accelerator-produced radionuclides and Ra-226, and may also include large activities of PET radionuclides used to produce PET radioactive drugs.

Discussion: The Radiation Protection Program that licensees are required to develop, document, and implement in accordance with 10 CFR 20.1101 must include provisions for safe use of licensed material. Licensees are responsible for developing, documenting, and implementing procedures to ensure the security and safe use of all licensed material from the time it arrives at their facilities until it is used, transferred, and/or disposed of. The written procedures should provide reasonable assurance that only appropriately trained personnel will handle and use licensed material without undue hazard to themselves, other workers, or members of the public.

The Radiation Protection Program must cover the uses of accelerator-produced radioactive materials and Ra-226, which are now included in the definition of byproduct material as a result of the EPAct.

In addition, licensees must develop, implement, and maintain procedures for protective measures to be taken by occupational workers to maintain their doses ALARA. Protective measures may include:

- Use of syringe shields and/or vial shields,

- Wearing laboratory coats and gloves when handling unsealed byproduct material, and

- Monitoring hands after handling unsealed byproduct material.

When producing PET radioactive drugs, protective measures may include remote manipulation of material in shielded hot cells and the use of remote handling tools in other production tasks.

Appendix T contains model procedures that provide one method for the safe use of unsealed licensed material. This Appendix addresses some elements needed for the production of PET radioactive drugs.

Response from Applicant: Provide the following statement:

"We have developed and will implement and maintain procedures for safe use of unsealed byproduct material that meet the requirements of 10 CFR 20.1101 and 10 CFR 20.1301."

8.26 ITEM 10: SPILL/CONTAMINATION PROCEDURES

Regulations: 10 CFR 19.11(a)(3), 10 CFR 20.1101, 10 CFR 20.1406, 10 CFR 20.2202, 10 CFR 20.2203, 10 CFR 30.32, 10 CFR 30.35(g), 10 CFR 30.50, 10 CFR 30.51, 10 CFR 35.27.

Part 35	Applicability
100	✓
200	✓
300	✓
400	✓*
500	✓*
600	
1000	✓

*If source does not meet sealed source definition in 10 CFR Part 35.

Criteria: Before using licensed material, the licensee must develop, document, and implement a Radiation Protection Program that includes proper response to spills of licensed material. Licensed material now also includes accelerator-produced radionuclides and Ra-226.

Discussion: The Radiation Protection Program that licensees are required to develop, document, and implement in accordance with 10 CFR 20.1101 must include provisions for responding to spills or other contamination events in order to prevent the spread of radioactive material. Appendix N contains model emergency response procedures, including model spill procedures. Spill procedures should address all types and forms of licensed material used and should be posted in restricted areas where licensed materials are used or stored. The instructions should specifically state the names and telephone numbers of persons to be notified (e.g., RSO, staff, State and local authorities, and NRC, when applicable). Additionally, the instructions should contain procedures for evacuation of the area, and containment of spills and other releases, as well as appropriate methods for reentering and decontaminating facilities (when necessary).

The provisions for responding to spills and other contamination events must cover any unique properties of accelerator-produced radionuclides or discrete sources of Ra-226 that the applicant possesses. These radioactive materials are now included in the definition of byproduct material as a result of the EPAct. When producing PET radioactive drugs, the procedures should also address spills or loss of control of curie quantities of material.

Response from Applicant: Provide the following statement:

"We have developed and will implement and maintain written procedures for safe response to spills of licensed material in accordance with 10 CFR 20.1101."

8.27 ITEM 10: INSTALLATION, MAINTENANCE, ADJUSTMENT, REPAIR, AND INSPECTION OF THERAPY DEVICES CONTAINING SEALED SOURCES

Part 35	Applicability
100	
200	
300	
400	
500	
600	✓
1000	✓

Regulations: 10 CFR 20.1101, 10 CFR 30.32, 10 CFR 30.34, 10 CFR 35.605, 10 CFR 35.655, 10 CFR 35.2605, 10 CFR 35.2655.

Criteria: In accordance with 10 CFR 35.605 and 10 CFR 35.655, licensees must ensure that therapy devices containing sealed sources are installed, maintained, adjusted, repaired, and inspected by persons specifically licensed to conduct these activities. The above activities should be conducted according to the manufacturers' written recommendations and instructions and according to the SSDR. In addition, 10 CFR 35.655 requires that teletherapy and GSR units be fully inspected and serviced during source replacement or at intervals not to exceed 5 years, whichever comes first, to ensure that the source exposure mechanism functions properly. Maintenance is necessary to ensure that the device functions as designed and source integrity is not compromised.

Discussion: Maintenance and repair includes installation, replacement, and relocation or removal of the sealed source(s) or therapy unit that contains a sealed source(s). Maintenance and repair also includes any adjustment involving any mechanism on the therapy device, treatment console, or interlocks that could expose the source(s), reduce the shielding around the source(s), affect the source drive controls, or compromise the radiation safety of the unit or the source(s).

The NRC requires that maintenance and repair (as defined above) be performed only by persons specifically licensed by NRC or an Agreement State to perform such services. Most licensee employees do not perform maintenance and repair because they do not have the specialized equipment and technical expertise to perform these activities. Applicants requesting authorization to possess and use LDR remote afterloaders should review 10 CFR 35.605 before responding to this item. Section 10 CFR 35.605 allows for an AMP to perform certain service activities with regard to LDR remote afterloader units.

Response from Applicant: No response is necessary if the licensee contracts with personnel who are licensed by NRC or an Agreement State to install, maintain, adjust, repair, and inspect the specific therapy device possessed by the licensee. However, if the applicant requests that an employee who is trained by the manufacturer be authorized to perform the aforementioned activities, the applicant must provide sufficient information to allow the NRC to evaluate and approve such authorization (see CFR 35.605 and 10 CFR 35.655). This should include the following:

- Name of the proposed employee and types of activities requested,

AND

- Description of the training and experience demonstrating that the proposed employee is qualified by training and experience for the use requested,

AND

- Copy of the manufacturer's training certification and an outline of the training in procedures to be followed.

Note: The applicant should specify only those installation, maintenance, inspection, adjustment, and repair functions, as described in a certificate or letter from the manufacturer of the device, that document the employee's training in the requested function(s).

8.28 ITEM 10: MINIMIZATION OF CONTAMINATION

Part 35	Applicability
100	✓
200	✓
300	✓
400	✓
500	✓
600	✓
1000	✓

Regulations: 10 CFR 20.1406 and 10 CFR 35.67.

Criteria: Applicants for new licenses must describe in the application how facility design and procedures for operation will minimize, to the extent practicable, contamination of the facility and the environment, facilitate eventual decommissioning, and minimize, to the extent practicable, the generation of radioactive waste.

Discussion: All applicants for new licenses need to consider the importance of designing and operating their facilities to minimize the amount of radioactive contamination generated at the site during its operating lifetime and to minimize the generation of radioactive waste during decontamination. This is especially important for licensed activities involving unsealed byproduct material. As described in Item 8.26, "Spill/Contamination Procedures," cleanup procedures should be implemented for contamination events. Recommended limits for acceptable levels of surface contamination in restricted and unrestricted areas are provided in Appendix R, Tables R.2 and R.3.

Sealed sources and devices that are approved by NRC or an Agreement State and located and used according to their SSDR certificates usually pose little risk of contamination. Leak tests performed as specified in the SSDR certificate should identify defective sources. Leaking sources must be immediately withdrawn from use and stored, repaired, or disposed of according to NRC requirements. These steps minimize the spread of contamination and reduce radioactive waste associated with decontamination efforts.

The NRC now has regulatory authority over sealed sources and devices containing accelerator-produced radioactive material and discrete sources of Ra-226 under the new definition of byproduct material resulting from the EPAct. There may be sources and devices containing this newly defined byproduct material that do not have SSDR certificates. These devices and sources are, however, subject to the standard leak test provisions included in materials licenses.

Response from Applicant: A response from applicants is not required under the following condition: the NRC will consider that the above criteria have been met if the information provided in the applicant's responses satisfy the criteria in Sections 8.15, 8.16, 8.21, 8.25, 8.27, and 8.29, on the topics: facility and equipment, facility diagram, Radiation Protection Program, and waste management.

8.29   ITEM 11: WASTE MANAGEMENT

Part 35	Applicability
100	✓
200	✓
300	✓
400	✓
500	✓
600	✓
1000	✓

Regulations: 10 CFR 20.1101, 10 CFR 20.1301, 10 CFR 20.1302, 10 CFR 20.1501, 10 CFR 20.1904, 10 CFR 20.2001-2007, 10 CFR 20.2102, 10 CFR 20.2103, 10 CFR 20.2107, 10 CFR 20.2108, 10 CFR 30.33(a)(2), 10 CFR 30.41, 10 CFR 30.51, 10 CFR 31.11; 10 CFR 35.92, 10 CFR 35.2092, 10 CFR 61.3, 10 CFR 71.5.

Criteria: Licensed materials must be disposed of in accordance with NRC requirements by:

• Transfer to an authorized recipient (10 CFR 30.41(b)),

• Decay-in-storage,

• Release in effluents within the limits in 10 CFR 20.1301, or

• As authorized under 10 CFR 20.2002 through 20.2005.

Discussion: The Radiation Protection Program that licensees are required to develop, document, and implement in accordance with 10 CFR 20.1101 must include provisions for waste disposal of licensed material. Licensed material now includes accelerator-produced radioactive material and discrete sources of Ra-226 as currently included in the new definition of byproduct material resulting from the EPAct. Appendix W contains model procedures that represent one way to provide for decay-in-storage and generator or other licensed material return. Applicants are reminded to take into account the following information when they develop procedures (as applicable):

- Except for material suitable for decay-in-storage and some animal carcasses handled by the licensee, solids are transferred to an authorized recipient licensed to receive such waste in accordance with 10 CFR 20.2001(b), 10 CFR 20.2006, or in applicable regulations in 10 CFR Parts 30 or 61. Follow the packaging instructions received from the transfer agent and the burial site operator. Keep the consignment sheet from the transfer agent as the record of disposal.

- When setting up a program for decay-in-storage, consider short-term and long-term storage. Consider designing long-term storage to allow for segregation of wastes with different half-lives (e.g., the use of multiple shielded containers) and use of containers with shielded covers to maintain occupational exposure at ALARA levels. Storage areas must be in a secure location. *Note:* Some short half-life radionuclide products (e.g., Tc-99m/Mo-99 generator columns and some yttrium-90 (Y-90) microspheres) contain long half-life contaminants that may preclude disposal by decay-in-storage.

- Waste from *in vitro* kits (except mock iodine-125) that are generally licensed under 10 CFR 31.11 is exempt from waste disposal regulations in 10 CFR Part 20, as set forth in 10 CFR 31.11(f). Radioactive labels should be defaced or removed. There is no need to keep any record of release or make any measurement.

- Consider the monitoring and control mechanisms in place to ensure compliance with the appropriate requirements regarding the release of material into air and water under 10 CFR 20.1302 and 20.2003, respectively.

  - Regulations for disposal in the sanitary sewer appear in 10 CFR 20.2003. Material must be readily soluble or dispersible in water. There are also monthly and annual limits, based on the total sanitary sewerage release of the facility. (Excreta from patients undergoing medical diagnosis or therapy are not subject to these limitations; see 10 CFR 20.2003(b).)

  - Limits on permissible concentrations in effluents to unrestricted areas are enumerated in Table II of Appendix B to 10 CFR Part 20. These limits apply at the boundary of the restricted area. If PET radioactive drugs are produced, the program should include methods of measuring, monitoring, and controlling effluent releases at all stages of production.

  - Liquid scintillation-counting media containing 1.85 kBq (0.05 µCi) per gram of H-3 or C-14 may be disposed of without regard to their radioactivity (10 CFR 20.2005(a)(1)).

- If applicants/licensees propose to treat or dispose of licensed material by incineration, they must comply with 10 CFR 20.2004. Contact the appropriate NRC Regional Office for guidance on treatment or disposal of material by incineration.

- Applicants that wish to use waste volume reduction operations (e.g., compactors) should provide a detailed description (as outlined below), along with their response to Item 8.16 (Facility Diagram):

  - A description of the compactor to demonstrate that it is designed to safely compact the waste generated (e.g., manufacturer's specifications, annotated sketches, photographs);

  - The types, quantities, and concentrations of the waste to be compacted;

- An analysis of the potential for airborne release of radioactive material during compaction activities;

- The location of the compactors in the waste processing area(s), as well as a description of the ventilation and filtering systems used in conjunction with the compactors, and procedures for monitoring filter blockage and exchange;

- Methods used to monitor worker breathing zones and/or exhaust systems;

- The types and frequencies of surveys that will be performed for contamination control in the compactor area;

- The instructions provided to compactor operators, including instructions for protective clothing, checks for proper functioning of equipment, and methods of handling uncompacted waste and examining containers for defects.

- "Empty" transport shield return: Applicants requesting authorization under 10 CFR 30.32(j) to produce PET radioactive drugs for noncommercial transfer to other medical use members in the consortium should request authorization to receive contaminated transport shields returned from consortium members. Individual consortium members are responsible for handling unused dosages, empty vials, and syringes under their own waste management program. (See Appendix AA.)

Nuclear pacemakers: Medical licensees are often the first to come into contact with plutonium-powered pacemakers or the first to be contacted by nursing homes and funeral homes when a patient with an implanted pacemaker dies. In such cases, and when the licensee is not responsible for control or disposal of the pacemaker, notify the NRC and attempt to contact the hospital where the pacemaker was implanted to arrange for explantation. The licensee that implanted the device is responsible for the follow-up, explantation, and return of the pacemaker to the manufacturer for proper disposal. The NRC Information Notice 98-12, "Licensees' Responsibilities Regarding Reporting and Follow-up Requirements for Nuclear-Powered Pacemakers," provides additional information.

Response from Applicant:

- Contact the appropriate NRC Regional Office for guidance on treatment or disposal of waste by incineration.

- See Appendix AA when requesting authorization to receive contaminated transport shields from consortium members.

- For other treatment or disposal of waste, provide the following statement:

  "We have developed and will implement and maintain written waste disposal procedures for licensed material, in accordance with 10 CFR 20.1101, that also meet the requirements of the applicable section of Subpart K to 10 CFR Part 20 and of 10 CFR 35.92."

8.30   ITEM 12: FEES

Regulation: 10 CFR 170.31.

On NRC Form 313, enter the appropriate fee category from 10 CFR 170.31 and the amount of the fee enclosed with the application.

Note: There is no fee category associated with the authorization under 10 CFR 30.32(j) for the production of PET radioactive drugs for noncommercial distribution to medical use consortium members.

Part 35	Applicability
100	✓
200	✓
300	✓
400	✓
500	✓
600	✓
1000	✓

8.31   ITEM 13: CERTIFICATION

Individuals acting in a private capacity are required to date and sign NRC Form 313. Otherwise, representatives of the corporation or legal entity filing the application should date and sign NRC Form 313. These representatives must be authorized to make binding commitments and to sign official documents on behalf of the applicant. An application for licensing a medical facility must be signed by the applicant's or licensee's management. The individual who signs the application should be identified by title of the office held. As discussed previously in Section 3, "Management Responsibility," signing the application acknowledges management's commitment and responsibilities for the Radiation Protection Program. Management includes the chief executive officer or other individual having the authority to manage, direct, or administer the licensee's activities, or those persons' delegate or delegates. The NRC will return all unsigned applications for proper signature.

Part 35	Applicability
100	✓
200	✓
300	✓
400	✓
500	✓
600	✓
1000	✓

Note: It is a criminal offense to make a willful false statement or representation on applications or correspondence (18 U.S.C. 1001).

PROGRAM-RELATED GUIDANCE – NO RESPONSE REQUIRED FROM APPLICANTS ON NRC FORM 313

The information provided in the following sections is included because it is a key element of a licensee's program and the information is provided as guidance to applicants in setting up their programs to satisfy regulatory requirements.

8.32 ITEM 8: SAFETY INSTRUCTION FOR INDIVIDUALS WORKING IN OR FREQUENTING RESTRICTED AREAS

Part 35	Applicability
100	✓
200	✓
300	✓
400	✓
500	✓
600	✓
1000	✓

Regulations: 10 CFR 19.12, 10 CFR 35.27, 10 CFR 35.310, 10 CFR 35.410, 10 CFR 35.610, 10 CFR 35.2310.

Criteria: Individuals working with or in the vicinity of licensed material must have adequate safety instructions as required by 10 CFR Parts 19 and 35. Licensed material now includes accelerator-produced radioactive material and discrete sources of Ra-226 as currently included in the new definition of byproduct material resulting from the EPAct. For individuals who, in the course of employment, are likely to receive in a year an occupational dose of radiation over 1 millisievert (mSv) (100 millirem (mrem)), the licensee must provide safety instructions as required by 10 CFR 19.12. Additional requirements for training in radiation safety for individuals involved with therapeutic treatment of patients are described in 10 CFR 35.310, 10 CFR 35.410, and 10 CFR 35.610. Under 10 CFR 35.27 the licensee's AUs and ANPs are required to provide safety instruction to all personnel using byproduct material under their supervision.

Discussion: The AUs, ANPs, AMPs, RSOs, and their supervised employees are most likely to receive doses in excess of 1 mSv (100 mrem) in a year. If an applicant produces PET radioactive drugs under 10 CFR 30.32(j), or prepares them under 10 CFR 35.100(b), 35.200(b), or 35.300(b), the employees making them or preparing them for noncommercial transfer to other consortium members are also likely to receive doses in excess of 1 mSv (100 mrem) in a year. Licensees also must evaluate potential radiation doses received by any individual working in or frequenting restricted areas. All individuals working with or around licensed materials, including the newly defined byproduct material, should receive safety instructions commensurate with their assigned duties, and if it is likely that they could receive doses over 1 mSv (100 mrem) in a year, they must receive instructions as specified by 10 CFR 19.12. For example, a licensee might determine that housekeeping staff, while not likely to receive doses over 1 mSv (100 mrem), should be informed of the nature of the licensed material and the meaning of the radiation symbol, and instructed not to touch the licensed material and to remain out of the room if the door to the licensed material storage location is open. Providing minimal instruction to ancillary staff (e.g., housekeeping, security) may assist in controlling abnormal events, such as loss of radioactive material.

In addition to safety instructions required by 10 CFR 19.12, and in accordance with 10 CFR 35.310, 10 CFR 35.410, and 10 CFR 35.610, the licensee must provide radiation safety instructions to personnel (e.g., nurses) caring for patients undergoing radiopharmaceutical therapy and/or implant therapy who cannot be released in accordance with 10 CFR 35.75. This safety instruction should be commensurate with the duties of the personnel and include safe handling, patient control, visitor control, contamination control, waste control, and notification of the RSO and the AU if the patient has a medical emergency or dies.

In accordance with 10 CFR 35.27(a), individuals working with licensed material (which includes the newly defined byproduct material) under the supervision of an AU must receive instructions on the licensee's written radiation protection procedures, written directive procedures, and NRC regulations and license conditions with respect to the use of byproduct material.

In accordance with 10 CFR 35.27(b), a licensee that permits the preparation of byproduct material, including the newly defined byproduct material, for medical use by an individual under the supervision of an ANP or an AU, as allowed by 10 CFR 35.11(b)(2), shall instruct supervised individuals in the preparation of byproduct material for medical use and require the individuals to follow their instructions, the licensee's written radiation protection procedures, the license conditions, and NRC regulations. Under 10 CFR 35.27(c), a licensee that permits supervised activities, under paragraphs 10 CFR 35.27(a) and (b), is responsible for the acts and omissions of the supervised individuals.

Appendix J provides a model training program that provides one way to satisfy the requirements referenced above. Appendix J does not address special considerations applicable to the production of PET radioactive drugs. Therefore, licensees producing and transferring these drugs must include in their training programs additional elements needed to satisfy the requirements.

Response from Applicant: No response is necessary.

8.33 PUBLIC DOSE

Part 35	Applicability
100	✓
200	✓
300	✓
400	✓
500	✓
600	✓
1000	✓

Regulations: 10 CFR 20.1003, 10 CFR 20.1101, 10 CFR 20.1301, 10 CFR 20.1302, 10 CFR 20.1801, 10 CFR 20.1802, 10 CFR 20.2107.

Criteria: Licensees must do the following:

- Ensure that licensed material will be used, transported, and stored in such a way that members of the public will not receive more than 1 mSv (100 mrem) in 1 year, and the dose in any unrestricted area will not exceed 0.02 mSv (2 mrem) in any 1 hour from licensed operations. Licensed material now includes accelerator-produced radioactive material and discrete sources of Ra-226 as currently included in the new definition of byproduct material resulting from the EPAct.

- Ensure that air emissions of radioactive materials to the environment will not result in exposures to individual members of the public in excess of 0.1 mSv (10 mrem) (TEDE) in 1 year from these emissions.

- Control and maintain constant surveillance of licensed material, including the newly defined byproduct material, that is not in storage and secure stored licensed material to prevent unauthorized access, removal, or use.

Discussion: "Member of the public" is defined in 10 CFR 20.1003 as "any individual except when that individual is receiving an occupational dose." Members of the public include persons who are not radiation workers. This includes workers who live, work, or may be near locations where licensed material, including the newly defined byproduct material, is used or stored and employees whose assigned duties do not include the use of licensed materials and who work in the vicinity where it is used or stored. "Public dose" is defined in 10 CFR 20.1003 as "the dose received by a member of the public from exposure to radiation and/or radioactive material released by a licensee, or to any other source of radiation under the control of a licensee." Public dose is controlled, in part, by ensuring that licensed material is secure (e.g., located in a locked area) to prevent unauthorized access or use by individuals coming into the area. Some medical use devices containing licensed material are usually restricted by controlling access to the keys needed to operate the devices and/or to keys to the locked storage area. Only AUs and personnel using byproduct material under their supervision should have access to these keys.

Typical unrestricted areas may include offices, shops, laboratories, areas outside buildings, property, and nonradioactive equipment storage areas. The licensee does not control access to these areas for purposes of controlling exposure to radiation or radioactive materials; however, the licensee may control access to these areas for other reasons, such as security.

For areas adjacent to facilities where licensed material, including the newly defined byproduct material, is used or stored, calculations or a combination of calculations and measurements (e.g., using an environmental TLD) are often used to show compliance.

The definition of "public dose" in 10 CFR 20.1003 does not include doses received due to exposure to patients released in accordance with 10 CFR 35.75. Dose to members of the public in waiting rooms was addressed in Informational Notice (IN) 94-09.[9] The provisions of 10 CFR 20.1301(a) should not be applied to radiation received by a member of the general public from patients released under 10 CFR 35.75. If a patient is released pursuant to 10 CFR 35.75, licensees are not required to limit the radiation dose to members of the public (e.g., visitors in a waiting room or individuals near a PET "quiet room") from a patient to 0.02 mSv (2 mrem) in any one hour. Patient waiting rooms and "quiet rooms" need only be controlled for those patients not meeting the release criteria in 10 CFR 35.75.

The regulations in 10 CFR 20.1301(c) allow licensees to permit visitors to a patient who cannot be released under 10 CFR 35.75 to receive a dose greater than 0.1 rem (1 mSv), provided the dose does not exceed 0.5 rem (5 mSv) and the AU has determined before the visit that it is appropriate.

In assessing the adequacy of facilities to control public dose, licensees should consider the design factors discussed under "Facility Diagram" in Section 8.16 and may find confirmatory surveys to be useful in assuring compliance with 10 CFR 20.1301.

The licensee must control emissions to air of all byproduct material, including the newly defined byproduct material, such that the individual member of the public likely to receive the highest TEDE does not exceed the constraint level in 10 CFR 20.2101 of 0.10 mSv (10 mrem) per year

---

[9] IN 94-09, "Release of Patients with Residual Radioactivity from Medical Treatment and Control of Areas Due to Presence of Patients Containing Radioactivity Following Implementation of Revised 10 CFR Part 20," dated February 1994.

from those emissions. If exceeded, the licensee must report this in accordance with 10 CFR 20.2203 and take prompt actions to ensure against recurrence.

Response from Applicant: No response required.

8.34 OPENING PACKAGES

Part 35	Applicability
100	✓
200	✓
300	✓
400	✓
500	✓
600	✓
1000	✓

Regulations: 10 CFR 20.1906 and 10 CFR 20.2103.

Criteria: Licensees must ensure that packages are opened safely and that the requirements of 10 CFR 20.1906 are met. Licensees must retain records of package surveys in accordance with 10 CFR 20.2103.

Discussion: Licensees must establish, maintain, and retain written procedures for safely opening packages to ensure that the monitoring requirements of 10 CFR 20.1906 are met and that radiation exposure to personnel coming near or in contact with the packages containing radioactive material are ALARA.

Appendix P contains model procedures that represent one method for safely opening packages containing radioactive materials. Applicants are reminded that 10 CFR 20.1906(b) requires, in part, that licensees monitor the external surfaces of a labeled package, including those containing the newly defined byproduct material, for radioactive contamination within 3 hours of receipt if it is received during normal working hours, or not later than 3 hours from the beginning of the next working day if it is received after working hours. If authorized under 10 CFR 30.32(j) for the production and noncommercial transfer of PET radioactive drugs, the package opening procedures should be followed when receiving "empty" radiation transport shields back from consortium members.

Response from Applicant: No response required.

8.35 PROCEDURES FOR ADMINISTRATIONS WHEN A WRITTEN DIRECTIVE IS REQUIRED

Part 35	Applicability
100	
200	
300	✓
400	✓
500	
600	✓
1000	✓

Regulations: 10 CFR 35.27, 10 CFR 35.40, 10 CFR 35.41, 10 CFR 35.2040, 10 CFR 35.2041.

Criteria: The requirements for written directives (WDs) are set forth in 10 CFR 35.40. Under 10 CFR 35.41, medical use licensees are required to develop, maintain, and implement written procedures to provide high confidence that licensed material is administered as directed by AUs. Licensed material now includes accelerator-produced radioactive material and discrete sources of Ra-226 as currently included in the new definition of byproduct material resulting from the EPAct.

Discussion: The procedures do not need to be submitted to NRC. This gives licensees the flexibility to revise the procedures to enhance effectiveness without obtaining NRC approval. Appendix S provides guidance on developing the procedures.

Response from Applicant: No response required.

8.36 RELEASE OF PATIENTS OR HUMAN RESEARCH SUBJECTS

Part 35	Applicability
100	
200	
300	✓
400	✓
500	
600	
1000	✓

Regulations: 10 CFR 35.75, 10 CFR 35.2075.

Criteria: Licensees may release from confinement patients or human research subjects (patients) who have been administered licensed material if the TEDE to any other individual from exposure to the released patient is not likely to exceed 5 mSv (0.5 rem). Licensees must provide radiation safety instructions to patients released (or to their parent or guardian) in accordance with 10 CFR 35.75(b). Licensed material now includes accelerator-produced radioactive material and discrete sources of Ra-226 as currently included in the new definition of byproduct material resulting from the EPAct.

Discussion: Under 10 CFR 35.75, the licensee is required to provide the released individual (patient) with instructions, including written instructions, on actions recommended to maintain doses to other individuals ALARA if the TEDE to any other individual is likely to exceed 1 mSv (0.1 rem). If the dose to a breast-feeding infant or a child could exceed 1 mSv (0.1 rem), assuming there was no interruption of breast-feeding, the instructions also shall include:

- Guidance on the interruption or discontinuation of breast-feeding, and

- Information on the potential consequences of failure to follow the guidance.

Appendix U provides guidance to the applicant on one way for determining when:

- The licensee may authorize the release of a patient who has been administered radiopharmaceuticals or who has been treated with implants containing radioactive material (Section 1), and

- Instructions to the patient are required by 10 CFR 35.75(b) (Section 2).

Appendix U lists activities for commonly used radionuclides and the corresponding dose rates with which a patient may be released in compliance with the dose limits in 10 CFR 35.75.

Response from Applicant: No response required.

8.37   MOBILE MEDICAL SERVICE

Regulations: 10 CFR 35.2, 10 CFR 35.12, 10 CFR 35.18, 10 CFR 35.80, 10 CFR 35.647, 10 CFR 35.2080, 10 CFR 35.2647, 10 CFR 71.5, 10 CFR 71.12, 10 CFR 71.13, 10 CFR 71.14, 10 CFR 71.37, 10 CFR 71.38; Subpart H of 10 CFR Part 71, 10 CFR 150.20, 49 CFR Parts 171-178.

Part 35	Applicability
100	✓
200	✓
300	✓
400	✓
500	✓
600	✓
1000	✓

Criteria: In addition to the requirements in 10 CFR 35.80 and 10 CFR 35.647, as applicable, mobile medical service licensees must comply with all other applicable regulations.

Discussion: Applicants for licensure of mobile medical services should review Sections 8.1 through 8.31 of this NUREG for information to be submitted as part of their applications; many of the requirements in these sections are relevant to the use of byproduct material, including the newly defined byproduct material, by mobile medical service providers, with details being dependent upon the scope of such programs. "Temporary job site" means a location, other than the specific location(s) of use authorized on the license, where mobile medical services are conducted. Mobile medical service licensees may transport licensed material and equipment into a client's building, or may bring patients into the transport (e.g., van). In either case, the van should be located on the client's property that is under the client's control. Mobile PET medical service licensees must consider a "quiet room" as an area of use if the patients in the "quiet room" cannot be released under the provisions of 10 CFR 35.75.

A self-contained mobile medical service involves a mobile treatment or administration facility that provides ready-to-deliver mobile medical services on arrival at a client's site. Companies providing transportation only will not be licensed for medical use under 10 CFR Part 35. Before using a remote afterloader for this type of service, the device should be installed in an appropriately shielded treatment room.

The general types of services provided as mobile medical services are:

•   Mobile medical services (byproduct material, trained personnel, and facility) that provide the device/facility (e.g., in-van use) and treatment of (or administration to) patients at the client site. These mobile medical service providers are responsible for all aspects of byproduct material use and authorized patient treatments (or administrations).

•   Mobile medical service providers (byproduct material and trained personnel) that provide transportation to and use of the byproduct material within the client's facility. These mobile medical service providers are also responsible for all aspects of byproduct material use and authorized patient treatments (or administrations).

Mobile medical service licensees must ensure that the criteria in 10 CFR 35.75 are met before releasing patients treated in their facilities.

Refer to Appendix V for additional guidance on information to provide in applications.

Note: Agreement State licensees that request reciprocity for activities conducted in non-Agreement States are subject to the general license provisions described in 10 CFR 150.20. This general license authorizes persons holding a specific license from an Agreement State to conduct the same activity in non-Agreement States if the specific license issued by the Agreement State does not limit the authorized activity to specific locations or installations. An NRC licensee who wishes to conduct operations at temporary job sites in an Agreement State should contact that State's Radiation Control Program Office for information about State regulations, including notification requirements, whether the AU meets the requirements to be an AU in that State, and if mobile medical services are allowed within the Agreement State through reciprocity. Therefore, to ensure compliance with Agreement State reciprocity requirements, an NRC licensee shall request authorization well in advance of scheduled work. In addition to the requirements specified in 10 CFR 150.20, applicants requesting a mobile medical service license should contact all States where they plan to conduct mobile medical services, to clarify requirements, including training and experience requirements for AUs, as well as requirements associated with an authorization to practice medicine within the State's jurisdiction.

Response from Applicant: No response required.

8.38 AUDIT PROGRAM

Regulations: 10 CFR 20.1101 and 10 CFR 20.2102.

Criteria: Under 10 CFR 20.1101, all licensees must annually review the content and implementation of the Radiation Protection Program. The review should ensure the following:

Part 35	Applicability
100	✓
200	✓
300	✓
400	✓
500	✓
600	✓
1000	✓

- Compliance with NRC and applicable DOT regulations and the terms and conditions of the license; and

- Occupational doses and doses to members of the public are ALARA (10 CFR 20.1101).

Discussion: The applicant should develop and implement procedures for the required review or audit of the Radiation Protection Program's content and implementation. Appendix L contains model procedures that are only a suggested guide and are one way to meet this requirement. Some sections of Appendix L may not be pertinent to every licensee or to each review or audit. For example, licensees do not need to address areas that do not apply to their activities, and activities that have not occurred since the last review or audit need not be reviewed at the next review or audit. Appendix L also addresses some aspects of the Radiation Safety Program audit items associated with the production of PET radioactive drugs and other nonmedical uses authorized on the license. Licensees engaged in these activities may need to supplement the audit items in Appendix L to address any additional regulatory requirements for nonmedical uses. Reviews or audits of the content and implementation of the Radiation Protection Program must be conducted at least annually.

The NRC encourages licensee management to conduct performance-based reviews by observing work in progress, interviewing staff about the Radiation Protection Program, and spot-checking required records. As part of their review programs, licensees should consider performing unannounced audits of authorized and supervised users to determine if, for example, Operating and Emergency Procedures are available and are being followed.

It is essential that once identified, violations and radiation safety concerns are corrected comprehensively and in a timely manner. The following three-step corrective action process has proven effective:

- Conduct a complete and thorough review of the circumstances that led to the violation.

- Identify the root cause of the violation.

- Take prompt and comprehensive corrective actions that will address the immediate concerns and prevent recurrence of the violation.

The NRC's goal is to encourage prompt identification and prompt, comprehensive correction of violations and deficiencies.

Response from Applicant: No response is necessary.

References: See the Notice of Availability on the inside front cover of this report to obtain copies of: NRC's Enforcement Policy, "General Statement of Policy and Procedures on NRC Enforcement Actions," and IN 96-28, "Suggested Guidance Relating to Development and Implementation of Corrective Action," dated May 1, 1996. The NRC's Enforcement Policy is also available on the Internet at the NRC's Web site, http://www.nrc.gov/reading-rm/doc-collections/nuregs/staff/sr1600.

8.39 OPERATING AND EMERGENCY PROCEDURES

Part 35	Applicability
100	✓
200	✓
300	✓
400	✓
500	✓
600	✓
1000	✓

Regulations: 10 CFR 19.11(a)(3), 10 CFR 20.1101, 10 CFR 20.1601, 10 CFR 20.1602, 10 CFR 20.1801, 10 CFR 20.1802, 10 CFR 20.1906, 10 CFR 20.2102, 10 CFR 20.2201-2203, 10 CFR 21.21, 10 CFR 30.50, 10 CFR 35.12, 10 CFR 35.41, 10 CFR 35.75, 10 CFR 35.310, 10 CFR 35.315, 10 CFR 35.404, 10 CFR 35.406, 10 CFR 35.410, 10 CFR 35.415, 10 CFR 35.610, 10 CFR 35.615, 10 CFR 35.3045, 10 CFR 35.3047, 10 CFR 35.3067.

Criteria: This section summarizes operating and emergency procedures. Many of these procedures are covered in greater detail in other sections of this document. The NRC now has regulatory authority over sealed sources and devices containing accelerator-produced radioactive material and discrete sources of Ra-226 under the new definition of byproduct material resulting from the EPAct.

The licensee must develop, implement, and maintain specific operating and emergency procedures containing the following elements:

- Instructions for opening packages containing licensed material (see Section 8.34);

- Instructions for using licensed material, operating therapy treatment devices, and performing routine maintenance on devices containing sealed sources, according to the manufacturer's written recommendations and instructions and in accordance with regulatory requirements (see Section 8.27). *Note:* There may be sources and devices containing the newly defined byproduct material that do not have SSDR certificates. If these legacy sources or devices have manufacturers' recommendations or instructions, they should be followed. If not, contact the appropriate NRC Regional Office for licensing guidance. These devices and sources are, however, subject to the standard leak test provisions included in materials licenses.

- Instructions for conducting area radiation level and contamination surveys (see Section 8.24);

- Instructions for administering licensed material in accordance with the WD (see Section 8.35);

- Steps to ensure that patient release is in accordance with 10 CFR 35.75 (see Section 8.36);

- Instructions for calibration of survey and dosage measuring instruments (see Sections 8.17 and 8.18);

- Periodic spot checks of therapy device units, sources, and treatment facilities (see Section 8.19);

- Instructions for radioactive waste management (see Section 8.29);

- Steps to take, and whom to contact (e.g., RSO, local officials), when the following has occurred: (a) leaking or damaged source, (b) device malfunction and/or damage, (c) licensed material spills, (d) theft or loss of licensed material, or (e) any other incidents involving licensed material (see Sections 8.26, 8.45);

- Steps for source retrieval and access control of damaged sealed source(s) and/or malfunctioning devices containing sealed source(s) (see Section 8.22); and

- Steps to take if a therapy patient undergoes emergency surgery or dies.

The licensee should consider the following:

- Making operating procedures, including emergency procedures, available to all users (e.g., post the procedures or the location of procedure storage);

- Maintaining a current copy of the procedures at each location of use (or, if this is not practicable, posting a notice describing the procedures, and stating where they may be examined).

- When developing the procedures described above, the licensee is reminded that 10 CFR 20.1101(b) requires that the licensee use, to the extent practical, procedures and engineering controls based on sound radiation protection principles to achieve occupational doses and doses to members of the public that are ALARA.

- When receiving and using byproduct material (which includes the newly defined byproduct material), the licensee is reminded that it must be licensed to possess the byproduct material and that the radioactive material must be secured (or controlled) and accounted for at all times.

Discussion: Sealed sources and unsealed byproduct material used for therapy can deliver significant doses in a short time. The same may be true for high-activity PET radionuclides used to produce PET radioactive drugs, if not shielded. Access control to high- and very-high-radiation areas and the security of licensed material are described in 10 CFR 20.1601, 10 CFR 20.1602, 10 CFR 20.1801, and 10 CFR 20.1802. Unauthorized access to licensed material, including the newly defined byproduct material, by untrained individuals could lead to a significant radiological hazard. Many licensees achieve access control by permitting only trained individuals to have access to licensed material (e.g., keys, lock combinations, security badges). Accountability of licensed material, including the newly defined byproduct material, may be ensured by conducting physical inventories, controlling receipt and disposal, and maintaining use records.

If a therapy patient undergoes emergency surgery or dies, it is necessary to ensure the safety of others attending the patient. As long as the patient's body remains unopened, the radiation received by anyone near it is due almost entirely to gamma rays. When an operation or autopsy is to be performed, there should be an increased awareness of the possible exposure of the hands and face to relatively intense beta radiation. Procedures for emergency surgery or autopsy can be found in Section 5.3 of NCRP Report No. 37, "Precautions in the Management of Patients Who Have Received Therapeutic Amounts of Radionuclides."

Applicants should develop emergency procedures that address a spectrum of incidents (e.g., major spills, leaking sources, medical events, interlock failures, stuck sources).

After its occurrence becomes known to the licensee, the NRC must be notified when an incident involving licensed material, including the newly defined byproduct material, occurs. Refer to the regulations (10 CFR 20.2201-20.2203, 10 CFR 30.50, 10 CFR 21.21, 10 CFR 35.3045, 10 CFR 35.3047, and 10 CFR 35.3067) for a description of when notifications are required.

Appendix N provides model procedures that are one method for responding to some types of emergencies. Applicants requesting authorization for licensed activities not addressed by the model procedures in Appendix N should develop operational and emergency procedures to address these other activities.

Response from Applicant: No response is necessary.

Reference: Copies of NCRP Report No. 37, "Precautions in the Management of Patients Who Have Received Therapeutic Amounts of Radionuclides"; NCRP Report No. 105, "Radiation Protection for Medical and Allied Health Personnel," 1989; and NCRP Report No. 107, "Implementation of the Principle of As Low As Reasonably Achievable (ALARA) for Medical and Dental Personnel," 1990, may be obtained from the National Council on Radiation Protection and Measurements, 7910 Woodmont Avenue, Suite 800, Bethesda, MD 20814-3095, or ordered electronically at http://www.ncrp.com.

8.40 MATERIAL RECEIPT AND ACCOUNTABILITY

Part 35	Applicability
100	✓
200	✓
300	✓
400	✓
500	✓
600	✓
1000	✓

Regulations: 10 CFR 20.1801, 10 CFR 20.1802, 10 CFR 20.1906, 10 CFR 20.2201, 10 CFR 30.35(g)(2), 10 CFR 30.41, 10 CFR 30.51, 10 CFR 35.67, 10 CFR 35.406.

Criteria: To maintain accountability of licensed material, licensees must do the following:

- Secure licensed material;

- Maintain records of receipt, transfer, and disposal of licensed material; and

- Conduct physical inventories at required frequencies to account for licensed material.

Licensed material now includes accelerator-produced radioactive material and discrete sources of Ra-226 as currently included in the new definition of byproduct material resulting from the EPAct.

Discussion: Licensed materials must be tracked from "cradle to grave," from receipt (from another licensee or from its own radionuclide production facility) to its eventual transfer/disposal in order to ensure accountability; to identify that licensed material is missing and document the last confirmed possession of the material when it is lost, stolen, or misplaced; and to ensure that possession limits listed on the license are not exceeded.

Response from Applicant: No response is necessary.

8.41 ORDERING AND RECEIVING

Part 35	Applicability
100	✓
200	✓
300	✓
400	✓
500	✓
600	✓
1000	✓

Regulations: 10 CFR 20.1801, 10 CFR 20.1802, 10 CFR 20.1906, 10 CFR 30.32(j), 10 CFR 30.51.

Criteria: The requirements for receiving packages containing licensed material are found in 10 CFR 20.1906. Licensed material now includes accelerator-produced radioactive material and discrete sources of Ra-226 as currently included in the new definition of byproduct material resulting from the EPAct. Additionally, the security of licensed material, required by 10 CFR 20.1801 and 10 CFR 20.1802, must be considered for all receiving areas. Under 10 CFR 30.51, licensees are required, in part, to maintain records showing the receipt of byproduct material.

Discussion: Licensees must ensure that the type and quantity of licensed material possessed, including the newly defined byproduct material, is in accordance with the license. Additionally, licensees must ensure that packages are secured and radiation exposure from packages is minimized.

When ordering PET radioactive drugs produced under the provisions of 10 CFR 30.32(j), the medical use licensee must be a member of the consortium. *Note*: Authorization under 10 CFR 30.32(j) for the production of PET radioactive drugs for noncommercial transfer to medical use licensee members in the consortium restricts the transfer of these drugs only to members of the consortium. Licensees with this authorization must ensure that the drugs produced under this provision are transferred only to consortium members. The definition of a consortium is found in 10 CFR 30.4. Members of the consortium are authorized to receive these PET radioactive drugs by provisions in 10 CFR 35.100(a), 35.200(a), and 35.300(a).

Appendix O contains model procedures that are one method for ordering and receiving licensed material. Applicants that request authorization to produce PET radioactive drugs for noncommercial transfer to other medical use consortium members may have to supplement the procedures in Appendix O by developing procedures for filling orders for these drugs from other consortium members to meet regulatory requirements.

Response from Applicant: No response is necessary.

8.42   SEALED SOURCE INVENTORY

Regulations: 10 CFR 20.1801, 10 CFR 20.1802, 10 CFR 30.51, 10 CFR 35.67, 10 CFR 35.406, 10 CFR 35.2067, 10 CFR 35.2406.

Part 35	Applicability
100	✓*
200	✓*
300	✓*
400	✓
500	✓
600	✓
1000	✓

* Sealed sources for calibration, transmission, and reference use (35.65).

Criteria: The NRC requires the licensee in possession of a sealed source or brachytherapy source to conduct a semi-annual physical inventory of all such sources in its possession.

Discussion: According to 10 CFR 35.67, the licensee must conduct a semi-annual physical inventory of all sealed sources and brachytherapy sources in its possession. Individual GSR sources are exempt from this physical inventory requirement, as stated in 10 CFR 35.67(g). However, under 10 CFR 30.51, the licensee must maintain records of GSR source receipt, transfer, and disposal to indicate the current inventory of sources at the licensee's facility.

Response from Applicant: No response is necessary.

8.43   RECORDS OF DOSAGES AND USE OF BRACHYTHERAPY SOURCE

Regulations: 10 CFR 30.51, 10 CFR 35.63, 10 CFR 35.204, 10 CFR 35.2063, 10 CFR 35.2204, 10 CFR 35.2406.

Part 35	Applicability
100	✓
200	✓
300	✓
400	✓
500	
600	
1000	✓

Criteria: Licensees must record the use of licensed material to reflect proper use and accountability. Records of use must be maintained for 3 years.

Discussion: Licensees are required to make and maintain records of each dosage and administration prior to medical use. The records must include:

- Radiopharmaceutical;

- Patient's or human research subject's name or identification number (if one has been assigned);

- Prescribed dosage, determined dosage, or a notation that the total activity is less than 1.1 MBq (30 µCi);

- Date and time of dosage determination; and

- Name of the individual who determined the dosage.

Dosage determination for unit dosages may be made either by direct measurement or by a decay correction based on the determination (e.g., measurement) made by the manufacturer or preparer licensed under 10 CFR 32.72 or equivalent Agreement State requirements or an NRC or Agreement State medical use licensee authorized under 10 CFR 30.32(j) to produce PET radioactive drugs for noncommercial transfer to consortium members.

See Appendix AA for requirements to measure dosages for applicants applying for authorization under 10 CFR 30.32(j) to produce PET radioactive drugs for consortium members.

If molybdenum concentration is measured under 10 CFR 35.204, records of molybdenum concentration must be made under 10 CFR 35.2204 and must include, for each measured elution of technetium-99m:

- Ratio of the measurements expressed as kBq (µCi) of molybdenum-99 per MBq (mCi) of technetium-99m,

- Date and time of the measurement, and

- Name of the individual who made the measurement.

If strontium-82 (SR-82) and strontium-85 (Sr-85) concentrations are measured under 10 CFR 35.204, records of Sr-82 and Sr-85 concentrations must be made under 10 CFR 35.2204 and must include for each measured elution of rubidium-82 (Rb-82):

- Ratio of the measurements expressed in KBq (µCi) of Sr-82 per MBq (mCi) of Rb-82 Chloride and KBq (µCi) of Sr-85 per MBq (mci) of Rb-82,

- Date and time of the measurement, and

- Name of the individual who made the measurement.

If the licensee uses manual brachytherapy sources, the following records of use must be kept:

- When temporary implant brachytherapy sources are removed from storage, a record will include the number and activity of sources removed, the time and date they were removed from storage, the location of use, and the name of the individual who removed them from storage.

- When temporary implant brachytherapy sources are returned to storage, a record will include the number and activity of sources returned, the time and date they were returned to storage, and the name of the individual who returned them to storage.

- For permanent implants, a record will be made and will include the number and activity of sources removed from storage, the date they were removed from storage, the name of the individual who removed them from storage, the number and activity of sources not implanted, the date they were returned to storage, the name of the individual who returned them to storage, and the number and activity of sources permanently implanted in the patient or human research subject.

Response from Applicant: No response is necessary.

8.44 RECORDKEEPING

Part 35	Applicability
100	✓
200	✓
300	✓
400	✓
500	✓
600	✓
1000	✓

Regulations: 10 CFR Part 20, Subpart L; 10 CFR 30.51; 10 CFR Part 35, Subpart L.

Criteria: Licensees must maintain records as provided in 10 CFR Part 20, Subpart L; 10 CFR 30.51; and 10 CFR Part 35, Subpart L.

Discussion: The licensee must maintain certain records to comply with NRC regulations, the conditions of the license, and commitments made in the license application and correspondence with NRC. Operating procedures should identify which individuals in the organization are responsible for maintaining which records.

A table of recordkeeping requirements appears in Appendix X.

Response from Applicant: No response is necessary.

8.45 REPORTING

Part 35	Applicability
100	✓
200	✓
300	✓
400	✓
500	✓
600	✓
1000	✓

Regulations: 10 CFR Part 20, Subpart M; 10 CFR 21.21; 10 CFR 30.50; 10 CFR Part 35, Subpart M.

Criteria: Licensees are required to report to NRC via telephone, written report, or both, in the event that the safety or security of byproduct material may be compromised. The specific events that require reporting are explained in Subpart M of Part 35,

Subpart M of Part 20; and in 10 CFR 21.21 and 30.50. The timing and type of report are specified within these parts.

Discussion: The NRC requires licensees to report incidents that might compromise the health and safety of patients, health care providers, or the public. Therefore, 10 CFR Parts 20, 21, 30, and 35 include provisions that describe reporting requirements associated with the medical use of byproduct material.

A table of reporting requirements appears in Appendix Y.

Response from Applicant: No response is necessary.

8.46 LEAK TESTS

Regulations: 10 CFR 20.1501, 10 CFR 20.2103, 10 CFR 30.53, 10 CFR 35.67, 10 CFR 35.2067, 10 CFR 35.3067.

Criteria: The NRC requires testing to determine if there is any radioactive leakage from sealed sources.

Part 35	Applicability
100	✓*
200	✓*
300	✓*
400	✓
500	✓
600	✓
1000	✓

*If possess sealed sources under 35.65

Discussion: Licensees must perform leak testing of sealed sources (e.g., calibration, transmission, and reference sources) or brachytherapy sources, in accordance with 10 CFR 35.67. The NRC now has regulatory authority over sealed sources and devices containing accelerator-produced radioactive material and discrete sources of Ra-226 under the new definition of byproduct material resulting from the EPAct. There may be sources and devices containing this newly defined byproduct material that do not have SSDR certificates. These devices and sources are, however, subject to the standard leak test provisions included in materials licenses.

Appendix Q provides model procedures that are one way to perform leak testing for sealed sources, including Ra-226 sealed sources. Under 10 CFR 35.67, licensees are required to perform leak tests at six-month intervals or at other intervals approved by NRC or an Agreement State and specified in the SSDR certificate and before first use unless accompanied by a certificate indicating that the test was performed within the past 6 months. The measurement of the leak test sample is a quantitative analysis requiring that instrumentation used to analyze the sample be capable of detecting 185 Bq (0.005 µCi) of radioactivity on the sample. Leak test samples should be collected at the most accessible area where contamination would accumulate if the sealed source were leaking.

The leak test may be performed in-house or by a contractor who is authorized by NRC or an Agreement State to perform leak tests as a service to other licensees.

The licensee or contractor does not need to leak-test sources if:

* Sources contain only byproduct material with a half-life of less than 30 days;

- Sources contain only byproduct material as a gas;

- Sources contain 3.7 MBq (100 μCi) or less of beta-emitting or gamma-emitting material, or 0.37 MBq (10 μCi) or less of alpha-emitting material;

- Sources contain Ir-192 seeds in nylon ribbon; or

- Sources are stored and not being used. The licensee, shall, however, test each such source for leakage before any use or transfer unless it has been leak-tested within 6 months before the date of use or transfer.

Response from Applicant: No response is necessary.

References: See the Notice of Availability on the inside front cover of this report to obtain a copy of NUREG-1556, Volume 18, "Consolidated Guidance About Materials Licenses: Program-Specific Guidance About Service Provider Licenses," dated November 2000.

8.47 SAFETY PROCEDURES FOR TREATMENTS WHEN PATIENTS ARE HOSPITALIZED

Part 35	Applicability
100	
200	
300	✓
400	✓
500	
600	✓
1000	✓

Regulations: 10 CFR 20.1101, 10 CFR 20.1301, 10 CFR 20.1501, 10 CFR 20.1801, 10 CFR 20.2103, 10 CFR 35.310, 10 CFR 35.315, 10 CFR 35.404, 10 CFR 35.410, 10 CFR 35.415, 10 CFR 35.604, 10 CFR 35.610, 10 CFR 35.615, 10 CFR 35.2404.

Criteria: Applicants must develop and implement procedures to ensure that access to therapy treatment rooms, and exposure rates from therapy treatments, are limited to maintain doses to occupational workers and members of the public within regulatory limits.

Discussion: Under 10 CFR 35.315, 10 CFR 35.415, and 10 CFR 35.615, licensees are required to take certain safety precautions for uses of byproduct material involving radiopharmaceutical therapy, manual brachytherapy, or remote afterloader brachytherapy involving patients who cannot be released in accordance with 10 CFR 35.75. Byproduct material now includes accelerator-produced radioactive material and discrete sources of Ra-226, as a result of the EPAct. This section of the guidance does not include guidance on safety procedures for teletherapy or GSR outpatient treatments. The precautions described below are provided to help ensure compliance with the exposure limits in 10 CFR Part 20.

Under 10 CFR 35.404(b) and 10 CFR 35.604(a), licensees are required to perform a radiation survey of the patient (and the remote afterloader unit) immediately after removing the last temporary implant source from the patient and prior to releasing the patient from licensee control. This is done to confirm that all sources have been removed and accounted for. When sources are placed within the patient's body, 10 CFR 35.615(e) requires that licensed activities be limited to treatments that allow for expeditious removal of a decoupled or jammed source.

In addition, applicants must take the following steps for patients who cannot be released under 10 CFR 35.75:

- Provide a room with a private sanitary facility for patients treated with a radiopharmaceutical therapy dosage (*Note:* 10 CFR 35.315(a) allows for a room shared with another radiopharmaceutical therapy patient);

- Provide a private room for patients implanted with brachytherapy sources (*Note:* 10 CFR 35.415 allows for a room shared with another brachytherapy patient);

- Visibly post a "Radioactive Materials" sign on the patient's room and note on the door or in the patient's chart indicating where and how long visitors may stay in the patient's room. (10 CFR 35.315 and 10 CFR 35.415);

- Either monitor material and items removed from the patient's room (e.g., patient linens, surgical dressings) with a radiation detection survey instrument set on its most sensitive scale with no interposed shielding to determine that their radioactivity cannot be distinguished from the natural background radiation level or handle them as radioactive waste (10 CFR 35.315 and 10 CFR 20.1501); and

- Notify the RSO, or his/her designee, and AU as soon as possible if the patient has a medical emergency or dies (10 CFR 35.315, 10 CFR 35.415, and 10 CFR 35.615).

Licensees are required to perform adequate surveys to evaluate the extent of radiation levels (10 CFR 20.1501). Therefore, licensees must evaluate the exposure rates around patients who are hospitalized in accordance with 10 CFR 35.75 following the dosage administration or implant (e.g., measured exposure rates, combination of measured and calculated exposure rates).

Licensees are required to secure licensed material in storage from unauthorized access or removal (10 CFR 20.1801). Access control and appropriate training of authorized personnel may prevent unauthorized removal of licensed material temporarily stored in the patient's room and unnecessary personnel exposures.

In order to control exposures to individuals in accordance with 10 CFR Part 20, the licensee should consider briefing patients on radiation safety procedures for confinement to bed, visitor control, identification of potential problems, notification of medical staff in the event of problems, and other items as applicable and consistent with good medical care.

Response from Applicant: No response is necessary.

8.48 TRANSPORTATION

Regulations: 10 CFR 20.1101, 10 CFR 30.32(j), 10 CFR 30.41, 10 CFR 30.51, 10 CFR 71.5, 10 CFR 71.9, 10 CFR 71.12, 10 CFR 71.13, 10 CFR 71.14, 10 CFR 71.37, 10 CFR 71.38, 10 CFR 71, Subpart H; 49 CFR Parts 171-178.

Part 35	Applicability
100	✓
200	✓
300	✓
400	✓
500	✓
600	✓
1000	✓

Criteria: Applicants who will prepare for shipment, ship, or transport radioactive materials, including radioactive waste, must develop, implement, and maintain safety programs for the transport of radioactive material to ensure compliance with NRC and DOT regulations.

Discussion: Most packages of licensed material for medical use contain quantities of radioactive material that require the use of Type A packages. Licensed material now includes accelerator-produced radioactive material and discrete sources of Ra-226 as currently included in the new definition of byproduct material resulting from the EPAct. Applicants requesting authorization under 10 CFR 30.32(j) to produce PET radioactive drugs for noncommercial transfer to consortium members should also review Appendix AA for requirements for providing information about the shielded radiation transport packages.

Many packages shipped by medical licensees (e.g., unused radiopharmaceutical dosages) frequently meet the "Limited Quantity" criteria described in 49 CFR 173.421 and are therefore excepted from certain DOT requirements, provided certain other less restrictive requirements are met (e.g., activity in the package is less than the limited quantity and the radiation level on the surface of the package does not exceed 0.005 mSv per hour (0.5 mrem per hour)).

The general license in 10 CFR 71.12, "General license: NRC-approved package," provides the authorization used by most licensees to transport, or to deliver to a carrier for transport, licensed material in a package for which a license, certificate of compliance, or other approval has been issued by NRC. This general license is subject to certain conditions. The requirements for transportation of licensed material are set forth in 10 CFR 71.5. The regulations in 10 CFR 71.9 exempt from the requirements in 10 CFR 71.5 any physician licensed by a State to dispense drugs in the practice of medicine, who is also licensed under 10 CFR Part 35 or the equivalent Agreement State regulations. This exemption applies to transport by the physician of licensed material for use in the practice of medicine.

Some medical use licensees (e.g., teletherapy or gamma stereotactic radiosurgery) may need to ship licensed material in Type B packages. The Type B package requirements for transporting or delivering the package to a carrier for transport are set forth in 10 CFR 71.12-71.14. These include registration as a user of the package and an NRC-approved quality assurance (QA) plan. For information about these QA plans, see Revision 1 of RG 7.10, "Establishing Quality Assurance Programs for Packaging Used in the Transport of Radioactive Material," dated June 1986. For further information about registering as a user of a package or submitting a QA program for review, contact NRC's Division of Spent Fuel Storage and Transportation by calling NRC toll-free at (800) 368-5642, extension 492-3300. For information about associated fees, contact NRC's OCFO by calling NRC toll-free at (800) 368-5642, extension 415-7544.

Some medical use licensees that ship radioactive material have chosen to transfer possession of radioactive materials to a manufacturer (or service licensee) with an NRC or Agreement State license, who then acts as the shipper. The manufacturer (or service-licensee), who is subject to the provisions of 10 CFR 71.12 or 10 CFR 71.14, as appropriate, then becomes responsible for proper packaging of the radioactive materials and compliance with NRC and DOT regulations. Licensees who do this must ensure that the manufacturer (or service licensee):

- Is authorized to possess the licensed material (see 10 CFR 30.41), and

- Actually takes possession of the licensed material under its license.

Licensees should also ensure that the manufacturer (or service licensee) is authorized to possess the material at temporary job sites (e.g., the licensee's facilities).

During an inspection, the NRC uses the provisions of 10 CFR 71.5 and a Memorandum of Understanding with DOT on the Transportation of Radioactive Material (signed June 6, 1979) to examine and enforce various DOT requirements applicable to medical use licensees. Appendix Z lists major DOT regulations that apply to medical licensees.

Response from Applicant: No response is needed from applicants during the licensing phase. However, before making shipments of licensed materials on its own in a Type B package, a licensee must have registered with NRC as a user of the package and obtained NRC's approval of its QA program. Transportation issues will be reviewed during inspection.

References:

- "A Review of Department of Transportation Regulations for Transportation of Radioactive Materials" can be obtained be calling DOT's Office of Hazardous Material Initiatives and Training at (202) 366-4425.

- See the Notice of Availability on the inside front cover of this report to obtain a copy of the Memorandum of Understanding with DOT on the Transportation of Radioactive Material, signed June 6, 1979; Revision 1, of RG 7.10, "Establishing Quality Assurance Programs for Packaging Used in the Transport of Radioactive Material," dated June 1986; and NUREG-1556, Volume 18, "Consolidated Guidance About Materials Licenses: Program-Specific Guidance About Service Provider Licenses."

9 AMENDMENTS AND RENEWALS TO A LICENSE

Part 35	Applicability
100	✓
200	✓
300	✓
400	✓
500	✓
600	✓
1000	✓

Regulations: 10 CFR 30.37, 10 CFR 30.38, 10 CFR 35.13.

The NRC now has regulatory authority over sealed sources and devices containing accelerator-produced radioactive material and discrete sources of Ra-226, under the new definition of byproduct material resulting from the EPAct. Licensees may need license amendments for such purposes as to authorize use of these materials, to revise their Radiation Safety Programs to meet new requirements, or to provide new facility diagrams. The NRC issued a waiver on August 31, 2005, that permitted licensees to continue to use the newly defined byproduct material until the waiver was terminated on August 8, 2009. Licensees in Government agencies, Federally recognized Indian tribes, Delaware, the District of Columbia, Puerto Rico, the U.S. Virgin Islands, Indiana, Wyoming, and Montana who possess and use accelerator-produced radioactive material or discrete sources of Ra-226, or both, may continue to use these materials for medical use or prepare PET radioactive drugs for noncommercial distribution to other consortium members until the date of NRC's final licensing determination, provided the licensee submits an amendment application within 6 months after November 30, 2007. Other licensees should check with the appropriate NRC Regional Office to determine when they have to submit their license amendments.

Licensees are responsible for applying for amendments to licenses and for keeping them up-to-date. Furthermore, to continue a license after its expiration date, the licensee must submit an application for a license renewal at least 30 days before the expiration date (10 CFR 2.109, 10 CFR 30.36(a)).

Under 10 CFR 35.13, a licensee is required to apply for and receive a license amendment before several activities can occur, including:

- Receipt or use of byproduct material for a type of use permitted by 10 CFR Part 35, but not authorized on the licensee's current Part 35 license;

- Permitting anyone to work as an AU for medical uses, AMP, or ANP, unless the individual meets one of the exceptions listed in 10 CFR 35.13(b) (information required to document training and experience may be provided on the appropriate NRC Form 313A series of forms for change or addition of AU for medical uses, AMP, ANP, or RSO);

- Changing the RSO;

- Receiving byproduct material in excess of the amount, or receiving radionuclides or forms different than, currently authorized on the NRC license;

- Changing an area or address of use identified in the application or on the license. This includes additions and relocations of areas where PET radionuclides are produced or additions or relocations of a radionuclide delivery line from the PET radionuclide production area to a 10 CFR 35.100 or 10 CFR 35.200 medical use area. However, other changes and additions to the 10 CFR 35.100 and 10 CFR 35.200 medical use area do not require a license amendment and can be made, provided NRC is notified as required by 10 CFR 35.14 within 30 days following the changes, and

- Revising procedures required by 10 CFR 35.610, 35.642, 35.643, and 35.645, when the revision reduces the level of radiation safety.

In case of a medical emergency requiring an expedited license amendment, contact the materials licensing staff at the appropriate NRC Regional Office.

For both renewal and amendment requests, applicants should do the following:

- Use the most recent guidance in preparing an amendment or renewal request,

- Submit in duplicate either an NRC Form 313 or a letter requesting an amendment or renewal, and

- Provide the license number.

10 APPLICATIONS FOR EXEMPTIONS

Part 35	Applicability
100	✓
200	✓
300	✓
400	✓
500	✓
600	✓
1000	✓

Regulations: 10 CFR 19.31, 10 CFR 20.2301, 10 CFR 30.11, 10 CFR 35.15, 10 CFR 35.19.

Criteria: Licensees may request exemptions to regulations. The licensee must demonstrate that the exemption is authorized by law, will not endanger life or property or the common defense and security, and is otherwise in the public interest.

Discussion: Various sections of NRC's regulations address requests for exemptions (e.g., 10 CFR 19.31, 10 CFR 20.2301, 10 CFR 30.11(a)). These regulations state that NRC may grant an exemption, acting on its own initiative or on an application from an interested person.

Exemptions are not intended to revise regulations, are not intended for large classes of licenses, and are generally limited to unique situations. Exemption requests should be accompanied by descriptions of the following:

- Exemption and justification of why it is needed.

- Proposed compensatory safety measures intended to provide a level of health and safety equivalent to the regulation for which the exemption is being requested.

- Alternative methods for complying with the regulation and why compliance with the existing regulations is not feasible.

Until the NRC has granted an exemption in writing, it expects strict compliance with all applicable regulations.

Type A broad-scope licensees are granted certain exemptions as described in 10 CFR 35.15.

11 TERMINATION OF ACTIVITIES

Regulations: 10 CFR 20.1401, 10 CFR 20.1402, 10 CFR 20.1403, 10 CFR 20.1404, 10 CFR 20.1405, 10 CFR 20.1406, 10 CFR 30.34(b), 10 CFR 30.35(g), 10 CFR 30.36, 10 CFR 30.51(f).

Part 35	Applicability
100	✓
200	✓
300	✓
400	✓
500	✓
600	✓
1000	✓

Criteria: Pursuant to the regulations described above, the licensee must do the following:

- Notify NRC, in writing, within 60 days of:

    - the expiration of its license;

    - a decision to permanently cease licensed activities at the entire site (regardless of contamination levels);

    - a decision to permanently cease licensed activities in any separate building or outdoor area if it contains residual radioactivity making it unsuitable for release according to NRC requirements (see Note);

    - no principal activities having been conducted at the entire site under the license for a period of 24 months; and

    - no principal activities having been conducted for a period of 24 months in any separate building or outdoor area if it contains residual radioactivity making it unsuitable for release according to NRC requirements (see Note).

    Note: Residual radioactivity includes that from accelerator-produced radionuclides and discrete sources of Ra-226, which are now included in the definition of byproduct material as a result of the EPAct.

- Submit a decommissioning plan, if required by 10 CFR 30.36(g);

- Conduct decommissioning, as required by 10 CFR 30.36(h) and (j); and

- Submit, to the appropriate NRC Regional Office, a completed NRC Form 314, "Certificate of Disposition of Materials," (or equivalent information) and demonstrate that the premises are suitable for release for unrestricted use (e.g., results of final survey).

- Before a license is terminated, the licensee must send the records important to decommissioning to the appropriate NRC Regional Office. If licensed activities are transferred or assigned in accordance with 10 CFR 30.34(b), transfer records important to decommissioning to the new licensee.

Discussion: Useful guidance and other aids related to decommissioning are:

- NUREG-1757, Volume 2, "Consolidated NMSS Decommissioning Guidance: Characterization, Survey, and Determination of Radiological Criteria," dated September 2003.

- NUREG/BR-0241, "NMSS Handbook for Decommissioning Fuel Cycle and Materials Licenses," dated March 1997, containing the current regulatory guidance concerning decommissioning of facilities and termination of licenses.

- Appendix B of NUREG/BR-0241 containing a comprehensive list of NRC's decommissioning regulations and guidance. NUREG-1727 contains a list of superseded guidance; however, due to ongoing revisions, applicants are encouraged to consult with NRC staff regarding updates of decommissioning guidance.

- NUREG-1575, "Multi-Agency Radiation Survey and Site Investigation Manual (MARSSIM)," dated December 1997, should be reviewed by licensees who have large facilities to decommission.

- An acceptable computer code for calculating screening values to demonstrate compliance with the unrestricted dose limits is DandD, Version 2.1.0, (McFadden and others, 2001).

- NUREG-1757, Volume 2, includes a table (Table H.1) of acceptable license termination screening values of common beta/gamma radionuclides for building surface contamination. NUREG-1757, Volume 2, also contains methods for conducting site-specific dose assessments for facilities with contamination levels above those in the table.

Response from Applicant: The applicant is not required to submit a response to NRC during the initial application. The licensee's obligations are to undertake the necessary decommissioning activities, to submit NRC Form 314 or equivalent information, and to perform any other actions as summarized in the "Criteria."

References:

- NRC Form 314, "Certificate of Disposition of Materials," is available at http://www.nrc.gov/reading-rm/doc-collections/forms/.

- McFadden, K., D.A. Brosseau, W.A. Beyeler, and C.D. Updegraff, "Residual Radioactive Contamination from Decommissioning - User's Manual DandD, Version 2.1," NUREG/CR-5512, Volume 2, U.S. Nuclear Regulatory Commission, Washington, DC, April 2001.

APPENDICES

APPENDICES A-H

FORMS AND SAMPLES

APPENDIX A

NRC Form 313
"Application for Materials License"

NRC FORM 313 (10-2005) 10 CFR 30, 32, 33, 34, 35, 36, 39, and 40	U.S. NUCLEAR REGULATORY COMMISSION	APPROVED BY OMB: NO. 3150-0120 EXPIRES: 10/31/2008
APPLICATION FOR MATERIALS LICENSE		Estimated burden per response to comply with this mandatory collection request: 4.4 hours. Submittal of the application is necessary to determine that the applicant is qualified and that adequate procedures exist to protect the public health and safety. Send comments regarding burden estimate to the Records and FOIA/Privacy Services Branch (T-5 F53), U.S. Nuclear Regulatory Commission, Washington, DC 20555-0001, or by internet e-mail to infocollects@nrc.gov, and to the Desk Officer, Office of Information and Regulatory Affairs, NEOB-10202, (3150-0120), Office of Management and Budget, Washington, DC 20503. If a means used to impose an information collection does not display a currently valid OMB control number, the NRC may not conduct or sponsor, and a person is not required to respond to, the information collection.

INSTRUCTIONS: SEE THE APPROPRIATE LICENSE APPLICATION GUIDE FOR DETAILED INSTRUCTIONS FOR COMPLETING APPLICATION. SEND TWO COPIES OF THE ENTIRE COMPLETED APPLICATION TO THE NRC OFFICE SPECIFIED BELOW.

APPLICATION FOR DISTRIBUTION OF EXEMPT PRODUCTS FILE APPLICATIONS WITH:

DIVISION OF INDUSTRIAL AND MEDICAL NUCLEAR SAFETY
OFFICE OF NUCLEAR MATERIALS SAFETY AND SAFEGUARDS
U.S. NUCLEAR REGULATORY COMMISSION
WASHINGTON, DC 20555-0001

ALL OTHER PERSONS FILE APPLICATIONS AS FOLLOWS:

IF YOU ARE LOCATED IN:

ALABAMA, CONNECTICUT, DELAWARE, DISTRICT OF COLUMBIA, FLORIDA, GEORGIA, KENTUCKY, MAINE, MARYLAND, MASSACHUSETTS, NEW HAMPSHIRE, NEW JERSEY, NEW YORK, NORTH CAROLINA, PENNSYLVANIA, PUERTO RICO, RHODE ISLAND, SOUTH CAROLINA, TENNESSEE, VERMONT, VIRGINIA, VIRGIN ISLANDS, OR WEST VIRGINIA, SEND APPLICATIONS TO:

LICENSING ASSISTANCE TEAM
DIVISION OF NUCLEAR MATERIALS SAFETY
U.S. NUCLEAR REGULATORY COMMISSION, REGION I
475 ALLENDALE ROAD
KING OF PRUSSIA, PA 19406-1415

IF YOU ARE LOCATED IN:

ILLINOIS, INDIANA, IOWA, MICHIGAN, MINNESOTA, MISSOURI, OHIO, OR WISCONSIN, SEND APPLICATIONS TO:

MATERIALS LICENSING BRANCH
U.S. NUCLEAR REGULATORY COMMISSION, REGION III
2443 WARRENVILLE ROAD, SUITE 210
LISLE, IL 60532-4352

ALASKA, ARIZONA, ARKANSAS, CALIFORNIA, COLORADO, HAWAII, IDAHO, KANSAS, LOUISIANA, MISSISSIPPI, MONTANA, NEBRASKA, NEVADA, NEW MEXICO, NORTH DAKOTA, OKLAHOMA, OREGON, PACIFIC TRUST TERRITORIES, SOUTH DAKOTA, TEXAS, UTAH, WASHINGTON, OR WYOMING, SEND APPLICATIONS TO:

NUCLEAR MATERIALS LICENSING BRANCH
U.S. NUCLEAR REGULATORY COMMISSION, REGION IV
611 RYAN PLAZA DRIVE, SUITE 400
ARLINGTON, TX 76011-4005

PERSONS LOCATED IN AGREEMENT STATES SEND APPLICATIONS TO THE U.S. NUCLEAR REGULATORY COMMISSION ONLY IF THEY WISH TO POSSESS AND USE LICENSED MATERIAL IN STATES SUBJECT TO U.S.NUCLEAR REGULATORY COMMISSION JURISDICTIONS.

1. THIS IS AN APPLICATION FOR *(Check appropriate item)* ☐ A. NEW LICENSE ☐ B. AMENDMENT TO LICENSE NUMBER _____ ☐ C. RENEWAL OF LICENSE NUMBER _____	2. NAME AND MAILING ADDRESS OF APPLICANT *(Include ZIP code)*
3. ADDRESS WHERE LICENSED MATERIAL WILL BE USED OR POSSESSED	4. NAME OF PERSON TO BE CONTACTED ABOUT THIS APPLICATION TELEPHONE NUMBER

SUBMIT ITEMS 5 THROUGH 11 ON 8-1/2 X 11" PAPER. THE TYPE AND SCOPE OF INFORMATION TO BE PROVIDED IS DESCRIBED IN THE LICENSE APPLICATION GUIDE.

5. RADIOACTIVE MATERIAL a. Element and mass number; b. chemical and/or physical form; and c. maximum amount which will be possessed at any one time.	6. PURPOSE(S) FOR WHICH LICENSED MATERIAL WILL BE USED.
7. INDIVIDUAL(S) RESPONSIBLE FOR RADIATION SAFETY PROGRAM AND THEIR TRAINING EXPERIENCE.	8. TRAINING FOR INDIVIDUALS WORKING IN OR FREQUENTING RESTRICTED AREAS.
9. FACILITIES AND EQUIPMENT.	10. RADIATION SAFETY PROGRAM.
11. WASTE MANAGEMENT.	12. LICENSE FEES *(See 10 CFR 170 and Section 170.31)* FEE CATEGORY AMOUNT ENCLOSED $

13. CERTIFICATION. *(Must be completed by applicant)* THE APPLICANT UNDERSTANDS THAT ALL STATEMENTS AND REPRESENTATIONS MADE IN THIS APPLICATION ARE BINDING UPON THE APPLICANT.

THE APPLICANT AND ANY OFFICIAL EXECUTING THIS CERTIFICATION ON BEHALF OF THE APPLICANT, NAMED IN ITEM 2, CERTIFY THAT THIS APPLICATION IS PREPARED IN CONFORMITY WITH TITLE 10, CODE OF FEDERAL REGULATIONS, PARTS 30, 32, 33, 34, 35, 36, 39, AND 40, AND THAT ALL INFORMATION CONTANED HEREIN IS TRUE AND CORRECT TO THE BEST OF THEIR KNOWLEDGE AND BELIEF.

WARNING: 18 U.S.C. SECTION 1001 ACT OF JUNE 25, 1948 62 STAT. 749 MAKES IT A CRIMINAL OFFENSE TO MAKE A WILLFULLY FALSE STATEMENT OR REPRESENTATION TO ANY DEPARTMENT OR AGENCY OF THE UNITED STATES AS TO ANY MATTER WITHIN ITS JURISDICTION.

CERTIFYING OFFICER -- TYPED/PRINTED NAME AND TITLE	SIGNATURE	DATE

FOR NRC USE ONLY

TYPE OF FEE	FEE LOG	FEE CATEGORY	AMOUNT RECEIVED $	CHECK NUMBER	COMMENTS
APPROVED BY				DATE	

APPENDIX B

NRC Form 313A Series
"Medical Use Training and Experience and Preceptor Attestation"

Note: The most current versions of these forms are found on NRC's public Web site at http://www.nrc.gov/materials/miau/med-use-toolkit.html (Medical Uses Toolkit).

NRC FORM 313A (RSO) (2-2007)	U.S. NUCLEAR REGULATORY COMMISSION	

RADIATION SAFETY OFFICER TRAINING AND EXPERIENCE AND PRECEPTOR ATTESTATION
[10 CFR 35.50]

APPROVED BY OMB: NO. 3150-0120
EXPIRES: 10/31/2008

Name of Proposed Radiation Safety Officer

Requested Authorization(s) *The license authorizes the following medical uses (check all that apply):*

☐ 35.100   ☐ 35.200   ☐ 35.300   ☐ 35.400   ☐ 35.500   ☐ 35.600 (remote afterloader)

☐ 35.600 (teletherapy)   ☐ 35.600 (gamma stereotactic radiosurgery)   ☐ 35.1000 (_____)

PART I -- TRAINING AND EXPERIENCE
(Select one of the four methods below)

*Training and Experience, including board certification, must have been obtained within the 7 years preceding the date of application or the individual must have obtained related continuing education and experience since the required training and experience was completed. Provide dates, duration, and description of continuing education and experience related to the uses checked above.

☐ **1. Board Certification**

  a. Provide a copy of the board certification.

  b. Use Table 3.c. to describe training in radiation safety, regulatory issues, and emergency procedures for all types of medical use on the license.

  c. Skip to and complete Part II Preceptor Attestation.

OR

☐ **2. Current Radiation Safety Officer Seeking Authorization to Be Recognized as a Radiation Safety Officer for the Additional Medical Uses Checked Above**

  a. Use the table in section 3.c. to describe training in radiation safety, regulatory issues, and emergency procedures for the additional types of medical use for which recognition as RSO is sought.

  b. Skip to and complete Part II Preceptor Attestation.

OR

☐ **3. Structured Educational Program for Proposed Radiation Safety Officer**

  a. Classroom and Laboratory Training

Description of Training	Location of Training	Clock Hours	Dates of Training*
Radiation physics and instrumentation			
Radiation protection			
Mathematics pertaining to the use and measurement of radioactivity			
Radiation biology			
Radiation dosimetry			
Total Hours of Training:			

3. <u>**Structured Educational Program for Proposed Radiation Safety Officer**</u> (continued)

   b. Supervised Radiation Safety Experience
   (If more than one supervising individual is necessary to document supervised work experience, provide multiple copies of this section.)

Description of Experience	Location of Training/ License or Permit Number of Facility	Dates of Training*
Shipping, receiving, and performing related radiation surveys		
Using and performing checks for proper operation of instruments used to determine the activity of dosages, survey meters, and instruments used to measure radionuclides		
Securing and controlling byproduct material		
Using administrative controls to avoid mistakes in administration of byproduct material		
Using procedures to prevent or minimize radioactive contamination and using proper decontamination procedures		
Using emergency procedures to control byproduct material		
Disposing of byproduct material		
Licensed Material Used (e.g., 35.100, 35.200, etc.)+ _____ _____ _____ _____		

+ Choose all applicable sections of 10 CFR Part 35 to describe radioisotopes and quantities used: 35.100, 35.200, 35.300, 35.400, 35.500, 35.600 remote afterloader units, 35.600 teletherapy units, 35.600 gamma stereotactic radiosurgery units, emerging technologies (provide list of devices).

NRC FORM 313A (RSO)	U.S. NUCLEAR REGULATORY COMMISSION
(2-2007)	

RADIATION SAFETY OFFICER TRAINING AND EXPERIENCE AND PRECEPTOR ATTESTATION (continued)

3. Structured Educational Program for Proposed Radiation Safety Officer (continued)

b. Supervised Radiation Safety Experience (continued)

(If more than one supervising individual is necessary to document supervised work experience, provide multiple copies of this section.)

Supervising Individual	License/Permit Number listing supervising individual as a Radiation Safety Officer

This license authorizes the following medical uses:

☐ 35.100 ☐ 35.200 ☐ 35.300 ☐ 35.400

☐ 35.500 ☐ 35.600 (remote afterloader) ☐ 35.600 (teletherapy)

☐ 35.600 (gamma stereotactic radiosurgery) ☐ 35.1000 (_____)

c. Describe training in radiation safety, regulatory issues, and emergency procedures for all types of medical use on the license.

Description of Training	Training Provided By	Dates of Training*
Radiation safety, regulatory issues, and emergency procedures for 35.100, 35.200, and 35.500 uses		
Radiation safety, regulatory issues, and emergency procedures for 35.300 uses		
Radiation safety, regulatory issues, and emergency procedures for 35.400 uses		
Radiation safety, regulatory issues, and emergency procedures for 35.600 - teletherapy uses		
Radiation safety, regulatory issues, and emergency procedures for 35.600 - remote afterloader uses		
Radiation safety, regulatory issues, and emergency procedures for 35.600 - gamma stereotactic radiosurgery uses		
Radiation safety, regulatory issues, and emergency procedures for 35.1000, specify use(s):		

PAGE 3

NRC FORM 313A (RSO)
(2-2007)

U.S. NUCLEAR REGULATORY COMMISSION

RADIATION SAFETY OFFICER TRAINING AND EXPERIENCE AND PRECEPTOR ATTESTATION (continued)

3. Structured Educational Program for Proposed Radiation Safety Officer (continued)

c. Training in radiation safety, regulatory issues, and emergency procedures for all types of medical use on the license (continued)

Supervising Individual *If training was provided by supervising RSO, AU, AMP, or ANP. (If more than one supervising individual is necessary to document supervised training, provide multiple copies of this page.)*	License/Permit Number listing supervising individual

License/Permit lists supervising individual as:

☐ Radiation Safety Officer   ☐ Authorized User   ☐ Authorized Nuclear Pharmacist

☐ Authorized Medical Physicist

Authorized as RSO, AU, ANP, or AMP for the following medical uses:

☐ 35.100   ☐ 35.200   ☐ 35.300   ☐ 35.400

☐ 35.500   ☐ 35.600 (remote afterloader)   ☐ 35.600 (teletherapy)

☐ 35.600 (gamma stereotactic radiosurgery)   ☐ 35.1000 (_____)

d. Skip to and complete Part II Preceptor Attestation.

OR

☐ **4. Authorized User, Authorized Medical Physicist, or Authorized Nuclear Pharmacist identified on the licensee's license**

a. Provide license number.

b. Use the table in section 3.c. to describe training in radiation safety, regulatory issues, and emergency procedures for all types of medical use on the license.

c. Skip to and complete Part II Preceptor Attestation.

PART II – PRECEPTOR ATTESTATION

Note: This part must be completed by the individual's preceptor. The preceptor does not have to be the supervising individual as long as the preceptor provides, directs, or verifies training and experience required. If more than one preceptor is necessary to document experience, obtain a separate preceptor statement from each.

First Section
Check one of the following:

☐ **1. Board Certification**

☐ I attest that _____ has satisfactorily completed the requirements in
                  Name of Proposed Radiation Safety Officer

10 CFR 35.50(a)(1)(i) and (a)(1)(ii); or 35.50 (a)(2)(i) and (a)(2)(ii); or 35.50(c)(1).

OR

☐ **2. Structured Educational Program for Proposed Radiation Safety Officers**

☐ I attest that _____ has satisfactorily completed a structural educational
                  Name of Proposed Radiation Safety Officer

program consisting of both 200 hours of classroom and laboratory training and one year of full-time radiation safety experience as required by 10 CFR 35.50(b)(1).

OR

PAGE 4

RADIATION SAFETY OFFICER TRAINING AND EXPERIENCE AND PRECEPTOR ATTESTATION (continued)

Preceptor Attestation (continued)

First Section (continued)
Check one of the following:

☐ **3. Additional Authorization as Radiation Safety Officer**

☐ I attest that _____ is an
Name of Proposed Radiation Safety Officer

☐ Authorized User   ☐ Authorized Nuclear Pharmacist

☐ Authorized Medical Physicist

identified on the Licensees license and has experience with the radiation safety aspects of similar type of use of byproduct material for which the individual has Radiation Safety Officer responsibilities

- -

AND

Second Section
Complete for all *(check all that apply)*:

☐ I attest that _____ has training in the radiation safety, regulatory issues, and
Name of ProposedRadiation Safety Officer
emergency procedures for the following types of use:

☐ 35.100

☐ 35.200

☐ 35.300   oral administration of less than or equal to 33 millicuries of sodium iodide I-131, for which a written directive is required

☐ 35.300   oral administration of greater than 33 millicuries of sodium iodide I-131

☐ 35.300   parenteral administration of any beta-emitter, or a photon-emitting radionuclide with a photon energy less than 150 keV for which a written directive is required

☐ 35.300   parenteral administration of any other radionuclide for which a written directive is required

☐ 35.400

☐ 35.500

☐ 35.600   remote afterloader units

☐ 35.600   teletherapy units

☐ 35.600   gamma stereotactic radiosurgery units

☐ 35.1000   emerging technologies, including:

_____

_____

_____

APPENDIX B

AND

Third Section
Complete for ALL

☐ I attest that _____ has achieved a level of radiation safety knowledge

Name of Proposed Radiation Safety Officer

sufficient to function independently as a Radiation Safety Officer for a medical use licensee.

- -

Fourth Section
Complete the following for Preceptor Attestation and signature

I am the Radiation Safety Officer for _____

Name of Facility

License/Permit Number: _____

Name of Preceptor	Signature	Telephone Number	Date

NRC FORM 313A (AMP) (10-2006)	U.S. NUCLEAR REGULATORY COMMISSION	
AUTHORIZED MEDICAL PHYSICIST TRAINING AND EXPERIENCE AND PRECEPTOR ATTESTATION **[10 CFR 35.51]**		APPROVED BY OMB: NO. 3150-0120 EXPIRES: 10/31/2008

Name of Proposed Authorized Medical Physicist

Requested Authorization(s)
(check all that apply)

☐ 35.400 Ophthalmic use of strontium-90   ☐ 35.600 Teletherapy unit(s)

☐ 35.600 Remote afterloader unit(s)   ☐ 35.600 Gamma stereotactic radiosurgery unit(s)

PART I -- TRAINING AND EXPERIENCE
(Select one of the three methods below)

*Training and Experience, including Board Certification, must have been obtained within the 7 years preceding the date of application or the individual must have obtained related continuing education and experience since the required training and experience was completed. Provide dates, duration, and description of continuing education and experience related to the uses checked above.

☐ **1. Board Certification**

    a. Provide a copy of the board certification.

    b. Go to the table in 3.c. and describe training provider and dates of training for each type of use for which authorization is sought.

    c. Skip to and complete Part II Preceptor Attestation.

☐ **2. Current Authorized Medical Physicist Seeking Additional Authorization for use(s) checked above**

    a. Go to the table in section 3.c. to document training for new device.

    b. Skip to and complete Part II Preceptor Attestation

☐ **3. Education, Training, and Experience for Proposed Authorized Medical Physicist**

    a. Education: Document master's or doctor's degree in physics, medical physics, other physical science, engineering, or applied mathematics from an accredited college or university.

Degree	Major Field
College or University	

    b. Supervised Full-Time Medical Physics Training and Work Experience in clinical radiation facilities that provide high-energy external beam therapy (photons and electrons with energies greater than or equal to 1 million electron volts) and brachytherapy services.

      ☐ Yes. Completed 1 year of full-time training in medical physics (for areas identified below) under the supervision of _____ who meets the requirements for an Authorized Medical Physicist.

AND

      ☐ Yes. Completed 1 year of full-time work experience in medical physics (for areas identified below) under the supervision of _____ who meets the requirements for an Authorized Medical Physicist.

NRC FORM 313A (AMP)
(10-2006)

U.S. NUCLEAR REGULATORY COMMISSION

AUTHORIZED MEDICAL PHYSICIST TRAINING AND EXPERIENCE AND PRECEPTOR ATTESTATION (continued)

3. **Education, Training, and Experience for Proposed Authorized Medical Physicist** (continued)

   b. Supervised Full-Time Medical Physics Training and Work Experience (continued)

     If more than one supervising individual is necessary to document supervised training, provide multiple copies of this page.

Description of Training/ Experience	Location of Training/License or Permit Number of Training Facility/Medical Devices Used+	Dates of Training*	Dates of Work Experience*
Medical Physics			
Performing sealed source leak tests and inventories			
Performing decay corrections			
Performing full calibration and periodic spot checks of external beam treatment unit(s)			
Performing full calibration and periodic spot checks of stereotactic radiosurgery unit(s)			
Performing full calibration and periodic spot checks of remote afterloading unit(s)			
Conducting radiation surveys around external beam treatment unit(s), stereotactic radiosurgery unit(s), remote after loading unit(s)			

Supervising Individual**   License/Permit Number listing supervising individual as an authorized Medical Physicist

for the following types of use:

☐ Remote afterloader unit(s)   ☐ Teletherapy unit(s)   ☐ Gamma stereotactic radiosurgery unit(s)

+   Training and work experience must be conducted in clinical radiation facilities that provide high-energy external beam therapy (photons and electrons with energies greater than or equal to 1 million electron volts) and brachytherapy services.

*   1 year of Full-time medical physics training and 1 year of full time work experience cannot be concurrent.

**   If the supervising medical physicist is not an authorized medical physicist, the licensee must submit evidence that the supervising medical physicist meets the training and experience requirements in 10 CFR 35.51 and 35.59 for the types of use for which the individual is seeking authorization.

NRC FORM 313A (AMP)
(10-2006)

U.S. NUCLEAR REGULATORY COMMISSION

AUTHORIZED MEDICAL PHYSICIST TRAINING AND EXPERIENCE AND PRECEPTOR ATTESTATION (continued)

3. Education, Training, and Experience for Proposed Authorized Medical Physicist (continued)

c. Describe training provider and dates of training for each type of use for which authorization is sought.

Description of Training	Training Provider and Dates		
	Remote Afterloader	Teletherapy	Gamma Stereotactic Radiosurgery
Hands-on device operation			
Safety procedures for the device use			
Clinical use of the device			
Treatment planning system operation			

Supervising Individual If training is provided by Supervising Medical Physicist, (If more than one supervising individual is necessary to document supervised training, provide multiple copies of this page.)	License/Permit Number listing supervising individual as an authorized Medical Physicist

for the following types of use:

☐ Remote afterloader unit(s)   ☐ Teletherapy unit(s)   ☐ Gamma stereotactic radiosurgery unit(s)

If Applicable:

Authorization Sought	Device	Training Provided By	Dates of Training
35.400 Ophthalmic Use of strontium-90			

d. Skip to and complete Part II Preceptor Attestation.

NRC FORM 313A (AMP)	U.S. NUCLEAR REGULATORY COMMISSION
(10-2006)	

AUTHORIZED MEDICAL PHYSICIST TRAINING AND EXPERIENCE AND PRECEPTOR ATTESTATION (continued)

PART II – PRECEPTOR ATTESTATION

Note: This part must be completed by the individual's preceptor. The preceptor does not have to be the supervising individual as long as the preceptor provides, directs, or verifies training and experience required. If more than one preceptor is necessary to document experience, obtain a separate preceptor statement from each.

First Section
Check one of the following:

1. **Board Certification**

   [ ] I attest that _____ has satisfactorily completed the requirements in
   Name of Proposed Authorized Medical Physicist
   10 CFR 35.51(a)(1) and (a)(2).

OR

2. **Education, Training, and Experience**

   [ ] I attest that _____ has satisfactorily completed the 1-year of full-time
   Name of Proposed Authorized Medical Physicist
   training in medical physics and an additional year of full-time work experience as required by 10 CFR 35.51(b)(1).

- -

AND

Second Section
Complete the following:

   [ ] I attest that _____ has training for the types of use for which authorization
   Name of Proposed Authorized Medical Physicist
   is sought that include hands-on device operation, safety procedures, clinical use, and the operation of a treatment planning system.

- -

AND

Third Section
Complete the following:

   [ ] I attest that _____ has achieved a level of competency sufficient to
   Name of Proposed Authorized Medical Physicist
   function independently as an Authorized Medical Physicist for the following:

   [ ] 35.400 Ophthalmic use of strontium-90   [ ] 35.600 Teletherapy unit(s)

   [ ] 35.600 Remote afterloader unit(s)   [ ] 35.600 Gamma stereotactic radiosurgery unit(s)

- -

AND

Fourth Section
Complete the following for preceptor attestation and signature:

   [ ] I meet the requirements in 10 CFR 35.51, or equivalent Agreement State requirements for Authorized Medical Physicist for the following:

   [ ] 35.400 Ophthalmic use of strontium-90   [ ] 35.600 Teletherapy unit(s)

   [ ] 35.600 Remote afterloader unit(s)   [ ] 35.600 Gamma stereotactic radiosurgery unit(s)

Name of Preceptor	Signature	Telephone Number	Date

License/Permit Number/Facility Name

PAGE 4

NRC FORM 313A (ANP) (10-2006)	U.S. NUCLEAR REGULATORY COMMISSION	APPROVED BY OMB: NO. 3150-0120 EXPIRES: 10/31/2008

AUTHORIZED NUCLEAR PHARMACIST TRAINING AND EXPERIENCE AND PRECEPTOR ATTESTATION
[10 CFR 35.55]

Name of Proposed Authorized Nuclear Pharmacist	State or Territory Where Licensed

PART I -- TRAINING AND EXPERIENCE
(Select one of the two methods below)

* Training and Experience, including board certification, must have been obtained within the 7 years preceding the date of application or the individual must have obtained related continuing education and experience since the required training and experience was completed. Provide dates, duration, and description of continuing education and experience related to the nuclear pharmacy uses.

☐ **1. Board Certification**

    a. Provide a copy of the board certification.

    b. Skip to and complete Part II Preceptor Attestation.

☐ **2. Structured Educational Program for Proposed Authorized Nuclear Pharmacist**

    a. Classroom and Laboratory Training.

Description of Training	Location of Training	Clock Hours	Dates of Training*
Radiation physics and instrumentation			
Radiation protection			
Mathematics pertaining to the use and measurement of radioactivity			
Chemistry of byproduct material for medical use			
Radiation biology			
Total Hours of Training:			

APPENDIX B

AUTHORIZED NUCLEAR PHARMACIST TRAINING AND EXPERIENCE
AND PRECEPTOR ATTESTATION (continued)

2. **Structured Educational Program for Proposed Authorized Nuclear Pharmacist** (continued)

   b. Supervised Practical Experience in a Nuclear Pharmacy.

Description of Experience	Location of Experience/License or Permit Number of Facility	Clock Hours	Dates of Experience*
Shipping, receiving, and performing related radiation surveys			
Using and performing checks for proper operation of instruments used to determine the activity of dosages, survey meters, and, if appropriate, instruments used to measure alpha- or beta-emitting radionuclides			
Calculating, assaying, and safely preparing dosages for patients or human research subjects			
Using administrative controls to avoid medical events in administration of byproduct material			
Using procedures to prevent or minimize radioactive contamination and using proper decontamination procedures			
Total Hours of Experience:			
Supervising Individual			

   c. Go to and complete Part II Preceptor Attestation.

PAGE 2

NRC FORM 313A (ANP) (10-2006)	U.S. NUCLEAR REGULATORY COMMISSION

AUTHORIZED NUCLEAR PHARMACIST TRAINING AND EXPERIENCE
AND PRECEPTOR ATTESTATION (continued)

PART II – PRECEPTOR ATTESTATION

Note: This part must be completed by the individual's preceptor. The preceptor does not have to be the supervising individual as long as the preceptor provides, directs, or verifies training and experience required. If more than one preceptor is necessary to document experience, obtain a separate preceptor statement from each.

First Section
Check one of the following:

Board Certification

☐ I attest that _____ has satisfactorily completed the requirements in
Name of Proposed Authorized Nuclear Pharmacist

10 CFR 35.55(a)(1), (a)(2), and (a)(3) and has achieved a level of competency sufficient to function independently as an authorized nuclear pharmacist.

OR

Structured Educational Program

☐ I attest that _____ has satisfactorily completed a 700-hour structured
Name of Proposed Authorized Nuclear Pharmacist

educational program consisting of both 200 hours of classroom and laboratory training, and practical experience in nuclear pharmacy, as required by 10 CFR 35.55(b)(1) and has achieved a level of competency sufficient to function independently as an authorized nuclear pharmacist.

- -

Second Section
Complete the following for preceptor attestation and signature:

I am an Authorized Nuclear Pharmacist for _____ ,
Nuclear Pharmacy or Medical Facility

_____
License/Permit Number

Name of Preceptor	Signature	Telephone Number	Date

NRC FORM 313A (AUD) (10-2007)	U.S. NUCLEAR REGULATORY COMMISSION	
AUTHORIZED USER TRAINING AND EXPERIENCE AND PRECEPTOR ATTESTATION **(for uses defined under 35.100, 35.200, and 35.500)** **[10 CFR 35.190, 35.290, and 35.590]**		**APPROVED BY OMB: NO. 3150-0120** **EXPIRES: 10/31/2008**

Name of Proposed Authorized User	State or Territory Where Licensed

Requested Authorization(s) *(check all that apply)*

☐ 35.100 Uptake, dilution, and excretion studies

☐ 35.200 Imaging and localization studies

☐ 35.500 Sealed sources for diagnosis (specify device _____)

PART I -- TRAINING AND EXPERIENCE
(Select one of the three methods below)

* Training and Experience, including board certification, must have been obtained within the 7 years preceding the date of application or the individual must have obtained related continuing education and experience since the required training and experience was completed. Provide dates, duration, and description of continuing education and experience related to the uses checked above.

☐ **1. Board Certification**

    a. Provide a copy of the board certification.

    b. If using only 35.500 materials, stop here. If using 35.100 and 35.200 materials, skip to and complete Part II Preceptor Attestation.

☐ **2. Current 35.390 Authorized User Seeking Additional 35.290 Authorization**

    a. Authorized user on Materials License _____ meeting 10 CFR 35.390 or equivalent Agreement State requirements seeking authorization for 35.290.

    b. Supervised Work Experience.
    (If more than one supervising individual is necessary to document supervised work experience, provide multiple copies of this section.)

Description of Experience	Location of Experience/License or Permit Number of Facility	Clock Hours	Dates of Experience*
Eluting generator systems appropriate for the preparation of radioactive drugs for imaging and localization studies, measuring and testing the eluate for radionuclidic purity, and processing the eluate with reagent kits to prepare labeled radioactive drugs			
Total Hours of Experience:			

Supervising Individual	License/Permit Number listing supervising individual as an authorized user

Supervisor meets the requirements below, or equivalent Agreement State requirements *(check all that apply)*.

    ☐ 35.290   ☐ 35.390 + generator experience in 32.290(c)(1)(ii)(G)

APPENDIX B

☐ 3. **Training and Experience for Proposed Authorized User**

    a. Classroom and Laboratory Training.

Description of Training	Location of Training	Clock Hours	Dates of Training*
Radiation physics and instrumentation			
Radiation protection			
Mathematics pertaining to the use and measurement of radioactivity			
Chemistry of byproduct material for medical use *(not required for 35.590)*			
Radiation biology			
Total Hours of Training:			

    b. Supervised Work Experience (completion of this table is not required for 35.590).
    (If more than one supervising individual is necessary to document supervised work experience, provide multiple copies of this section.)

Supervised Work Experience		Total Hours of Experience:	
Description of Experience Must Include:	Location of Experience/License or Permit Number of Facility	Confirm	Dates of Experience*
Ordering, receiving, and unpacking radioactive materials safely and performing the related radiation surveys		☐ Yes ☐ No	
Performing quality control procedures on instruments used to determine the activity of dosages and performing checks for proper operation of survey meters		☐ Yes ☐ No	

NRC FORM 313A (AUD)
(10-2007)

U.S. NUCLEAR REGULATORY COMMISSION

AUTHORIZED USER TRAINING AND EXPERIENCE AND PRECEPTOR ATTESTATION (continued)

3. Training and Experience for Proposed Authorized User (continued)

b. Supervised Work Experience. (continued)

Description of Experience Must Include:	Location of Experience/License or Permit Number of Facility	Confirm	Dates of Experience*
Calculating, measuring, and safely preparing patient or human research subject dosages		☐ Yes ☐ No	
Using administrative controls to prevent a medical event involving the use of unsealed byproduct material		☐ Yes ☐ No	
Using procedures to contain spilled byproduct material safely and using proper decontamination procedures		☐ Yes ☐ No	
Administering dosages of radioactive drugs to patients or human research subjects		☐ Yes ☐ No	
Eluting generator systems appropriate for the preparation of radioactive drugs for imaging and localization studies, measuring and testing the eluate for radionuclidic purity, and processing the eluate with reagent kits to prepare labeled radioactive drugs		☐ Yes ☐ No	

Supervising Individual	License/Permit Number listing supervising individual as an authorized user

Supervisor meets the requirements below, or equivalent Agreement State requirements *(check one)*.

☐ 35.190   ☐ 35.290   ☐ 35.390   ☐ 35.390 + generator experience in 35.290(c)(1)(ii)(G)

c. For 35.590 only, provide documentation of training on use of the device.

Device	Type of Training	Location and Dates

d. For 35.500 uses only, stop here. For 35.100 and 35.200 uses, skip to and complete Part II Preceptor Attestation.

NRC FORM 313A (AUD)	U.S. NUCLEAR REGULATORY COMMISSION
(10-2007) AUTHORIZED USER TRAINING AND EXPERIENCE AND PRECEPTOR ATTESTATION (continued)	

PART II – PRECEPTOR ATTESTATION

Note: This part must be completed by the individual's preceptor. The preceptor does not have to be the supervising individual as long as the preceptor provides, directs, or verifies training and experience required. If more than one preceptor is necessary to document experience, obtain a separate preceptor statement from each. (Not required to meet training requirements in 35.590)

By checking the boxes below, the preceptor is attesting that the individual has knowledge to fulfill the duties of the position sought and not attesting to the individual's "general clinical competency."

First Section
Check one of the following for each use requested:

For 35.190

Board Certification

☐ I attest that _____ has satisfactorily completed the requirements in
Name of Proposed Authorized User

10 CFR 35.190(a)(1) and has achieved a level of competency sufficient to function independently as an authorized user for the medical uses authorized under 10 CFR 35.100.

OR

Training and Experience

☐ I attest that _____ has satisfactorily completed the 60 hours of training and
Name of Proposed Authorized User

experience, including a minimum of 8 hours of classroom and laboratory training, required by 10 CFR 35.190(c)(1), and has achieved a level of competency sufficient to function independently as an authorized user for the medical uses authorized under 10 CFR 35.100.

For 35.290

Board Certification

☐ I attest that _____ has satisfactorily completed the requirements in
Name of Proposed Authorized User

10 CFR 35.290(a)(1) and has achieved a level of competency sufficient to function independently as an authorized user for the medical uses authorized under 10 CFR 35.100 and 35.200.

OR

Training and Experience

☐ I attest that _____ has satisfactorily completed the 700 hours of training
Name of Proposed Authorized User

and experience, including a minimum of 80 hours of classroom and laboratory training, required by 10 CFR 35.290(c)(1), and has achieved a level of competency sufficient to function independently as an authorized user for the medical uses authorized under 10 CFR 35.100 and 35.200.

Second Section
Complete the following for preceptor attestation and signature:

☐ I meet the requirements below, or equivalent Agreement State requirements, as an authorized user for:

☐ 35.190　　☐ 35.290　　☐ 35.390　　☐ 35.390 + generator experience

Name of Preceptor	Signature	Telephone Number	Date

License/Permit Number/Facility Name

PAGE 4

NRC FORM 313A (AUT) (10-2007)	U.S. NUCLEAR REGULATORY COMMISSION	
AUTHORIZED USER TRAINING AND EXPERIENCE **AND PRECEPTOR ATTESTATION** **(for uses defined under 35.300)** **[10 CFR 35.390, 35.392, 35.394, and 35.396]**		APPROVED BY OMB: NO. 3150-0120 EXPIRES: 10/31/2008

Name of Proposed Authorized User	State or Territory Where Licensed

Requested Authorization(s) *(check all that apply)*:

☐ 35.300 Use of unsealed byproduct material for which a written directive is required

OR

☐ 35.300 Oral administration of sodium iodide I-131 requiring a written directive in quantities less than or equal to 1.22 gigabecquerels (33 millicuries)

☐ 35.300 Oral administration of sodium iodide I-131 requiring a written directive in quantities greater than 1.22 gigabecquerels (33 millicuries)

☐ 35.300 Parenteral administration of any beta-emitter, or photon-emitting radionuclide with a photon energy less than 150 keV for which a written directive is required

☐ 35.300 Parenteral administration of any other radionuclide for which a written directive is required

PART I -- TRAINING AND EXPERIENCE
(Select one of the three methods below)

* Training and Experience, including board certification, must have been obtained within the 7 years preceding the date of application or the individual must have related continuing education and experience since the required training and experience was completed. Provide dates, duration, and description of continuing education and experience related to the uses checked above.

☐ 1. **Board Certification**

   a. Provide a copy of the board certification.

   b. For 35.390, provide documentation on supervised clinical case experience. The table in section 3.c. may be used to document this experience.

   c. For 35.396, provide documentation on classroom and laboratory training, supervised work experience, and supervised clinical case experience. The tables in sections 3.a., 3.b., and 3.c. may be used to document this experience.

   d. Skip to and complete Part II Preceptor Attestation.

☐ 2. **Current 35.300, 35.400, or 35.600 Authorized User Seeking Additional Authorization**

   a. Authorized User on Materials License _____ under the requirements below or equivalent Agreement State requirements *(check all that apply)*:

      ☐ 35.390   ☐ 35.392   ☐ 35.394   ☐ 35.490   ☐ 35.690

   b. If currently authorized for a subset of clinical uses under 35.300, provide documentation on additional required supervised case experience. The table in section 3.c. may be used to document this experience. Also provide completed Part II Preceptor Attestation.

   c. If currently authorized under 35.490 or 35.690 and requesting authorization for 35.396, provide documentation on classroom and laboratory training, supervised work experience, and supervised clinical case experience. The tables in sections 3.a., 3.b., and 3.c. may be used to document this experience. Also provide completed Part II Preceptor Attestation.

NRC FORM 313A (AUT)
(10-2007)

U.S. NUCLEAR REGULATORY COMMISSION

AUTHORIZED USER TRAINING AND EXPERIENCE AND PRECEPTOR ATTESTATION (continued)

☐ **3. Training and Experience for Proposed Authorized User**

a. Classroom and Laboratory Training ☐ 35.390 ☐ 35.392 ☐ 35.394 ☐ 35.396

Description of Training	Location of Training	Clock Hours	Dates of Training*
Radiation physics and instrumentation			
Radiation protection			
Mathematics pertaining to the use and measurement of radioactivity			
Chemistry of byproduct material for medical use			
Radiation biology			
Total Hours of Training:			

b. Supervised Work Experience ☐ 35.390 ☐ 35.392 ☐ 35.394 ☐ 35.396

If more than one supervising individual is necessary to document supervised training, provide multiple copies of this page.

Supervised Work Experience		**Total Hours of Experience:**	
Description of Experience Must Include:	Location of Experience/License or Permit Number of Facility	Confirm	Dates of Experience*
Ordering, receiving, and unpacking radioactive materials safely and performing the related radiation surveys		☐ Yes ☐ No	
Performing quality control procedures on instruments used to determine the activity of dosages and performing checks for proper operation of survey meters		☐ Yes ☐ No	
Calculating, measuring, and safely preparing patient or human research subject dosages		☐ Yes ☐ No	
Using administrative controls to prevent a medical event involving the use of unsealed byproduct material		☐ Yes ☐ No	
Using procedures to contain spilled byproduct material safely and using proper decontamination procedures		☐ Yes ☐ No	

NRC FORM 313A (AUT) (10-2007)	U.S. NUCLEAR REGULATORY COMMISSION

AUTHORIZED USER TRAINING AND EXPERIENCE AND PRECEPTOR ATTESTATION (continued)

3. Training and Experience for Proposed Authorized User (continued)

b. Supervised Work Experience (continued)

Supervising Individual	License/Permit Number listing supervising individual as an authorized user

Supervising individual meets the requirements below, or equivalent Agreement State requirements *(check all that apply)***:

- [ ] 35.390
- [ ] 35.392
- [ ] 35.394
- [ ] 35.396

With experience administering dosages of:

- [ ] Oral NaI-131 requiring a written directive in quantities less than or equal to 1.22 gigabecquerels (33 millicuries)
- [ ] Oral NaI-131 in quantities greater than 1.22 gigabecquerels (33 millicuries)
- [ ] Parenteral administration of beta-emitter, or photon-emitting radionuclide with a photon energy less than 150 keV requiring a written directive is required
- [ ] Parenteral administration of any other radionuclide requiring a written directive

** Supervising Authorized User must have experience in administering dosages in the same dosage category or categories as the individual requesting authorized user status.

c. Supervised Clinical Case Experience

If more than one supervising individual is necessary to document supervised work experience, provide multiple copies of this page.

Description of Experience	Number of Cases Involving Personal Participation	Location of Experience/License or Permit Number of Facility	Dates of Experience*
Oral administration of sodium iodide I-131 requiring a written directive in quantities less than or equal to 1.22 gigabecquerels (33 millicuries)			
Oral administration of sodium iodide I-131 requiring a written directive in quantities greater than 1.22 gigabecquerels (33 millicuries)			
Parenteral administration of any beta-emitter, or photon-emitting radionuclide with a photon energy less than 150 keV for which a written directive is required			
Parenteral administration of any other radionuclide for which a written directive is required _____ (List radionuclides)			

NRC FORM 313A (AUT)
(10-2007)

U.S. NUCLEAR REGULATORY COMMISSION

AUTHORIZED USER TRAINING AND EXPERIENCE AND PRECEPTOR ATTESTATION (continued)

3. Training and Experience for Proposed Authorized User (continued)

c. Supervised Clinical Case Experience (continued)

Supervising Individual	License/Permit Number listing supervising individual as an authorized user

Supervising individual meets the requirements below, or equivalent Agreement State requirements *(check all that apply)***:

☐ 35.390   With experience administering dosages of:

☐ 35.392

☐ 35.394   ☐ Oral NaI-131 requiring a written directive in quantities less than or equal to 1.22 gigabecquerels (33 millicuries)

☐ 35.396   ☐ Oral NaI-131 in quantities greater than 1.22 gigabecquerels (33 millicuries)

           ☐ Parenteral administration of beta-emitter, or photon-emitting radionuclide with a photon energy less than 150 keV requiring a written directive is required

           ☐ Parenteral administration of any other radionuclide requiring a written directive

**   Supervising Authorized User must have experience in administering dosages in the same dosage category or categories as the individual requesting authorized user status.

d. Provide completed Part II Preceptor Attestation.

PART II – PRECEPTOR ATTESTATION

Note: This part must be completed by the individual's preceptor. The preceptor does not have to be the supervising individual as long as the preceptor provides, directs, or verifies training and experience required. If more than one preceptor is necessary to document experience, obtain a separate preceptor statement from each.

By checking the boxes below, the preceptor is attesting that the individual has knowledge to fulfill the duties of the position sought and not attesting to the individual's "general clinical competency."

First Section
Check one of the following for each requested authorization:

For 35.390:

Board Certification

☐ I attest that _____ has satisfactorily completed the training and experience
             Name of Proposed Authorized User

requirements in 35.390(a)(1).

OR

Training and Experience

☐ I attest that _____ has satisfactorily completed the 700 hours of training
             Name of Proposed Authorized User

and experience, including a minimum of 200 hours of classroom and laboratory training, as required by 10 CFR 35.390 (b)(1).

PAGE 4

NRC FORM 313A (AUT) (10-2007)	U.S. NUCLEAR REGULATORY COMMISSION

AUTHORIZED USER TRAINING AND EXPERIENCE AND PRECEPTOR ATTESTATION (continued)

Preceptor Attestation (continued)

First Section (continued)

For 35.392 (Identical Attestation Statement Regardless of Training and Experience Pathway):

☐ I attest that _____ has satisfactorily completed the 80 hours of classroom
　　　　　　　　　　 Name of Proposed Authorized User

and laboratory training, as required by 10 CFR 35.392(c)(1), and the supervised work and clinical case experience required in 35.392(c)(2).

For 35.394 (Identical Attestation Statement Regardless of Training and Experience Pathway):

☐ I attest that _____ has satisfactorily completed the 80 hours of classroom
　　　　　　　　　　 Name of Proposed Authorized User

and laboratory training, as required by 10 CFR 35.394 (c)(1), and the supervised work and clinical case experience required in 35.394(c)(2).

- -

Second Section

☐ I attest that _____ has satisfactorily completed the required clinical case
　　　　　　　　　　 Name of Proposed Authorized User

experience required in 35.390(b)(1)(ii)G listed below:

☐ Oral NaI-131 requiring a written directive in quantities less than or equal to 1.22 gigabecquerels (33 millicuries)

☐ Oral NaI-131 in quantities greater than 1.22 gigabecquerels (33 millicuries)

☐ Parenteral administration of beta-emitter, or photon-emitting radionuclide with a photon energy less than 150 keV requiring a written directive is required

☐ Parenteral administration of any other radionuclide requiring a written directive

- -

Third Section

☐ I attest that _____ has satisfactorily achieved a level of competency to
　　　　　　　　　　 Name of Proposed Authorized User

function independently as an authorized user for:

☐ Oral NaI-131 requiring a written directive in quantities less than or equal to 1.22 gigabecquerels (33 millicuries)

☐ Oral NaI-131 in quantities greater than 1.22 gigabecquerels (33 millicuries)

☐ Parenteral administration of beta-emitter, or photon-emitting radionuclide with a photon energy less than 150 keV requiring a written directive is required

☐ Parenteral administration of any other radionuclide requiring a written directive

NRC FORM 313A (AUT) (10-2007)	U.S. NUCLEAR REGULATORY COMMISSION

AUTHORIZED USER TRAINING AND EXPERIENCE AND PRECEPTOR ATTESTATION (continued)

Fourth Section

For 35.396:

Current 35.490 or 35.690 authorized user:

☐ I attest that _____ is an authorized user under 10 CFR 35.490 or 35.690

Name of Proposed Authorized User

or equivalent Agreement State requirements, has satisfactorily completed the 80 hours of classroom and laboratory training, as required by 10 CFR 35.396 (d)(1), and the supervised work and clinical case experience required by 35.396(d)(2), and has achieved a level of competency sufficient to function independently as an authorized user for:

☐ Parenteral administration of any beta-emitter, or photon-emitting radionuclide with a photon energy less than 150 keV for which a written directive is required

☐ Parenteral administration of any other radionuclide for which a written directive is required

OR

Board Certification:

☐ I attest that _____ has satisfactorily completed the board certification

Name of Proposed Authorized User

requirements of 35.396(c), has satisfactorily completed the 80 hours of classroom and laboratory training required by 10 CFR 35.396 (d)(1) and the supervised work and clinical case experience required by 35.396(d)(2), and has achieved a level of competency sufficient to function independently as an authorized user for:

☐ Parenteral administration of any beta-emitter, or photon-emitting radionuclide with a photon energy less than 150 keV for which a written directive is required

☐ Parenteral adminstration of any other radionuclide for which a written directive is required

- -

Fifth Section
Complete the following for preceptor attestation and signature:

☐ I meet the requirements below, or equivalent Agreement State requirements, as an authorized user for:

☐ 35.390   ☐ 35.392   ☐ 35.394   ☐ 35.396

☐ I have experience administering dosages in the following categories for which the proposed Authorized User is requesting authorization.

☐ Oral NaI-131 requiring a written directive in quantities less than or equal to 1.22 gigabecquerels (33 millicuries)

☐ Oral NaI-131 in quantities greater than 1.22 gigabecquerels (33 millicuries)

☐ Parenteral administration of beta-emitter, or photon-emitting radionuclide with a photon energy less than 150 keV requiring a written directive is required

☐ Parenteral administration of any other radionuclide requiring a written directive

Name of Preceptor	Signature	Telephone Number	Date

License/Permit Number/Facility Name

NRC FORM 313A (AUS)
(10-2007)

U.S. NUCLEAR REGULATORY COMMISSION

AUTHORIZED USER TRAINING AND EXPERIENCE AND PRECEPTOR ATTESTATION
(for uses defined under 35.400 and 35.600)
[10 CFR 35.490, 35.491, and 35.690]

APPROVED BY OMB: NO. 3150-0120
EXPIRES: 10/31/2008

Name of Proposed Authorized User	State or Territory Where Licensed

Requested Authorization(s) (check all that apply)	☐ 35.400 Manual brachytherapy sources	☐ 35.600 Teletherapy unit(s)
	☐ 35.400 Ophthalmic use of strontium-90	☐ 35.600 Gamma stereotactic radiosurgery unit(s)
	☐ 35.600 Remote afterloader unit(s)	

PART I -- TRAINING AND EXPERIENCE
(Select one of the three methods below)

* Training and Experience, including Board Certification, must have been obtained within the 7 years preceding the date of application or the individual must have obtained related continuing education and experience since the required training and experience was completed. Provide dates, duration, and description of continuing education and experience related to the uses checked above.

☐ **1. Board Certification**

    a. Provide a copy of the board certification.

    b. For 35.600, go to the table in 3.e. and describe training provider and dates of training for each type of use for which authorization is sought.

    c. Skip to and complete Part II Preceptor Attestation.

☐ **2. Current 35.600 Authorized User Requesting Additional Authorization for 35.600 Use(s) Checked Above**

    a. Go to the table in section 3.e. to document training for new device.

    b. Skip to and complete Part II Preceptor Attestation.

☐ **3. Training and Experience for Proposed Authorized User**

    a. Classroom and Laboratory Training ☐ 35.490 ☐ 35.491 ☐ 35.690

Description of Training	Location of Training	Clock Hours	Dates of Training*
Radiation physics and instrumentation			
Radiation protection			
Mathematics pertaining to the use and measurement of radioactivity			
Radiation biology			
Total Hours of Training:			

3. Training and Experience for Proposed Authorized User (continued)

b. Supervised Work and Clinical Experience for 10 CFR 35.490 *(If more than one supervising individual is necessary to document supervised work experience, provide multiple copies of this page.)*

Supervised Work Experience		Total Hours of Experience:	
Description of Experience Must Include:	Location of Experience/License or Permit Number of Facility	Confirm	Dates of Experience*
Ordering, receiving, and unpacking radioactive materials safely and performing the related radiation surveys		☐ Yes ☐ No	
Checking survey meters for proper operation		☐ Yes ☐ No	
Preparing, implanting, and safely removing brachytherapy sources		☐ Yes ☐ No	
Maintaining running inventories of material on hand		☐ Yes ☐ No	
Using administrative controls to prevent a medical event involving the use of byproduct material		☐ Yes ☐ No	
Using emergency procedures to control byproduct material		☐ Yes ☐ No	

Clinical experience in radiation oncology as part of an approved formal training program	Location of Experience/License or Permit Number of Facility	Dates of Experience*
Approved by: ☐ Residency Review Committee for Radiation Oncology of the ACGME ☐ Royal College of Physicians and Surgeons of Canada ☐ Committee on Postdoctoral Training of the American Osteopathic Association		
Supervising Individual	License/Permit Number listing supervising individual as an Authorized User	

NRC FORM 313A (AUS)
(10-2007)

U.S. NUCLEAR REGULATORY COMMISSION

AUTHORIZED USER TRAINING AND EXPERIENCE AND PRECEPTOR ATTESTATION (continued)

3. Training and Experience for Proposed Authorized User (continued)

c. Supervised Clinical Experience for 10 CFR 35.491

Description of Experience	Location of Experience/License or Permit Number of Facility	Clock Hours	Dates of Experience*
Use of strontium-90 for ophthalmic treatment, including: examination of each individual to be treated; calculation of the dose to be administered; administration of the dose; and follow up and review of each individual's case history			
Supervising Individual	License/Permit Number listing supervising individual as an Authorized User		

d. Supervised Work and Clinical Experience for 10 CFR 35.690

☐ Remote afterloader unit(s)   ☐ Teletherapy unit(s)   ☐ Gamma stereotactic radiosurgery unit(s)

Supervised Work Experience		Total Hours of Experience:		
Description of Experience Must Include:	Location of Experience/License or Permit Number of Facility		Confirm	Dates of Experience*
Reviewing full calibration measurements and periodic spot-checks			☐ Yes ☐ No	
Preparing treatment plans and calculating treatment doses and times			☐ Yes ☐ No	
Using administrative controls to prevent a medical event involving the use of byproduct material			☐ Yes ☐ No	
Implementing emergency procedures to be followed in the event of the abnormal operation of the medical unit or console			☐ Yes ☐ No	
Checking and using survey meters			☐ Yes ☐ No	
Selecting the proper dose and how it is to be administered			☐ Yes ☐ No	

APPENDIX B

3. <u>**Training and Experience for Proposed Authorized User**</u>   (continued)

   d. Supervised Work and Clinical Experience for 10 CFR 35.690 (continued)

Clinical experience in radiation oncology as part of an approved formal training program	Location of Experience/License or Permit Number of Facility	Dates of Experience*
Approved by: ☐ Residency Review Committee for Radiation Oncology of the ACGME ☐ Royal College of Physicians and Surgeons of Canada ☐ Committee on Postdoctoral Training of the American Osteopathic Association		
Supervising Individual	License/Permit Number listing supervising individual as an Authorized User	

   e. For 35.600, describe training provider and dates of training for each type of use for which authorization is sought.

Description of Training	Training Provider and Dates		
	Remote Afterloader	Teletherapy	Gamma Stereotactic Radiosurgery
Device operation			
Safety procedures for the device use			
Clinical use of the device			
Supervising Individual. *If training provided by Supervising Individual (If more than one supervising individual is necessary to document supervised work experience, provide multiple copies of this page.)*	License/Permit Number listing supervising individual as an Authorized User		
Authorized for the following types of use: ☐ Remote afterloader unit(s) ☐ Teletherapy unit(s) ☐ Gamma stereotactic radiosurgery unit(s)			

   f. Provide completed Part II Preceptor Attestation.

NRC FORM 313A (AUS)
(10-2007)

U.S. NUCLEAR REGULATORY COMMISSION

AUTHORIZED USER TRAINING AND EXPERIENCE AND PRECEPTOR ATTESTATION (continued)

PART II – PRECEPTOR ATTESTATION

Note: This part must be completed by the individual's preceptor. The preceptor does not have to be the supervising individual as long as the preceptor provides, directs, or verifies training and experience required. If more than one preceptor is necessary to document experience, obtain a separate preceptor statement from each.

By checking the boxes below, the preceptor is attesting that the individual has knowledge to fulfill the duties of the position sought and not attesting to the individual's "general clinical competency."

First Section
Check one of the following for each requested authorization:

For 35.490:

Board Certification

☐ I attest that _____ has satisfactorily completed the requirements in
Name of Proposed Authorized User

35.490(a)(1) and has achieved a level of competency sufficient to function independently as an authorized user of manual brachytherapy sources for the medical uses authorized under 10 CFR 35.400.

OR

Training and Experience

☐ I attest that _____ has satisfactorily completed the 200 hours of
Name of Proposed Authorized User

classroom and laboratory training, 500 hours of supervised work experience, and 3 years of supervised clinical experience in radiation oncology, as required by 10 CFR 35.490(b)(1) and (b)(2), and has achieved a level of competency sufficient to function independently as an authorized user of manual brachytherapy sources for the medical uses authorized under 10 CFR 35.400.

For 35.491:

☐ I attest that _____ has satisfactorily completed the 24 hours of
Name of Proposed Authorized User

classroom and laboratory training applicable to the medical use of strontium-90 for ophthalmic radiotherapy, has used strontium-90 for ophthalmic treatment of 5 individuals, as required by 10 CFR 35.491(b), and has achieved a level of competency sufficient to function independently as an authorized user of strontium-90 for ophthalmic use.

- -

Second Section

For 35.690:

Board Certification

☐ I attest that _____ has satisfactorily completed the requirements in
Name of Proposed Authorized User
35.690(a)(1).

OR

Training and Experience

☐ I attest that _____ has satisfactorily completed 200 hours of classroom
Name of Proposed Authorized User

and laboratory training, 500 hours of supervised work experience, and 3 years of supervised clinical experience in radiation therapy, as required by 10 CFR 35.690(b)(1) and (b)(2).

AND

NRC FORM 313A (AUS)	U.S. NUCLEAR REGULATORY COMMISSION
(10-2007)	

AUTHORIZED USER TRAINING AND EXPERIENCE AND PRECEPTOR ATTESTATION (continued)

Preceptor Attestation (continued)

Third Section

For 35.690: (continued)

[ ] I attest that _____ has received training required in 35.690(c) for device

Name of Proposed Authorized User

operation, safety procedures, and clinical use for the type(s) of use for which authorization is sought, as checked below.

[ ] Remote afterloader unit(s) [ ] Teletherapy unit(s) [ ] Gamma stereotactic radiosurgery unit(s)

- -

AND

Fourth Section

[ ] I attest that _____ has achieved a level of competency sufficient to

Name of Proposed Authorized User

achieve a level of competency sufficient to function independently as an authorized user for:

[ ] Remote afterloader unit(s) [ ] Teletherapy unit(s) [ ] Gamma stereotactic radiosurgery unit(s)

- -

Fifth Section

Complete the following for preceptor attestation and signature:

[ ] I meet the requirements in 10 CFR 35.490, 35.491, 35.690, or equivalent Agreement State requirements, as an authorized user for:

[ ] 35.400 Manual brachytherapy sources [ ] 35.600 Teletherapy unit(s)

[ ] 35.400 Ophthalmic use of strontium-90 [ ] 35.600 Gamma stereotactic radiosurgery unit(s)

[ ] 35.600 Remote afterloader unit(s)

Name of Preceptor	Signature	Telephone Number	Date

License/Permit Number/Facility Name

PAGE 6

APPENDIX C

License Application Checklists

License Application Checklists

This Appendix contains checklists that may be used to assist in organizing an application. It addresses information a medical use licensee needs to provide for authorization to produce PET radioactive drugs for noncommercial transfer to consortium members. See Appendix AA for additional information.

Table C.1, Applicability Table, may be used to determine if particular information must be provided or if "N/A" (not applicable) may be the response to each item that follows. To determine those items to which applicants must respond, "highlight" the columns under the categories of materials requested in Item 5 (e.g., 10 CFR 35.300, 35.400). If any "Y" beside an item is highlighted, applicants must provide detailed information in response to that item. If the letters "N/A" are highlighted, applicants may respond "N/A" on their applications. If any "N" beside an item is highlighted, no information in response is required, but NRC regulations that apply to the given category apply to that type of license. If any "P" beside an item is highlighted, applicants should provide a commitment as described in the section referenced in the body of this document. If any "G" beside an item is highlighted, see subsequent sections for required responses. "APP" indicates that this document contains an appendix that addresses the item.

Section #	Topic	35.100/200	35.300	35.400	35.500	35.600	35.1000	APP
	Table C.1 Applicability Table							
8.5	Unsealed Byproduct Material – Uptake, Dilution, Excretion, Imaging, and Localization Studies	Y						
8.5	Unsealed Byproduct Material – Written Directive Required		Y					
8.5	Manual Brachytherapy			Y				
8.5	Sealed Sources for Diagnosis				Y			
8.5	Teletherapy Units					Y		
8.5	Remote Afterloader Units					Y		
8.5	Gamma Stereotactic Radiosurgery Units					Y		
8.5	Other Medical Uses						Y	
8.6	Sealed Sources and Devices	N	N	Y	Y	Y	Y	
8.7	Discrete Source of Ra-226 (Other than sealed sources)	Y	Y	N	N	N	Y	
8.8	Financial Assurance Determination	Y	Y	Y	Y	Y	Y	
8.9	Purpose(s) for Which Licensed Material Will Be Used	Y	Y	Y	Y	Y	Y	
8.10	Training and Experience	G	G	G	G	G	G	
8.11	Radiation Safety Officer	Y	Y	Y	Y	Y	Y	I, D
8.12	Authorized User(s) (AUs)	Y	Y	Y	Y	Y	Y	D
8.13	Authorized Nuclear Pharmacist (ANP)	Y	Y	N/A	N/A	N/A	Y	D
8.14	Authorized Medical Physicist (AMP)	N/A	N/A	Y*	N/A	Y	Y	D
8.15	Facilities and Equipment	G	G	G	G	G	G	
8.16	Facility Diagram	Y	Y	Y	Y	Y	Y	
8.17	Radiation Monitoring Instruments	Y, P	Y, P	Y, P	Y, P	Y, P	Y, P	K
8.18	Dose Calibrator and Other Equipment	P	P	N/A	N/A	N/A	P	
8.19	Therapy Unit - Calibration and Use	N/A	N/A	N	N/A	Y	N	
8.20	Other Equipment and Facilities	N	N	N	N	Y	N	
8.21	Radiation Protection Program	G	G	G	G	G	G	
8.22	Safety Procedures and Instructions	N/A	N/A	N/A	N/A	Y	N/A	
8.23	Occupational Dose	P	P	P	P	P	P	M

Section #	Topic	35.100/200	35.300	35.400	35.500	35.600	35.1000	APP
8.24	Area Surveys	P	P	P	P	P	P	R
8.25	Safe Use of Unsealed Licensed Material	P	P	N/A	N/A	N/A	P	T
8.26	Spill/Contamination Procedures	P	P	P	N/A	N/A	P	N
8.27	Service of Therapy Devices Containing Sealed Sources	N/A	N/A	N/A	N/A	Y	Y	
8.28	Minimization of Contamination	N	N	N	N	N	N	
8.29	Waste Management	P	P	P	P	P	P	W
8.30	Fees	Y	Y	Y	Y	Y	Y	
8.31	Certification	Y	Y	Y	Y	Y	Y	
8.32	Safety Instruction for Individuals in Restricted Areas	N	N	N	N	N	N	J
8.33	Public Dose	N	N	N	N	N	N	
8.34	Opening Packages	N	N	N	N	N	N	
8.35	Written Directive Procedures	N/A	N	N	N/A	N	N	S
8.36	Release of Patients or Human Research Subjects	N	N	N	N/A	N/A	N	U
8.37	Mobile Medical Service	N	N	N	N	N	N	V
8.38	Audit Program	N	N	N	N	N	N	L
8.39	Operating and Emergency Procedures	N	N	N	N	N	N	N
8.40	Material Receipt and Accountability	N	N	N	N	N	N	
8.41	Ordering and Receiving	N	N	N	N	N	N	O
8.42	Sealed Source Inventory	N	N	N	N	N	N	
8.43	Records of Dosages and Use of Brachytherapy Source	N	N	N	N	N	N	
8.44	Recordkeeping	N	N	N	N	N	N	X
8.45	Reporting	N	N	N	N	N	N	Y
8.46	Leak Tests	N	N	N	N	N	N	Q
8.47	Safety Procedures for Treatments when Patients are Hospitalized	N/A	N	N	N/A	N**	N	
8.48	Transportation	N	N	N	N	N	N	Z

Table C.1 Applicability Table

* Y beside item 8.13 for use under 35.400 applies to Sr-90 only.
** N/A for teletherapy and gamma stereotactic radiosurgery outpatient treatments.

Table C.2 outlines the detailed responses that may be made to Items 5 and 6 on Form 313 for the type of radioactive material requested and the purposes for which it will be used. For example, if the applicant is seeking a license for unsealed byproduct material under 10 CFR 35.100 or 35.200, then the applicant should check the "yes" column next to 10 CFR 35.100 and 35.200 in Table C.2. The table then indicates appropriate responses for that type of use. An applicant may copy the checklist and include it in the license application.

The applicant should review the guidance in Section 5.2 and mark security-related information appropriately.

Note: The NRC now has regulatory authority for accelerator-produced radioactive material and discrete sources of Ra-226, as a result of the EPAct. Uses of these materials are added to Table C.2.

Table C.2 Items 5 and 6 on NRC Form 313: Radioactive Material and Use
(If using this checklist, check applicable rows and fill in details, and attach copy of checklist to the application.)

☐ Yes ☐ No	This response includes security-related sensitive information (see Section 5.2) which is included in Attachment ____ and marked "Security-related information – withhold under 10 CFR 2.390"			
Yes	**Radionuclide**	**Form or Manufacturer/ Model No.**	**Maximum Quantity**	**Purpose of Use**
	Any byproduct material permitted by 10 CFR 35.100	Any	As needed	Any uptake, dilution, and excretion study permitted by 10 CFR 35.100.
	Any byproduct material permitted by 10 CFR 35.200	Any	As needed	Any imaging and localization study permitted by 10 CFR 35.200.
	F-18	Any	____ curies	Production of PET radioactive drugs under 10 CFR 30.32(j).
	O-15	Any	____ curies	Production of PET radioactive drugs under 10 CFR 30.32(j).
	C-11	Any	____ curies	Production of PET radioactive drugs under 10 CFR 30.32(j).
	Any byproduct material permitted by 10 CFR 35.300	Any	____ millicuries	Any radiopharmaceutical therapy procedure permitted by 10 CFR 35.300.
	Iodine-131	Any	___ millicuries	Administration of I-131 sodium iodide.
	Byproduct material permitted by 10 CFR 35.400 (Radionuclide _____)	Sealed source or device (Manufacturer _____, Model No._____)	___ millicuries	Any brachytherapy procedure permitted by 10 CFR 35.400.
	Byproduct material permitted by 10 CFR 35.400 (Radionuclide _____)	Sealed source or device (Manufacturer _____, Model No._____)	___ millicuries	Any brachytherapy procedure permitted by 10 CFR 35.400.
	Byproduct material permitted by 10 CFR 35.400 (Radionuclide _____)	Sealed source or device (Manufacturer _____, Model No._____)	___ millicuries	Any brachytherapy procedure permitted by 10 CFR 35.400.
	Byproduct material permitted by 10 CFR 35.400 (Radionuclide _____)	Sealed source or device (Manufacturer _____, Model No._____)	___ millicuries	Any brachytherapy procedure permitted by 10 CFR 35.400.

Table C.2 Items 5 and 6 on NRC Form 313: Radioactive Material and Use

(If using this checklist, check applicable rows and fill in details, and attach copy of checklist to the application.)

Yes	Radionuclide	Form or Manufacturer/ Model No.	Maximum Quantity	Purpose of Use
	Strontium-90	Sealed source or device (Manufacturer _____, Model No._____)	___millicuries	Treatment of superficial eye conditions using an applicator distributed pursuant to 10 CFR 32.74 and permitted by 10 CFR 35.400.
	Byproduct material permitted by 10 CFR 35.500 Check all that apply: ☐ Gd-153; ☐ I-125; ☐ Other, describe	Sealed source or device (Manufacturer _____, Model No._____)	___curies per source and ___curies total	Diagnostic medical use of sealed sources permitted by 10 CFR 35.500 in compatible devices registered pursuant to 10 CFR 30.32(g).
	Iridium-192	Sealed source or device (Manufacturer _____, Model No._____)	___curies per source and ___curies total	One source for medical use permitted by 10 CFR 35.600, in a Manufacturer _____, Model No. _____ remote afterloading brachytherapy device. One source in its shipping container as necessary for replacement of the source in the remote afterloader device.
	Cobalt-60	Sealed source or device (Manufacturer _____, Model No._____)	___curies per source and ___curies total	One source for medical use permitted by 10 CFR 35.600, in a Manufacturer _____, Model No. _____ teletherapy unit. One source in its shipping container as necessary for replacement of the source in the teletherapy unit.
	Cobalt-60	Sealed source or device (Manufacturer _____, Model No._____)	___curies per source and ___curies total	For medical use permitted by 10 CFR 35.600, in a Manufacturer _____, Model No. _____ stereotactic radiosurgery device. Sources in the shipping container as necessary for replacement of the sources in the stereotactic

Table C.2 Items 5 and 6 on NRC Form 313: Radioactive Material and Use *(If using this checklist, check applicable rows and fill in details, and attach copy of checklist to the application.)*				
Yes	**Radionuclide**	**Form or Manufacturer/ Model No.**	**Maximum Quantity**	**Purpose of Use**
				radiosurgery device.
	Any byproduct material under 10 CFR 31.11	Prepackaged kits	___millicuries	*In vitro* studies.
	Depleted uranium	Metal	___kilograms	Shielding in a teletherapy unit.
	Depleted uranium	Metal	___kilograms	Shielding in a linear accelerator.
	Any radionuclide in excess of 30 millicuries for use in calibration, transmission, and reference sources. (List radionuclide: _____)	Sealed source or device (Manufacturer _____, Model No._____)	___millicuries	For use in a Manufacturer _____, Model No. _____ for calibration and checking of licensee's survey instruments.
	Americium-241	Sealed source or device (Manufacturer _____, Model No._____)	___millicuries per source and ___millicuries total	Use as an anatomical marker.
	Plutonium (principal radionuclide Pu-238)	Sealed sources	___millicuries per source and ___grams total	As a component of Manufacturer _____, Model No. _____ nuclear-powered cardiac pacemakers for clinical evaluation in accordance with manufacturer's protocol dated _____. This authorization includes: follow-up, explantation, recovery, disposal, and implantation.
	Other	Form or Manufacturer/Model No. _____	___millicuries	Purpose of use _____

Table C.3 contains a checklist that may be used to identify the attached documents that the
applicant is supplying for items for which a response is required. For example, an applicant may
fill in the name of the Radiation Safety Officer in Table C.3 and then check the boxes indicating
which documents pertaining to the RSO are being included in the license application. An
applicant may copy the checklist and include it in the license application.

Table C.3	Items 7 through 11 on NRC Form 313: Training & Experience, Facilities & Equipment, Radiation Protection Program, and Waste Disposal
(Check all applicable rows and fill in details and attach a copy of the checklist to the application or provide information separately.)	

Item Number and Title	Suggested Response	Check box to indicate material included in application
Item 7: Radiation Safety Officer Name:	*For an individual previously identified as an RSO on an NRC or Agreement State license or permit:*	
	Previous license number (if issued by the NRC), or a copy of a license (if issued by an Agreement State), or a copy of a permit (if issued by an NRC master materials licensee) on which the individual was specifically named as the RSO.	☐
	For an individual qualifying under 10 CFR 35.57(a)(3):	
	Documentation that the individual was: • the RSO for only the medical uses of accelerator-produced radioactive material or discrete sources of Ra-226 included in the definition of byproduct material as a result of the EPAct; • the RSO for the medical uses of these materials before or during the effective period of NRC's waiver of August 31, 2005.	☐
	For an individual qualifying under 10 CFR 35.50(a):	
	Copy of certification by a specialty board whose certification process has been recognized[10] by NRC or an Agreement State under 10 CFR 35.50(a). **AND**	☐
	Description of the training and experience specified in 10 CFR 35.50(e) demonstrating that the proposed RSO is qualified by training in radiation safety, regulatory issues, and emergency procedures as applicable to the types of use for which the applicant seeks approval of an individual to serve as RSO. **AND**	☐
	Written attestation, signed by a preceptor RSO, that the individual has satisfactorily completed training in and experience required for certification, as well as training in radiation safety, regulatory issues, and emergency procedures for the types of use for which the licensee seeks approval, and has achieved a level of radiation safety knowledge sufficient to function independently as an RSO. **AND**	☐
	If applicable, description of recent related continuing education and experience as required by 10 CFR 35.59.	☐

---

[10]The names of board certifications that have been recognized by the NRC or an Agreement State are posted on the NRC's Web site http://www.nrc.gov/materials/miau/med-use-toolkit.html.

Table C.3 Items 7 through 11 on NRC Form 313: Training & Experience, Facilities & Equipment, Radiation Protection Program, and Waste Disposal

(Check all applicable rows and fill in details and attach a copy of the checklist to the application or provide information separately.)

Item Number and Title	Suggested Response	Check box to indicate material included in application
	For an individual qualifying under 10 CFR 35.50(b):	
	Description of the training and experience specified in 10 CFR 35.50(b) demonstrating that the proposed RSO is qualified by training and experience as applicable to the types of use for which the applicant seeks approval of an individual to serve as RSO. AND	❐
	Description of the training and experience specified in 10 CFR 35.50(e) demonstrating that the proposed RSO is qualified by training in radiation safety, regulatory issues, and emergency procedures as applicable to the types of use for which the applicant seeks approval of an individual to serve as RSO. AND	❐
	Written attestation, signed by a preceptor RSO, that the individual has satisfactorily completed the required training and experience specified in 10 CFR 35.50(b), as well as the training in radiation safety, regulatory issues, and emergency procedures for the types of use for which the licensee seeks approval, and has achieved a level of radiation safety knowledge sufficient to function independently as an RSO. AND	❐
	If applicable, description of recent related continuing education and experience as required by 10 CFR 35.59.	❐
	For an individual qualifying under 10 CFR 35.50(c)(1):	
	Copy of the certification(s) as a medical physicist by a board whose certification process has been recognized[11] by the NRC or an Agreement State under 10 CFR 35.51(a) and description of the experience specified in 10 CFR 35.50(c)(1) demonstrating that the proposed RSO is qualified by experience as applicable to the types of use for which the applicant seeks approval of an individual to serve as RSO. AND	❐
	Description of the training and experience specified in 10 CFR 35.50(e) demonstrating that the proposed RSO is qualified by training in radiation safety, regulatory issues, and emergency procedures as applicable to the types of use for which the applicant seeks approval of an individual to serve as RSO. AND	❐

---

[11] The names of board certifications that have been recognized by the NRC or an Agreement State are posted on the NRC's Web site http://www.nrc.gov/materials/miau/med-use-toolkit.html.

Table C.3 Items 7 through 11 on NRC Form 313: Training & Experience, Facilities & Equipment, Radiation Protection Program, and Waste Disposal *(Check all applicable rows and fill in details and attach a copy of the checklist to the application or provide information separately.)*		
Item Number and Title	**Suggested Response**	**Check box to indicate material included in application**
	Written attestation, signed by a preceptor RSO, that the individual has satisfactorily completed the required training and experience specified for certification, as well as training in radiation safety, regulatory issues, and emergency procedures for the types of use for which the licensee seeks approval, and has achieved a level of radiation safety knowledge sufficient to function independently as an RSO. **AND**	❒
	If applicable, description of recent related continuing education and experience as required by 10 CFR 35.59.	❒
	For an individual qualifying under 10 CFR 35.50(c)(2):	
	Copy of the licensee's license indicating that the individual is an AU, AMP, or ANP identified on the licensee's license and has experience with radiation safety aspects of similar types of use of byproduct material for which the applicant seeks approval of an individual to serve as RSO. **AND**	❒
	Description of the training and experience specified in 10 CFR 35.50(e) demonstrating that the proposed RSO is qualified by training in radiation safety, regulatory issues, and emergency procedures as applicable to the types of use for which the applicant seeks approval of an individual to serve as RSO. **AND**	❒
	Written attestation, signed by a preceptor RSO, that the individual has satisfactorily completed the requirements in 10 CFR 35.50(c)(2), as well as training in radiation safety, regulatory issues, and emergency procedures for the types of use for which the licensee seeks approval, and has achieved a level of radiation safety knowledge sufficient to function independently as an RSO. **AND**	❒
	If applicable, description of recent related continuing education and experience as required by 10 CFR 35.59.	❒

Table C.3	Items 7 through 11 on NRC Form 313: Training & Experience, Facilities & Equipment, Radiation Protection Program, and Waste Disposal *(Check all applicable rows and fill in details and attach a copy of the checklist to the application or provide information separately.)*	
Item Number and Title	**Suggested Response**	**Check box to indicate material included in application**
Item 7: Authorized Users for medical uses:	*For an individual previously identified as an AU on an NRC or Agreement State license or permit:*	
Name(s), (including license number authorizing practice of medicine, podiatry, or dentistry if not provided previously or in attachment); Requested uses for each individual	Previous license number (if issued by the NRC), or a copy of the license (if issued by an Agreement State), or a copy of a permit issued by an NRC master materials licensee, or a copy of a permit issued by an NRC or Agreement State broad-scope licensee, or a copy of a permit issued by an NRC Master Materials License broad-scope permittee on which the physician, dentist, or podiatrist was specifically named as an AU for the uses requested.	❑
	For an AU requesting authorization for an additional medical use:	
	Description of the additional training and experience to demonstrate the AU is also qualified for the new medical uses requested (e.g., training and experience needed to meet the requirements in 10 CFR 35.290 (b), 35.396, 35.390(b)(1)(ii)(G), or 35.690(c)). **AND**	❑
	A preceptor attestation, if required (e.g., attestation is required to meet the requirements in 10 CFR 35.396, 35.390(b)(1)(ii)(G), or 35.690(c)).	
	For an individual qualifying under 10 CFR 35.57(b)(3):	
	Documentation that the physician, podiatrist, or dentist: • used only accelerator-produced radioactive materials, or discrete sources of Ra-226, or both, for medical uses before or during the effective period of NRC's waiver of August 31, 2005; and • used these materials for the same medical uses requested.	❑
	For an individual qualifying under 10 CFR Part 35, Subparts D, E, F, G, and/or H, who is board-certified:	
	Copy of the certification(s) by a specialty board(s) whose certification process has been recognized[12] by the NRC under 10 CFR Part 35, Subpart D, E, F, G, or H, as applicable to the use requested. **AND**	❑

---

[12]The names of board certifications that have been recognized by the NRC or an Agreement State are posted on the NRC's Web site http://www.nrc.gov/materials/miau/med-use-toolkit.html.

Table C.3   Items 7 through 11 on NRC Form 313: Training & Experience, Facilities & Equipment, Radiation Protection Program, and Waste Disposal
(Check all applicable rows and fill in details and attach a copy of the checklist to the application or provide information separately.)

Item Number and Title	Suggested Response	Check box to indicate material included in application
	For an individual with a board certification recognized under 10 CFR 35.390, a description of the supervised work experience administering dosages of radioactive drugs required in 10 CFR 35.390(b)(1)(ii)(G) demonstrating that the proposed AU is qualified for the types of administrations for which authorization is sought; AND	❏
	For an individual with a board certification recognized under 10 CFR 35.390 for medical uses described in 10 CFR 35.200, a description of the supervised work experience eluting generator systems required in 10 CFR 35.290(c)(1)(ii)(G) demonstrating the proposed AU is also qualified for imaging and localization medical uses; AND	❏
	For an individual with a board certification recognized under 10 CFR 35.490 or 35.690 seeking authorization under 10 CFR 35.396(d), a description of the classroom and laboratory training and supervised work experience required to demonstrate qualifications for administering parenteral administrations of unsealed byproduct material requiring a written directive; AND	❏
	For an individual seeking authorization under 10 CFR Part 35, Subpart H, description of the training specified in 10 CFR 35.690(c) demonstrating that the proposed AU is qualified for the type(s) of use for which authorization is sought; AND	❏
	Written attestation, signed by a preceptor physician AU, that the training and experience specified for certification, as well as the clinical casework, or training and experience required by 10 CFR 35.396(d), or training for 10 CFR 35.600 types of use, if appropriate, have been satisfactorily completed and that a level of competency sufficient to function independently as an AU for the medical uses authorized has been achieved; AND	❏
	If applicable, description of recent related continuing education and experience as required by 10 CFR 35.59.	❏

Table C.3	Items 7 through 11 on NRC Form 313: Training & Experience, Facilities & Equipment, Radiation Protection Program, and Waste Disposal	
	(Check all applicable rows and fill in details and attach a copy of the checklist to the application or provide information separately.)	

Item Number and Title	Suggested Response	Check box to indicate material included in application
	For an individual qualifying under 10 CFR Part 35, Subparts D, E, F, G, and/or H, who is not board-certified:	
	A description of the training and experience identified in 10 CFR Part 35, Subparts D, E, F, G, and H, demonstrating that the proposed AU is qualified by training and experience for the use(s) requested. AND	☐
	For an individual seeking authorization under 10 CFR Part 35, Subpart H, description of the training specified in 10 CFR 35.690 (c) demonstrating that the proposed AU is qualified for the type(s) of use for which authorization is sought. AND	☐
	Written attestation, signed by a preceptor physician AU, that the above training and experience have been satisfactorily completed and that a level of competency sufficient to function independently as an AU for the medical uses authorized has been achieved. AND	☐
	If applicable, description of recent related continuing education and experience as required by 10 CFR 35.59.	☐
Item 7: Authorized Nuclear Pharmacists Name(s) and license to practice pharmacy:	*For an individual previously identified as an ANP on an NRC or Agreement State license or permit:*	
	Previous license number (if issued by the NRC), or a copy of the license (if issued by an Agreement State), or a copy of a permit issued by an NRC master materials licensee, or a copy of a permit issued by an NRC or Agreement State broad-scope licensee, or a copy of a permit issued by an NRC Master Materials License broad-scope permittee on which the individual was specifically named ANP.	☐
	For an individual qualifying under 10 CFR 35.57(a)(3):	
	Documentation that the nuclear pharmacist: • used only accelerator-produced radioactive materials or discrete sources of Ra-226, or both, in the practice of nuclear pharmacy before or during the effective period of NRC's waiver of August 31, 2005; and • used these materials for the same uses requested.	☐

Table C.3 Items 7 through 11 on NRC Form 313: Training & Experience, Facilities & Equipment, Radiation Protection Program, and Waste Disposal

(Check all applicable rows and fill in details and attach a copy of the checklist to the application or provide information separately.)

Item Number and Title	Suggested Response	Check box to indicate material included in application
	For an individual qualifying under 10 CFR 35.55(a):	
	Copy of the certification(s) of the specialty board whose certification process has been recognized[13] under 10 CFR 35.55(a). AND	❐
	Written attestation, signed by a preceptor ANP, that training and experience required for certification have been satisfactorily completed and that a level of competency sufficient to function independently as an ANP has been achieved. AND	❐
	If applicable, description of recent related continuing education and experience as required by 10 CFR 35.59.	❐
	For an individual qualifying under 10 CFR 35.55(b):	
	Description of the training and experience specified in 10 CFR 35.55(b) demonstrating that the proposed ANP is qualified by training and experience. AND	❐
	Written attestation, signed by a preceptor ANP, that the above training and experience have been satisfactorily completed and that a level of competency sufficient to function independently as an ANP has been achieved. AND	❐
	If applicable, description of recent related continuing education and experience as required by 10 CFR 35.59.	❐
Item 7: Authorized Medical Physicists Name(s):	*For an individual previously identified as an AMP on an NRC or Agreement State license or permit:*	
	Previous license number (if issued by the NRC), or a copy of the license (if issued by an Agreement State), or a copy of a permit issued by an NRC master materials licensee, or a copy of a permit issued by an NRC or Agreement State broad-scope licensee, or a copy of a permit issued by an NRC Master Materials License broad-scope permittee on which the individual was specifically named an AMP for the uses requested.	❐

---

[13]The names of board certifications that have been recognized by the NRC or an Agreement State are posted on the NRC's Web site http://www.nrc.gov/materials/miau/med-use-toolkit.html.

Table C.3	Items 7 through 11 on NRC Form 313: Training & Experience, Facilities & Equipment, Radiation Protection Program, and Waste Disposal

(Check all applicable rows and fill in details and attach a copy of the checklist to the application or provide information separately.)

Item Number and Title	Suggested Response	Check box to indicate material included in application
	For an individual qualifying under 10 CFR 35.57(a)(3):	
	Documentation that the medical physicist: • used only accelerator-produced radioactive material, discrete sources of Ra-226, or both, for medical uses before or during the effective period of NRC's waiver of August 31, 2005; and • used these materials for the same medical uses requested.	☐
	For an individual qualifying under 10 CFR 35.51(a):	
	Copy of the certification(s) of the specialty board(s) whose certification process has been recognized[14] under 10 CFR 35.51(a). **AND**	☐
	Description of the training and experience specified in 10 CFR 35.51(c) demonstrating that the proposed AMP is qualified by training in the types of use for which he or she is requesting AMP status, including hands-on device operation, safety procedures, clinical use, and operation of a treatment planning system. **AND**	☐
	Written attestation, signed by a preceptor AMP, that the required training and experience required for certification, as well as the training and experience specified in 10 CFR 35.51(c) have been satisfactorily completed, and that a level of competency sufficient to function independently as an AMP has been achieved. **AND**	☐
	If applicable, description of recent related continuing education and experience as required by 10 CFR 35.59.	☐
	For an individual qualifying under 10 CFR 35.51(b):	
	Description of the training and experience demonstrating that the proposed AMP is qualified by training and experience identified in 10 CFR 35.51(b)(1) for the uses requested. **AND**	☐

---

[14]The names of board certifications that have been recognized by the NRC or an Agreement State are posted on the NRC's Web site http://www.nrc.gov/materials/miau/med-use-toolkit.html.

Table C.3	Items 7 through 11 on NRC Form 313: Training & Experience, Facilities & Equipment, Radiation Protection Program, and Waste Disposal	
	(Check all applicable rows and fill in details and attach a copy of the checklist to the application or provide information separately.)	
Item Number and Title	**Suggested Response**	**Check box to indicate material included in application**
	Description of the training and experience specified in 10 CFR 35.51(c) demonstrating that the proposed AMP is qualified by training in the types of use for which he or she is requesting AMP status, including hands-on device operation, safety procedures, clinical use, and operation of a treatment planning system.	❒
	AND	
	Written attestation, signed by a preceptor AMP, that the required training and experience have been satisfactorily completed and that a level of competency sufficient to function independently as an AMP has been achieved.	❒
	AND	
	If applicable, description of recent related continuing education and experience as required by 10 CFR 35.59.	❒
Item 7: Authorized User for nonmedical uses	*Note:* For purposes of this section of the table, the term "authorized user" is used to mean individuals authorized for the nonmedical uses described. See Sections 8.11 and 8.12.	
Name(s):	*For an individual previously authorized for nonmedical use on an NRC or Agreement State license or permit:*	
Requested types, quantities, and nonmedical uses for each individual	Previous license number (if issued by the NRC), or a copy of the license (if issued by an Agreement State), or a copy of a permit issued by an NRC master materials licensee, or a copy of a permit issued by an NRC or Agreement State broad-scope licensee, or a copy of a permit issued by an NRC Master Materials License broad-scope permittee on which the individual was specifically named an AU for the types, quantities, and uses requested.	❒
	For individuals qualifying under 10 CFR 30.33(a)(3):	
	Documentation of the individual's training and experience demonstrating that the individual is qualified to use the types and quantities of licensed materials for the requested uses.	❒
Item 9: Facility Diagram	A diagram is enclosed that describes the facilities and identifies activities conducted in all contiguous areas surrounding the area(s) of use. The following information is included:	❒
	• Guidance in Section 5.2 was reviewed and security-related sensitive information provided is marked accordingly.	❒
	• Drawings should be to scale, indicating the scale used.	❒

Table C.3	Items 7 through 11 on NRC Form 313: Training & Experience, Facilities & Equipment, Radiation Protection Program, and Waste Disposal *(Check all applicable rows and fill in details and attach a copy of the checklist to the application or provide information separately.)*	
Item Number and Title	**Suggested Response**	**Check box to indicate material included in application**
	• Location, room numbers, and principal use of each room or area where byproduct material is prepared, used or stored, location of direct transfer delivery tubes from a PET radionuclide/radioactive drug production facility or production area of PET radioactive drugs under 10 CFR 30.32(j), and areas where higher energy gamma- emitting radionuclides (e.g., PET radionuclides) are used;	❏
	• Location, room numbers, and principal use of each adjacent room (e.g., office, file, toilet, closet, hallway), including areas above, beside, and below therapy treatment rooms, indicating whether the room is a restricted or unrestricted area as defined in 10 CFR 20.1003; and	❏
	• Provide shielding calculations and include information about the type, thickness, and density of any necessary shielding to enable independent verification of shielding calculations, including a description of any portable shields used (e.g., shielding of proposed patient rooms used for implant therapy, including the dimensions of any portable shield, if one is used; source storage safe).	❏
	In addition to the above, for teletherapy and GSR facilities, applicants should provide the directions of primary beam usage for teletherapy units and, in the case of an isocentric unit, the plane of beam rotation.	❏
Item 9: Radiation Monitoring Instruments	A statement that: "Radiation monitoring instruments will be calibrated by a person qualified to perform survey meter calibrations." **AND/OR**	❏
	A statement that: "We have developed and will implement and maintain written survey meter calibration procedures in accordance with the requirements in 10 CFR 20.1501 and that meet the requirements of 10 CFR 35.61." **AND**	❏
	A description of the instrumentation (e.g., gamma counter, solid state detector, portable or stationary count rate meter, portable or stationary dose rate or exposure rate meter, single or multichannel analyzer, liquid scintillation counter, proportional counter) that will be used to perform required surveys. **AND**	❏
	A statement that: "We reserve the right to upgrade our survey instruments as necessary as long as they are adequate to measure the type and level of radiation for which they are used."	❏
Item 9: Dose Calibrator and Other Dosage Measuring Equipment	A statement that: "Equipment used to measure dosages will be calibrated in accordance with nationally recognized standards or the manufacturer's instructions."	❏

Item Number and Title	Suggested Response	Check box to indicate material included in application
	When administering dosages of alpha-emitting unsealed byproduct material in other than unit dosages made by a manufacturer or preparer licensed under 10 CFR 32.72 or 10 CFR 30.32(j),	
	• A statement that: "Dosages will be determined by relying on the provider's dose label for measurement of the radioactivity and a combination of volumetric measurement and mathematical calculation."	❏
	OR	
	• We are providing a description of the dosage measurement equipment, the nationally recognized calibration standard (or manufacturer's calibration instructions), and dosage measurement procedures.	❏
Item 9: Therapy Unit - Calibration and Use	We are providing the procedures required by 10 CFR 35.642, 10 CFR 35.643, and 10 CFR 35.645, if applicable to the license application.	❏
Item 9: Other Equipment and Facilities	Guidance in Section 5.2 was reviewed and security-related information provided is marked accordingly.	❏
	Attached is a description, identified as Attachment 9.4, of additional facilities and equipment.	❏
	For manual brachytherapy facilities, we are providing a description of the emergency response equipment.	❏
	For PET radionuclide use, PET radioactive drug production, and radiopharmaceutical therapy programs, we are providing a description of the additional facilities and equipment for these uses.	❏
	For teletherapy, GSR, and remote afterloader facilities, we are providing a description of the following:	
	• Warning systems and restricted area controls (e.g., locks, signs, warning lights and alarms, interlock systems) for each therapy treatment room;	❏
	• Area radiation monitoring equipment;	❏
	• Viewing and intercom systems (except for LDR units);	❏
	• Steps that will be taken to ensure that no two units can be operated simultaneously, if other radiation-producing equipment (e.g., linear accelerator, X-ray machine) is in the treatment room;	❏
	• Methods to ensure that whenever the device is not in use or is unattended, the console keys will be inaccessible to unauthorized persons; and	❏
	• Emergency response equipment.	❏

Table C.3 Items 7 through 11 on NRC Form 313: Training & Experience, Facilities & Equipment, Radiation Protection Program, and Waste Disposal
(Check all applicable rows and fill in details and attach a copy of the checklist to the application or provide information separately.)

Table C.3 Items 7 through 11 on NRC Form 313: Training & Experience, Facilities & Equipment, Radiation Protection Program, and Waste Disposal *(Check all applicable rows and fill in details and attach a copy of the checklist to the application or provide information separately.)*		
Item Number and Title	**Suggested Response**	**Check box to indicate material included in application**
Item 10: Safety Procedures and Instructions	Attached are procedures required by 10 CFR 35.610.	❒
	Guidance in Section 5.2 was reviewed and security-related sensitive information provided is marked accordingly.	❒
Item 10: Occupational Dose	A statement that: "Either we will perform a prospective evaluation demonstrating that unmonitored individuals are not likely to receive, in 1 year, a radiation dose in excess of 10% of the allowable limits in 10 CFR Part 20 or we will provide dosimetry that meets the requirements listed under 'Criteria' in NUREG-1556, Vol. 9, Rev. 1, 'Consolidated Guidance About Materials Licenses: Program-Specific Guidance About Medical Use Licenses.'" **OR**	❒
	A description of an alternative method for demonstrating compliance with the referenced regulations.	❒
Item 10: Area Surveys	A statement that: "We have developed and will implement and maintain written procedures for area surveys in accordance with 10 CFR 20.1101 that meet the requirements of 10 CFR 20.1501 and 10 CFR 35.70."	❒
Item 10: Safe Use of Unsealed Licensed Material	A statement that: "We have developed and will implement and maintain procedures for safe use of unsealed byproduct material that meet the requirements of 10 CFR 20.1101 and 10 CFR 20.1301."	❒
Item 10: Spill/Contamination Procedures	A statement that: "We have developed and will implement and maintain written procedures for safe response to spills of licensed material in accordance with 10 CFR 20.1101."	❒
Item 10: Installation, Maintenance, Adjustment, Repair, and Inspection of Therapy Devices Containing Sealed Sources	Name of the proposed employee and types of activities requested: _____ **AND**	❒
	Description of the training and experience demonstrating that the proposed employee is qualified by training and experience for the use requested. **AND**	❒
	Copy of the manufacturer's training certification and an outline of the training in procedures to be followed.	❒
Item 10: Minimization of Contamination	A response is not required under the following condition: the NRC will consider that the above criteria have been met if the information provided in applicant's responses satisfy the criteria in Sections 8.15, 8.16, 8.21, 8.25, 8.27, and 8.29, on the topics: facilities and equipment, facility diagram, Radiation Protection Program, safety program, and waste management.	N/A

Table C.3 Items 7 through 11 on NRC Form 313: Training & Experience, Facilities & Equipment, Radiation Protection Program, and Waste Disposal *(Check all applicable rows and fill in details and attach a copy of the checklist to the application or provide information separately.)*		
Item Number and Title	**Suggested Response**	**Check box to indicate material included in application**
Item 11: Waste Management	A statement that: "We have developed and will implement and maintain written waste disposal procedures for licensed material in accordance with 10 CFR 20.1101, that also meet the requirements of the applicable section of 10 CFR Part 20, Subpart K, and of 10 CFR 35.92."	☐
	Attached is a description of the radioactive waste incinerator facility and related portions of the Radiation Safety Program (10 CFR 20.2004).	☐
	Attached is a request to receive potentially contaminated radiation transport shields from consortium members receiving PET radioactive drugs noncommercially transferred under 10 CFR 30.32(j) authorization.	☐

APPENDIX D

Documentation of Training and Experience to Identify Individuals on a License as Authorized User, Radiation Safety Officer, Authorized Medical Physicist, or Authorized Nuclear Pharmacist

Note: The most current guidance is found on NRC's public Web site at http://www.nrc.gov/materials/miau/med-use-toolkit.html (Medical Uses Toolkit).

Documentation of Training and Experience to Identify Individuals on a License as Authorized User, Radiation Safety Officer, Authorized Medical Physicist, or Authorized Nuclear Pharmacist

I. Experienced Authorized Users, Authorized Medical Physicists, Authorized Nuclear Pharmacists, or Radiation Safety Officer

An applicant or licensee who is adding an experienced authorized user (AU) for medical uses, authorized medical physicist (AMP), authorized nuclear pharmacist (ANP), or Radiation Safety Officer (RSO) to its medical use license or application only needs to provide evidence that the individual is listed on a medical use license issued by the NRC or Agreement State, a permit issued by an NRC master materials licensee, a permit issued by an NRC or Agreement State broad-scope licensee, or a permit issued by an NRC master material broad-scope permittee, provided that the individual is authorized for the same types of use(s) requested in the application under review, and the individual meets the recentness of training criteria described in 10 CFR 35.59. When adding an experienced ANP to the license, the applicant also may provide evidence that the individual is listed on an NRC or Agreement State commercial nuclear pharmacy license or identified as an ANP by a commercial nuclear pharmacy authorized to identify ANPs. For individuals who have been previously authorized by, but not listed on, the commercial nuclear pharmacy license, medical broad-scope license, or Master Materials License medical broad-scope permit, the applicant should submit either verification of previous authorizations granted or evidence of acceptable training and experience.

II. Experienced Physicians, Podiatrists, Dentists, Nuclear Pharmacists, Medical Physicists, and Radiation Safety Officers Who Only Used Accelerator-Produced Nuclear Materials, or Discrete Sources of Radium-226, or Both, for Medical or Nuclear Pharmacy Uses.

In implementing the EPAct, the NRC "grandfathered" physicians, podiatrists, dentists, medical physicists, and nuclear pharmacists that used only accelerator-produced radioactive materials, discrete sources of radium-226 (Ra-226), or both, for medical or nuclear pharmacy uses, before or under the NRC waiver of August 31, 2005, when using these materials for the same uses. These individuals, as well as individuals that performed RSO duties only for uses of accelerator-produced radionuclides or discrete sources of Ra-226 at medical or nuclear pharmacy facilities before or during the effective period of the waiver, do not have to meet the requirements of 10 CFR 35.59, or the training and experience requirements in 10 CFR Part 35, Subparts B, D, E, F, and G.

The applicant or licensee that is adding one of these experienced individuals to its medical use license should document that the individual used only accelerator-produced radionuclides, or discrete sources of Ra-226, or both, for medical or nuclear pharmacy uses before or during the effective period of the waiver and that the materials were used for the same uses requested. This documentation may be, but is not restricted to, evidence that the individual was listed on an Agreement State or non-Agreement State license or permit authorizing these materials for the requested uses.

III. Applications that Include Individuals for New Authorized User, Authorized Medical Physicist, Authorized Nuclear Pharmacist or Radiation Safety Officer Recognition by NRC

Applicants should submit the appropriate completed form in the NRC Form 313A series to show that the individuals meet the correct training and experience criteria in 10 CFR Part 35, Subparts B, D, E, F, G, and H. For the applicant's convenience, the NRC Form 313A series has been separated into six separate forms. The forms are NRC FORM 313A (RSO) for the Radiation Safety Officer; NRC FORM 313A (AMP) for the authorized medical physicist; NRC FORM 313A (ANP) for the authorized nuclear pharmacist; NRC FORM 313A (AUD) for the authorized user of the medical uses included in 10 CFR 35.100, 35.200, and/or 35.500; NRC FORM 313A (AUT) for the authorized user for the medical use included in 10 CFR 35.300; and NRC FORM 313A (AUS) for the authorized user for the medical uses included in 10 CFR 35.400 and/or 35.600.

There are two primary training and experience routes to qualify an individual as a new AU, AMP, ANP, or RSO. The first is by means of certification by a board recognized by NRC and listed on the NRC Web site as provided in 10 CFR 35.50(a), 35.51(a), 35.55(a), 35.190(a), 35.290(a), 35.390(a), 35.392(a), 35.394(a), 35.490(a), 35.590(a), or 35.690(a). Preceptor attestations must also be submitted for all individuals to qualify under 10 CFR Part 35, Subparts B and D through H. Additional training may also need to be documented for RSOs, AMPs, and AUs under 10 CFR 35.600.

The second route is by meeting the structured educational program, supervised work experience, and preceptor attestation requirements in 10 CFR Part 35, Subparts B, D, E, F, G, and H. In some cases there may be additional training and experience routes for recognized AUs, ANPs, AMPs, or RSOs to seek additional authorizations.

IV. Recentness of Training

The required training and experience, including board certification, described in 10 CFR Part 35 must be obtained within the 7 years preceding the date of the application, or the individual must document having had related continuing education, retraining, and experience since obtaining the required training and experience. Examples of acceptable continuing education and experience for physicians include the following:

- Successful completion of classroom and laboratory review courses that include radiation safety practices relative to the proposed type of authorized medical use,

- Practical and laboratory experience with patient procedures using radioactive material for the same use(s) for which the applicant is requesting authorization,

- Practical and laboratory experience under the supervision of an AU at the same or another licensed facility that is authorized for the same use(s) for which the applicant is requesting authorization, and

- For therapy devices, experience with the therapy unit and/or comparable linear accelerator experience and completion of an in-service review of operating and emergency procedures relative to the therapy unit to be used by the applicant.

V. General Instructions and Guidance for Filling Out NRC Form 313A Series

If the applicant is proposing an individual for more than one type of authorization, the applicant may need to either submit multiple forms in the NRC Form 313A series or fill out some sections more than once. For example, an applicant that requests a physician be authorized for 10 CFR 35.200 and 10 CFR 35.300 medical uses and as the RSO, should provide three completed NRC Form 313A series forms (i.e., NRC Form 313A (RSO), NRC Form 313A (AUD) and NRC Form 313A (AUT)). Also, if the applicant requests that a physician be authorized for both high dose-rate remote afterloading and gamma stereotactic radiosurgery under 10 CFR 35.600, only one form, NRC Form 313A (AUS) needs to be completed, but one part (i.e., "Supervised Work and Clinical Experience") must be filled out twice.

To identify an Agreement State license, provide a copy of the license. To identify a Master Materials License permit, provide a copy of the permit. To identify an individual (i.e., supervising individual or preceptor) who is authorized under a broad-scope license or broad-scope permit of a Master Materials License, provide a copy of the permit issued by the broad-scope licensee/permittee. Alternatively, provide a statement signed by the Radiation Safety Officer or chairperson of the Radiation Safety Committee similar to the following: "_____(name of supervising individual or preceptor) is authorized under _____(name of licensee/permittee) broad-scope license number_____ to use_____(materials) during _____(time frame)."

INTRODUCTORY INFORMATION

Name of individual

Provide the individual's complete name so that NRC can distinguish the training and experience received from that received by others with a similar name.

Note: Do not include personal or private information (e.g., date of birth, Social Security Number, home address, personal telephone number) as part of your qualification documentation.

State or territory where licensed

The NRC requires physicians, dentists, podiatrists, and pharmacists to be licensed by a State or territory of the United States, the District of Columbia, or the Commonwealth of Puerto Rico to prescribe drugs in the practice of medicine, as well as licensed in the practice of dentistry, podiatry, or pharmacy, respectively (see definitions of "physician", "dentist", "podiatrist", and "pharmacist" in 10 CFR 35.2).

Requested Authorization(s).

Check all authorizations that apply and fill in the blanks as provided.

Part I. Training and Experience

There are always multiple pathways provided for each training and experience section. Select the applicable one.

Item 1.   Board Certification

The applicant or licensee may use this pathway if the proposed new authorized individual is certified by a board recognized by NRC (to confirm that NRC recognizes that board's certifications, see NRC's Web site http://www.nrc.gov/materials/miau/med-use-toolkit.html.

Note: An individual that is board-eligible will not be considered for this pathway until the individual is actually board-certified. Further, individuals holding other board certifications will also not be considered for this pathway.

The applicant or licensee will need to provide a copy of the board certification and other documentation of training, experience, or clinical casework as indicated on the specific form of the NRC Form 313A series.

All applicants under this pathway (except for 10 CFR 35.500 uses) must submit a completed Part II Preceptor Attestation.

Item 2.   Current Authorized Individuals Seeking Additional Authorizations

Provide the information requested for training, experience, or clinical casework as indicated on the specific form of the NRC Form 313A series. (*Note:* This section does not include individuals who are authorized only on foreign licenses.)

All applicants under this pathway must submit a completed Part II Preceptor Attestation.

Item 3.   Alternate Pathway for Training and Experience for Proposed New Authorized Individuals

This pathway is used for those individuals not listed on the license as authorized individuals, who do not meet the requirements for the board certification pathway.

The regulatory requirements refer to two categories of training: (a) classroom and laboratory training, and (b) supervised work experience. All hours credited to classroom and laboratory training must relate directly to radiation safety and safe handling of byproduct material and be allocated to one of the topics in the regulations. Each hour of training involving performance of radiation safety tasks or hands-on use of byproduct material may be credited to either (a) classroom and laboratory training, or (b) supervised work experience. Note that a single hour of training may only be counted once and may not be credited to both of these categories.

The proposed authorized individual may receive the required classroom and laboratory training, supervised work experience, and clinical casework at a single training facility or at multiple training facilities; therefore, space is provided to identify each location and date of training or experience. The date should be provided in the month/day/year (mm/dd/yyyy) format.

The specific number of hours needed for each training and supervised work experience element will depend upon the type of approval sought. Under the "classroom and laboratory training," provide the number of clock hours spent on each of the topics listed in the regulatory requirements.

The proposed authorized individual may obtain the required "classroom and laboratory training" in any number of settings, locations, and educational situations. For example, at some medical teaching/university institutions, a course may be provided for that particular need and taught in consecutive days. In other training programs, the period may be a semester or quarter as part of the formal curriculum. Also, the classroom and laboratory training may be obtained using a variety of other instructional methods. Therefore, the NRC will broadly interpret "classroom and laboratory training" to include various types of instruction, including online training, as long as it meets the specific clock hour requirements and the subject matter relates to radiation safety and safe handling of byproduct material for the uses requested.

Under the "supervised work experience" sections of the forms, provide only the total number of hours of supervised work experience and check the boxes for each of the topics listed in the regulatory requirements to confirm that the listed subject areas were included in the supervised work experience.

The "supervised work experience" for physicians must include, but is not limited to, the subject areas listed in the applicable training and experience requirements. The NRC recognizes that physicians in training will not dedicate all of their supervised work experience time specifically to the subject areas listed in the regulatory requirements and will be attending to other clinical activities involving the medical use of byproduct material (e.g., reviewing case histories or interpreting scans). Hours spent on these other duties not directly related to radiation safety or hands-on use of byproduct material, even though not specifically required by the NRC, may be credited to the supervised work experience category but not to the classroom and laboratory training category.

For nuclear pharmacists, under the "supervised practical experience" section, provide the number of clock hours for each topic. The supervised practical experience topics for the nuclear pharmacists include all the basic elements in the practice of nuclear pharmacy. Therefore, all the hours of supervised experience are allocated to these topics.

Note: If the proposed new authorized individual had more than one supervisor, provide the information requested for each supervising individual.

Part II. Preceptor Attestation

The NRC defines the term "preceptor" in 10 CFR 35.2, "Definitions," to mean "an individual who provides, directs, or verifies training and experience required for an individual to become an AU, an AMP, an ANP, or an RSO." While the supervising individual for the work experience may also be the preceptor, the preceptor does not have to be the supervising individual as long as the preceptor directs or verifies the training and experience required. The preceptor must attest in writing regarding the training and experience of any individual to serve as an authorized individual and attest that the individual has satisfactorily completed the appropriate training and experience requirements and has achieved a level of competency or a level of radiation safety knowledge sufficient to function independently. The preceptor language in NRC Forms 313A (AUD), 313A (AUT), and 313A (AUS) does not require an attestation of general clinical competency but requires sufficient attestation to demonstrate that the individual has the knowledge to fulfill the duties of the position for which the attestation is sought. The preceptor also has to meet specific requirements.

The NRC may require supervised work experience conducted under the supervision of an authorized individual in a licensed material use program. In this case, a supervisor is an individual who provides frequent direction, instruction, and direct oversight of the student as the student completes the required work experience in the use of byproduct material.

Supervision may occur at various licensed facilities, from a large teaching university hospital to a small private practice.

The NRC Form 313A series Part II - Preceptor Attestation has multiple sections. The preceptor must complete an attestation of the proposed user's training, experience, and competency to function independently, as well as provide information concerning his/her own qualifications and sign the attestation. Because there are a number of different pathways to obtain the required training and experience for different authorized individuals, specific instructions are provided below for each form in the NRC 313A series.

VI. RADIATION SAFETY OFFICER - Specific Instructions and Guidance for Filling Out NRC Form 313A (RSO)

See Section V, "General Instructions and Guidance for Filling out NRC Form 313A Series," for additional clarification on providing information about an individual's status on an Agreement State license, medical broad-scope license, or Master Materials License permit.

Part I. Training and Experience - select one of four methods below:

Item 1.   Board Certification

Provide the requested information (i.e., a copy of the board certification, documentation of specific radiation safety training for all types of use on the license, and a completed preceptor attestation). As indicated on the form, additional information is needed if the board certification or radiation safety training was completed more than 7 years ago.

Specific radiation safety training for each type of use on the license may be supervised by an RSO, AMP, ANP, or AU who is authorized for that type of use. Specific information regarding the supervising individual only needs to be provided in the table in 3.c if the training was provided by an RSO, AMP, ANP, or AU. If more than one supervising individual provided the training, identify each supervising individual by name and provide his/her qualifications.

Item 2.   Current Radiation Safety Officer Seeking Authorization to Be Recognized as a Radiation Safety Officer for the Additional Medical Use(s) Checked Above.

Provide the requested information (i.e., documentation of specific radiation safety training (complete the table in 3.c) and a completed preceptor attestation in Part II). As indicated on the form, additional information is needed if the specific radiation safety training was completed more than 7 years ago.

Specific radiation safety training for each type of use on the license may be supervised by an RSO, AMP, ANP, or AU who is authorized for that type of use. Specific information regarding the supervising individual only needs to be provided in the table in 3.c if the training was provided by an RSO, AMP, ANP, or AU. If more than one supervising individual provided the training, identify each supervising individual by name and provide his/her qualifications.

Item 3.   Structured Educational Program for Proposed New Radiation Safety Officer

As indicated on the form, additional information is needed if the training, supervised radiation safety experience, and specific radiation safety training was completed more than 7 years ago.

Submit a completed Section 3.a.

Submit a completed Section 3.b. The individual must have completed 1 year of full-time radiation safety experience under the supervision of an RSO. This is documented in Section 3.b by providing the ranges of dates for supervised radiation safety experience. If there was more than one supervising individual, identify each supervising individual by name and provide his/her qualifications.

Provide the requested information (i.e., documentation of specific radiation safety training for each use on the license (complete the table in 3.c)). Specific radiation safety training for each type of use on the license may be supervised by an RSO, AMP, ANP, or AU who is authorized for that type of use. Specific information regarding the supervising individual only needs to be provided in the table in 3.c if the training was provided by an RSO, AMP, ANP, or AU. If more than one supervising individual provided the training, identify each supervising individual by name and provide his/her qualifications.

Submit a completed Preceptor Attestation in Part II.

Item 4.   Authorized User, Authorized Medical Physicist, or Authorized Nuclear Pharmacist Identified on the Licensee's License

Provide the requested information (i.e., the license number and documentation of specific radiation safety training for each use on the license (complete the table in 3.c)). As indicated on the form, additional information is needed if the specific radiation safety training was completed more than 7 years ago.

Specific radiation safety training for each type of use on the license may be supervised by an RSO, AMP, ANP, or AU who is authorized for that type of use. Specific information regarding the supervising individual only needs to be provided in the table in 3.c if the training was provided by an RSO, AMP, ANP, or AU. If more than one supervising individual provided the training, identify each supervising individual by name and provide his/her qualifications.

Part II. Preceptor Attestation

The Preceptor Attestation page has four sections.

The attestation for the new proposed RSO's training or identification on the license as an AU, AMP, or ANP is in the first section.

The attestation for the specific radiation safety training is in the second section.

The attestation for the individual's competency to function independently as an RSO for a medical use license is in the third section.

The fourth and final section requests specific information about the preceptor's authorization as an RSO on a medical use license in addition to the preceptor's signature.

The preceptor for a new proposed RSO must fill out all four sections.

The preceptor for an RSO seeking authorization to be recognized as an RSO for the additional medical use(s) must fill out the second, third, and fourth sections.

VII.   AUTHORIZED MEDICAL PHYSICIST - Specific Instructions and Guidance for Filling Out NRC Form 313A (AMP)

See Section V, "General Instructions and Guidance for Filling Out NRC Form 313A Series," for additional clarification on providing information about an individual's status on an Agreement State license, medical broad-scope license, or Master Materials License permit.

Part I. Training and Experience - select one of the three methods below:

Item 1.   Board Certification

Provide the requested information (i.e., a copy of the board certification, documentation of device-specific training in the table in 3.c, and a completed Preceptor Attestation). As indicated on the form, additional information is needed if the board certification or device-specific training was completed more than 7 years ago.

Device-specific training may be provided by the vendor or a supervising medical physicist authorized for the requested type of use. Specific information regarding the supervising individual only needs to be provided in the table in 3.c if the training was provided by an AMP. If more than one supervising individual provided the training, identify each supervising individual by name and provide his/her qualifications.

Item 2. Current Authorized Medical Physicist Seeking Additional Uses(s) Checked above

Provide the requested information (i.e., documentation of device-specific training (complete the table in 3.c) and complete the Preceptor Attestation in Part II). As indicated on the form, additional information is needed if the device-specific training was completed more than 7 years ago.

Device-specific training may be provided by the vendor or a supervising medical physicist authorized for the requested type of use. Specific information regarding the supervising individual only needs to be provided in the table in 3.c if the training was provided by an AMP. If more than one supervising medical physicist provided the training, identify each supervising individual by name and provide his/her qualifications.

Item 3. Training and Experience for Proposed Authorized Medical Physicist

As indicated on the form, additional information is needed if the degree, training, and/or work experience was completed more than 7 years ago.

Submit a completed Section 3.a. Submit documentation of a graduate degree (for example, a copy of a diploma or transcript from an accredited college or university).

Submit a completed Section 3.b. The individual must have completed 1 year of full-time training in medical physics and an additional year of full-time work experience, which cannot be concurrent. This is documented in Section 3.b by providing the ranges of dates for training and work experience.

If the proposed AMP had more than one supervisor, provide the information requested in Section 3.b for each supervising individual. If the supervising individual is not an AMP, the applicant must provide documentation that the supervising individual meets the requirements in 10 CFR 35.51 and 10 CFR 35.59.

Submit a completed Section 3.c for each specific device for which the applicant is requesting authorization.

Device-specific training may be provided by the vendor or a supervising medical physicist authorized for the requested type of use. Specific information regarding the supervising individual only needs to be provided in the table in 3.c if the training was provided by an AMP.

If more than one supervising medical physicist provided the training, identify each supervising individual by name and provide his/her qualifications.

Submit a completed Preceptor Attestation in Part II.

Part II. Preceptor Attestation

The Preceptor Attestation page has four sections.

The attestation to the proposed AMP's training is in the first section.

The attestation for the device-specific training is in the second section.

The attestation of the individual's competency to function independently as an AMP for the specific devices requested by the applicant is in the third section.

The fourth and final section requests specific information about the preceptor's authorizations to use licensed material, in addition to the preceptor's signature.

The preceptor for a proposed new AMP must fill out all four sections of this page. The preceptor for an AMP seeking additional authorizations must complete the last three sections.

VIII. AUTHORIZED NUCLEAR PHARMACIST - Specific Instructions and Guidance for Filling Out NRC Form 313A (ANP)

See Section V, "General Instructions and Guidance for Filling out NRC Form 313A Series," for additional clarification on providing information about an individual's status on an Agreement State license, medical broad-scope license, or Master Materials License permit.

Part I. Training and Experience - select one of the two methods below:

Item 1. Board Certification

Provide the requested information (i.e., a copy of the board certification and a completed Preceptor Attestation). As indicated on the form, additional information is needed if the board certification occurred more than 7 years ago.

Item 2. Structured Educational Program for a Proposed Authorized Nuclear Pharmacist

As indicated on the form, additional information is needed if the training and/or supervised practical experience was completed more than 7 years ago.

Submit completed Sections 2.a and 2.b. If the proposed new nuclear pharmacist had more than one supervisor, provide the name of each supervising individual in Section 2.b.

Submit a completed Preceptor Attestation.

Part II. Preceptor Attestation

The Preceptor Attestation page has two sections. The preceptor must select either the board certification or the structured educational program when filling out the first section on this page.

The second and final section of the page requests specific information about the preceptor's authorization to use licensed material, in addition to the preceptor's signature.

IX. 10 CFR 35.100, 35.200, AND 35.500 AUTHORIZED USERS - Specific Instructions and Guidance for Filling Out NRC Form 313A (AUD)

See Section V, "General Instructions and Guidance for Filling out NRC Form 313A Series," for additional clarification on providing information about an individual's status on an Agreement State license, medical broad-scope license, or Master Materials License permit.

Part I. Training and Experience - select one of the three methods below:

Item 1. Board Certification

Provide the requested information (i.e., a copy of the board certification and a completed Preceptor Attestation). As indicated on the form, additional information is needed if the board certification occurred more than 7 years ago.

Item 2. Current 35.390 Authorized User Seeking Additional 10 CFR 35.290 Authorization

(a) Fill in the blank in Section 2.a with the current license number on which the proposed user is listed.

(b) Provide a description of the proposed user's experience that meets the requirements of 10 CFR 35.290 (c)(1)(ii)(G) as shown in the table in 2.b. As indicated on the form, additional information is needed if this experience was obtained more than 7 years ago.

List each supervising individual by name and include the license showing the supervising individual as an AU.

Item 3. Training and Experience for Proposed Authorized Users

As indicated on the form, additional information is needed if the training and/or work experience was completed more than 7 years ago.

Note: Providing the training and experience information required under 10 CFR 35.290 will allow the individual to be authorized to use materials permitted by both 10 CFR 35.100 and 10 CFR 35.200.

Submit a completed Section 3.a for each proposed authorized use.

Submit a completed Section 3.b, except for 10 CFR 35.500 uses. If the proposed user had more than one supervisor, provide the information requested in Section 3.b for each supervising individual.

Submit a completed Section 3.c for 10 CFR 35.500 uses.

Submit a completed Preceptor Attestation, except for 10 CFR 35.500 uses.

Part II. Preceptor Attestation

The Preceptor Attestation page has two sections.

The attestations for training and experience requirements in 10 CFR 35.190 and 10 CFR 35.290 are found in the first section.

The second and final section requests specific information about the preceptor's authorization(s) to use licensed material, in addition to the preceptor's signature.

The preceptor must fill out both sections.

Note: The attestation to the proposed user's training and competency to function independently under 10 CFR 35.190 covers the use of material permitted by 10 CFR 35.100 only. The attestation for the proposed user's training and competency to function independently under 10 CFR 35.290 will allow the individual to be authorized to use material permitted by both 10 CFR 35.100 and 10 CFR 35.200.

X.   35.300 AUTHORIZED USER - Specific Instructions and Guidance for Filling Out NRC Form 313A (AUT)

See Section V, "General Instructions and Guidance for Filling out NRC Form 313A Series," for additional clarification on providing information about an individual's status on an Agreement State license, medical broad-scope license, or Master Materials License permit.

Part I. Training and Experience - select one of the three methods below:

Item 1.   Board Certification

If the applicant is a nuclear medicine physician, radiologist, or radiation oncologist with a board certification listed under 10 CFR 35.300 on NRC's Web site, provide the requested information (i.e., a copy of the board certification, documentation of supervised clinical experience (complete the table in section 3.c), and a completed Preceptor Attestation). As indicated on the form, additional information is needed if the board certification or supervised clinical experience occurred more than 7 years ago. List each supervising individual by name and include the license showing the supervising individual as an AU.

If the applicant is a radiation oncologist whose board certification is not listed under 10 CFR 35.300 on NRC's Web site, provide the requested information (i.e., a copy of the board certification listed under either 10 CFR 35.400 or 10 CFR 35.600 on NRC's Web site, documentation of training and supervised work experience with unsealed materials requiring a written directive (complete the tables in Sections 3.a and 3.b), documentation of supervised clinical experience (complete the table in Section 3.c), and a completed Preceptor Attestation). As indicated on the form, additional information is needed if the board certification, training, and supervised work experience or clinical experience occurred more than 7 years ago. List each supervising individual by name and include the license showing the supervising individual as an AU.

Item 2. Current 10 CFR 35.300, 10 CFR 35.400, or 10 CFR 35.600 Authorized User Seeking Additional Authorization

Submit a completed Section 2.a, listing the license number and the user's current authorizations.

If the applicant is currently authorized for a subset of clinical uses under 10 CFR 35.300, submit the requested information (i.e., complete the table in Section 3.c to document the new supervised clinical case experience and the completed Preceptor Attestation). As indicated on the form, additional information is needed if the clinical case experience occurred more than 7 years ago. List each supervising individual by name and include the license showing the supervising individual as an AU.

If the applicant is currently authorized under 10 CFR 35.490 or 10 CFR 35.690 and meets the requirements in 10 CFR 35.396, submit the requested information (i.e., documentation of training and supervised work experience with unsealed materials requiring a written directive (complete the tables in Sections 3.a and 3.b), documentation of supervised clinical experience (complete the table in Section 3.c), and a completed Preceptor Attestation)). As indicated on the form, additional information is needed if the training and supervised work experience or clinical experience occurred more than 7 years ago. List each supervising individual by name and include the license showing the supervising individual as an AU.

Item 3. Training and Experience for Proposed Authorized Users

As indicated on the form, additional information is needed if the degree, training, and/or work experience was completed more than 7 years ago.

Submit a completed Section 3.a.

Submit a completed Section 3.b. List each supervising individual by name and include the license number showing the supervising individual as an AU.

Submit a completed Section 3.c for each requested authorization. List each supervising individual by name and include the license number showing the supervising individual as an AU.

Submit a completed Preceptor Attestation in Part II.

Part II. Preceptor Attestation

The Preceptor Attestation page has five sections.

The attestations for training and experience requirements in 10 CFR 35.390, 10 CFR 35.392, and 10 CFR 35.394 are in the first section.

The attestation for supervised clinical experience is in the second section.

The attestations for competency to function independently as an AU for specific uses is in the third section.

The attestation for training and experience requirements and competency to function independently for a radiation oncologist meeting the requirements in 10 CFR 35.396 is in the fourth section.

The fifth and final section requests specific information about the preceptor's authorization(s) to use licensed material, in addition to the preceptor's signature.

There are seven possible categories of individuals seeking AU status under this form. Follow the instructions for the applicable category.

The preceptor for a proposed AU who is a nuclear medicine physician, radiologist, or radiation oncologist with a board certification listed under 10 CFR 35.390 on NRC's Web site must complete the first, second, third, and fifth sections.

The preceptor for a proposed AU for all the uses listed in 10 CFR 35.390(b)(1)(ii)(G) who is a radiation oncologist with a board certification that is not listed under 10 CFR 35.390 on NRC's Web site must complete the first, second, third, and fifth sections.

The preceptor for a proposed AU for 10 CFR 35.390(b)(1)(ii)(G)(iii) and (iv) uses who is a radiation oncologist with a board certification listed under 10 CFR 35.490 or 10 CFR 35.690 on NRC's Web site must complete the fourth and fifth sections.

The preceptor for an AU who is currently authorized for a subset of clinical uses under 10 CFR 35.300 must complete the second, third, and fifth sections of this part, except for an AU meeting the criteria in 10 CFR 35.392 seeking to meet the training and experience requirements under 10 CFR 35.394.

The preceptor for an AU meeting the criteria in 10 CFR 35.392 seeking to meet the training and experience requirements under 10 CFR 35.394 must complete the first, second, third, and fifth sections.

The preceptor for an AU currently authorized under 10 CFR 35.490 or 10 CFR 35.690 and meeting the requirements in 10 CFR 35.396 must complete the fourth, and fifth sections.

The preceptor for a proposed new AU must complete the first, second, third and fifth sections.

XI. 35.400 AND 35.600 AUTHORIZED USERS - Specific Instructions and Guidance for Filling Out NRC Form 313A (AUS)

See Section V, "General Instructions and Guidance for Filling out NRC Form 313A Series," for additional clarification on providing information about an individual's status on an Agreement State license, medical broad-scope license, or Master Materials License permit.

Part I. Training and Experience - select one of the three methods below:

Item 1. Board Certification

Provide the requested information (i.e., a copy of the board certification) for 10 CFR 35.600 uses, documentation of device-specific training in the table in 3.e, and for all uses, a completed Preceptor Attestation. As indicated on the form, additional information is needed if the board certification or device-specific training was completed more than 7 years ago.

Device-specific training may be provided by the vendor for new users, or either a supervising AU or an AMP authorized for the requested type of use. Specific information regarding the supervising individual only needs to be provided in the table in 3.e if the training was provided by an AU or AMP. If more than one supervising individual provided the training, identify each supervising individual by name and provide his/her qualifications.

Item 2. Current 10 CFR 35.600 Authorized User Requesting Additional Authorization for 10 CFR 35.600 Use(s) Checked Above

Provide the requested information (i.e., documentation of device-specific training (complete the table in 3.e)) and a completed Preceptor Attestation in Part II. As indicated on the form, additional information is needed if the device-specific training was completed more than 7 years ago.

Device-specific training may be provided by the vendor, a supervising AU, or an AMP authorized for the requested type of use. Specific information regarding the supervising individual only needs to be provided in the table in 3.e if the training was provided by an AU or AMP. If more than one supervising individual provided the training, identify each supervising individual by name and provide his/her qualifications.

Item 3. Training and Experience for Proposed Authorized Users

As indicated on the form, additional information is needed if the training, residency program, supervised work, and clinical experience were completed more than 7 years ago.

Submit a completed Section 3.a for each requested use.

Submit a completed Section 3.b if applying for 10 CFR 35.400 uses. However, Section 3.b does not have to be completed when only applying for use of strontium-90 for ophthalmic use. If more than one supervising AU provided the supervised work and clinical experience, identify each supervising individual by name and provide his/her qualifications.

Submit a completed Section 3.c if only applying for use of strontium-90 for ophthalmic use. If more than one supervising AU provided the supervised clinical experience, identify each supervising individual by name and provide his/her qualifications.

Submit a completed Section 3.d for each requested 10 CFR 35.600 use. If more than one supervising AU provided the supervised work and clinical experience, identify each supervising individual by name and provide his/her qualifications.

Submit a completed Section 3.e for each specific 10 CFR 35.600 device for which the applicant is requesting authorization.

Device-specific training may be provided by the vendor, a supervising AU, or an AMP authorized for the requested type of use. Specific information regarding the supervising individual only needs to be provided in the table in 3.e if the training was provided by an AU or AMP. If more than one supervising individual provided the training, identify each supervising individual by name and provide his/her qualifications.

Submit a completed Preceptor Attestation in Part II.

Part II. Preceptor Attestation

The Preceptor Attestation part has five sections.

The attestation to the training and individual's competency for 10 CFR 35.400 uses or strontium-90 eye applicator use is in the first section.

The attestation to the training for the proposed AU for 10 CFR 35.600 uses is in the second section.

The attestation for the 10 CFR 35.600 device-specific training is in the third section.

The attestation of the individual's competency to function independently as an AU for the specific 10 CFR 35.600 devices requested by the applicant is in the fourth section.

The fifth and final section requests specific information about the preceptor's authorization(s) to use licensed material, in addition to the preceptor's signature.

The preceptor for a 10 CFR 35.400 proposed AU must fill out the first and fifth sections.

The preceptor for a 10 CFR 35.600 proposed AU must fill out the second, third, fourth and fifth sections.

The preceptor for an AU seeking additional 10 CFR 35.600 authorizations must complete the third, fourth, and fifth sections.

APPENDIX E

Sample License Application

Sample License Application

This Appendix includes the following sample forms:

- Sample Form 313, "Application for Materials License,"

- Sample Form 313A (AUD), "Authorized User Training and Experience and Preceptor Attestation,"

- Attachment 1, "Table E.1 Sample Submission: Table C.2 Completed,"

- Attachment 2, "Table E.2 Sample Submission: Table C.3 Completed," and

- Attachment 3, "Figure E.1 Sample License Application: Facility Diagram" (referenced in Attachment 1).

Sample Form 313
"Application for Materials License"

NRC FORM 313 U.S. NUCLEAR REGULATORY COMMISSION	APPROVED BY OMB: NO. 3150-0120 EXPIRES: 10/31/2008
(10-2006) 10 CFR 30, 32, 33 34, 35, 36, 39 and 40 **APPLICATION FOR MATERIALS LICENSE**	Estimated burden per response to comply with this mandatory information collection request, 7.4 hours. Submittal of the application is necessary to determine that the applicant is qualified and that adequate procedures exist to protect the public health and safety. Send comments regarding burden estimate to the Records Management Branch (T-6 E6), U.S. Nuclear Regulatory Commission, Washington, DC 20555-0001, or by internet e-mail to bjs1@nrc.gov, and to the Desk Officer, Office of Information and Regulatory Affairs, NEOB-10202, (3150-0120), Office of Management and Budget, Washington, DC 20503. If a means used to impose an information collection does not display a currently valid OMB control number, NRC may not conduct or sponsor, and a person is not required to respond to, the information collection.

INSTRUCTIONS: SEE THE APPROPRIATE LICENSE APPLICATION GUIDE FOR DETAILED INSTRUCTIONS FOR COMPLETING APPLICATION. SEND TWO COPIES OF THE ENTIRE COMPLETED APPLICATION TO THE NRC OFFICE SPECIFIED BELOW.

APPLICATION FOR DISTRIBUTION OF EXEMPT PRODUCTS FILE APPLICATIONS WITH:	IF YOU ARE LOCATED IN:
DIVISION OF INDUSTRIAL AND MEDICAL NUCLEAR SAFETY OFFICE OF NUCLEAR MATERIALS SAFETY AND SAFEGUARDS U.S. NUCLEAR REGULATORY COMMISSION WASHINGTON, DC 20555-001 **ALL OTHER PERSONS FILE APPLICATIONS AS FOLLOWS:** IF YOU ARE LOCATED IN: CONNECTICUT, DELAWARE, DISTRICT OF COLUMBIA, MAINE, MARYLAND, MASSACHUSETTS, NEW HAMPSHIRE, NEW JERSEY, NEW YORK, PENNSYLVANIA, RHODE ISLAND, OR VERMONT, SEND APPLICATIONS TO: LICENSING ASSISTANT SECTION NUCLEAR MATERIALS SAFETY BRANCH U.S. NUCLEAR REGULATORY COMMISSION, REGION I 475 ALLENDALE ROAD KING OF PRUSSIA, PA 19406-1415 ALABAMA, FLORIDA, GEORGIA, KENTUCKY, NORTH CAROLINA, PUERTO RICO, SOUTH CAROLINA, TENNESSEE, VIRGINIA, VIRGIN ISLANDS, OR WEST VIRGINIA,	ILLINOIS, INDIANA, IOWA, MICHIGAN, MINNESOTA, MISSOURI, OHIO, OR WISCONSIN, SEND APPLICATIONS TO: MATERIALS LICENSING SECTION U.S. NUCLEAR REGULATORY COMMISSION, REGION III 801 WARRENVILLE RD. LISLE, IL 60532-4351 ALASKA, ARIZONA, ARKANSAS, CALIFORNIA, COLORADO, HAWAII, IDAHO, KANSAS, LOUISIANA, MISSISSIPPI, MONTANA, NEBRASKA, NEVADA, NEW MEXICO, NORTH DAKOTA, OKLAHOMA, OREGON, PACIFIC TRUST TERRITORIES, SOUTH DAKOTA, TEXAS, UTAH, WASHINGTON, OR WYOMING, SEND APPLICATIONS TO: NUCLEAR MATERIALS LICENSING SECTION U.S. NUCLEAR REGULATORY COMMISSION, REGION IV 611 RYAN PLAZA DRIVE, SUITE 400 ARLINGTON, TX 76011-8084

PERSONS LOCATED IN AGREEMENT STATES SEND APPLICATIONS TO THE U.S. NUCLEAR REGULATORY COMMISSION ONLY IF THEY WISH TO POSSESS AND USE LICENSED MATERIAL IN STATES SUBJECT TO U.S. NUCLEAR REGULATORY COMMISSION JURISDICTIONS.

1. THIS IS AN APPLICATION FOR (Check appropriate item) [X] A. NEW LICENSE [ ] B. AMENDMENT TO LICENSE NUMBER _____ [ ] C. RENEWAL OF LICENSE NUMBER _____	2. NAME AND MAILING ADDRESS OF APPLICANT (Include Zip code) **Dr. Noe Directive** **Suite 112** **2 Physician Circle Parkway** **Anytown, WV 02201**
3. ADDRESS(ES) WHERE LICENSED MATERIAL WILL BE USED OR POSSESSED **Attached document contains security-related sensitive information**	4. NAME OF PERSON TO BE CONTACTED ABOUT THIS APPLICATION **Noe Directive, MD** TELEPHONE NUMBER **(123) 456-7890**

SUBMIT ITEMS 5 THROUGH 11 ON 8-1/2 X 11" PAPER. THE TYPE AND SCOPE OF INFORMATION TO BE PROVIDED IS DESCRIBED IN THE LICENSE APPLICATION GUIDE.

5. RADIOACTIVE MATERIAL a. Element and mass number; b. chemical and/or physical form; and c. maximum amount which will be possessed at any one time. **See Attachment 1**	6. PURPOSE(S) FOR WHICH LICENSED MATERIAL WILL BE USED. **See Attachment 1**
7. INDIVIDUAL(S) RESPONSIBLE FOR RADIATION SAFETY PROGRAM AND THEIR TRAINING EXPERIENCE **See Attachment 2**	8. TRAINING FOR INDIVIDUALS WORKING IN OR FREQUENTING RESTRICTED AREAS. **See Attachment 2**
9. FACILITIES AND EQUIPMENT. **See Attachment 2**	10. RADIATION SAFETY PROGRAM. **See Attachment 2**
11. WASTE MANAGEMENT. **See Attachment 2**	12. LICENSEE FEES (See 10 CFR 170 and Section 170.31) FEE CATEGORY 7C AMOUNT ENCLOSED $D, DDD.CC

13. CERTIFICATION. (Must be completed by applicant) THE APPLICANT UNDERSTANDS THAT ALL STATEMENTS AND REPRESENTATIONS MADE IN THIS APPLICATION ARE BINDING UPON THE APPLICANT.
THE APPLICANT AND ANY OFFICIAL EXECUTING THIS CERTIFICATION ON BEHALF OF THE APPLICANT, NAMED IN ITEM 2, CERTIFY THAT THIS APPLICATION IS PREPARED IN CONFORMITY WITH TITLE 10, CODE OF FEDERAL REGULATIONS, PARTS 30, , 32, 33, 34, 35, 36, 39, AND 40, AND THAT ALL INFORMATION CONTAINED HEREIN IS TRUE AND CORRECT TO THE BEST OF THEIR KNOWLEDGE AND BELIEF.
WARNING: 18 U.S.C. SECTION 1001 ACT OF JUNE 25, 1948 62 STAT. 749 MAKES IT A CRIMINAL OFFENSE TO MAKE A WILLFULLY FALSE STATEMENT OR REPRESENTATION TO ANY DEPARTMENT OR AGENCY OF THE UNITED STATES AS TO ANY MATTER WITHIN ITS JURISDICTION.

CERTIFYING OFFICER – TYPED/PRINTED NAME AND TITLE **Noe Directive, MD - President**	SIGNATURE *Noe Directive*	DATE **April 11, 2007**

FOR NRC USE ONLY

TYPE OF FEE	FEE LOG	FEE CATEGORY	AMOUNT RECEIVED $	CHECK NUMBER	COMMENTS
APPROVED BY			DATE		

NRC FORM 313 (10-2006) PRINTED ON RECYCLED PAPER

Sample Form 313A (AUD)
"Authorized User Training and Experience and Preceptor Attestation"

NRC FORM 313A (AUD) (3-2007)	U.S. NUCLEAR REGULATORY COMMISSION	
AUTHORIZED USER TRAINING AND EXPERIENCE **AND PRECEPTOR ATTESTATION** **(for uses defined under 35.100, 35.200, and 35.500)** **[10 CFR 35.190, 35.290, and 35.590]**		APPROVED BY OMB: NO. 3150-0120 EXPIRES: 10/31/2008

Name of Proposed Authorized User	State or Territory Where Licensed
Noe Directive, MD	**West Virginia Medical License WV-MDXXYY**

Requested Authorization(s) (*check all that apply*)

☒ 35.100 Uptake, dilution, and excretion studies

☒ 35.200 Imaging and localization studies

☐ 35.500 Sealed sources for diagnosis (specify device _____)

PART 1 - - TRAINING AND EXPERIENCE
(Select one of the three methods below)

* Training and Experience, including board certification, must have been obtained within the 7 years preceding the date of application or the individual must have obtained related continuing education and experience since the required training and experience was completed. Provide dates, duration, and description of continuing education and experience related to the uses checked above.

☐ 1. **Board Certification**
   a. Provide a copy of the board certification
   b. If using only 35.500 materials, stop here. If using 35.100 and 35.200 materials, skip to and complete Part II Preceptor Attestation.

☐ 2. **Current 35.390 Authorized User Seeking Additional 35.290 Authorization**
   a. Authorized user on Materials License _____ meeting 10 CFR 35.390 or equivalent Agreement State requirements seeking authorization for 35.290.
   b. Supervised Work Experience.
   (If more than one supervising individual is necessary to document supervised work experience, provide multiple copies of this section).

Description of Experience	Location of Experience/License or Permit Number of Facility	Clock Hours	Dates of Experience*
Eluting generator systems appropriate for the preparation of radioactive drugs for imaging and localization studies, measuring and testing the eluate for radionuclidic purity, and processing the eluate with reagent kits to prepare labeled radioactive drugs			
Total Hours of Experience:			
Supervising Individual	License/Permit Number listing supervising individual as an authorized user		

Supervisor meets the requirements below, or equivalent Agreement State requirements *(check all that apply)*.

☐ 35.290   ☐ 35.390 + generator experience in 32.290(c)(1)(ii)(G)

APPENDIX E

☒ 3.   **Training and Experience for Proposed Authorized User**

a.   Classroom and Laboratory Training.

Description of Training	Location of Training	Clock Hours	Dates of Training*
Radiation physics and instrumentation	**Radiation 200 for Diagnostic Physicians Sample Medical School Anytown, WV**	50	**July 1 to Aug 15, 2006**
Radiation protection	**Radiation 200 for Diagnostic Physicians Sample Medical School Anytown, WV**	50	**July 1 to Aug 15, 2006**
Mathematics pertaining to the use and measurement of radioactivity	**Radiation 200 for Diagnostic Physicians Sample Medical School Anytown, WV**	50	**July 1 to Aug 15, 2006**
Chemistry of byproduct material for medical use *(not required for 35.590)*	**Radiation 200 for Diagnostic Physicians Sample Medical School Anytown, WV**	50	**July 1 to Aug 15, 2006**
Radiation biology	**Radiation 200 for Diagnostic Physicians Sample Medical School Anytown, WV**	50	**July 1 to Aug 15, 2006**
	Total Hours of Training:	250	

b.   Supervised Work Experience (completion of this table is not required for 35.590).
(If more than one supervising individual is necessary to document supervised work experience, provide multiple copies of this section).

Supervised Work Experience		**Total Hours of Experience: 500**	
Description of Experience Must Include:	Location of Experience/License or Permit Number of Facility	Confirm	Dates of Experience*
Ordering, receiving, and unpacking radioactive materials safely and performing the related radiation surveys	**Sample Medical Institution Limited 1234 Main Street Anytown, WV 02120**	☒ Yes ☐ No	**August 2006 to March 2007**
Performing quality control procedures on instruments used to determine the activity of dosages and performing checks for proper operation of survey meters	**Sample Medical Institution Limited 1234 Main Street Anytown, WV 02120**	☒ Yes ☐ No	**August 2006 to March 2007**

NRC FORM 313A (AUD)
(3-2007)

U.S. NUCLEAR REGULATORY COMMISSION

AUTHORIZED USER TRAINING AND EXPERIENCE AND PRECEPTOR ATTESTATION (continued)

3. Training and Experience for Proposed Authorized User (continued)

b. Supervised Work Experience. (continued)

Description of Experience Must Include:	Location of Experience/License or Permit Number of Facility	Confirm	Dates of Experience*
Calculating, measuring, and safely preparing patient or human research subject dosages	**Sample Medical Institution Limited 1234 Main Street Anytown, WV 02120**	☒ Yes ☐ No	**August 2006 to March 2007**
Using administrative controls to prevent a medical event involving the use of unsealed byproduct material	**Sample Medical Institution Limited 1234 Main Street Anytown, WV 02120**	☒ Yes ☐ No	**August 2006 to March 2007**
Using procedures to contain spilled byproduct material safely and using proper decontamination procedures	**Sample Medical Institution Limited 1234 Main Street Anytown, WV 02120**	☒ Yes ☐ No	**August 2006 to March 2007**
Administering dosages of radioactive drugs to patients or human research subjects	**Sample Medical Institution Limited 1234 Main Street Anytown, WV 02120**	☒ Yes ☐ No	**August 2006 to March 2007**
Eluting generator systems appropriate for the preparation of radioactive drugs for imaging and localization studies, measuring and testing the eluate for radionuclidic purity, and processing the eluate with reagent kits to prepare labeled radioactive drugs		☐ Yes ☒ No	

Supervising Individual	License/Permit Number listing supervising individual as an authorized user
Thomas Group, D.O.	**99-02120-01**

Supervisor meets the requirements below, or equivalent Agreement State requirements *(check one)*.

☐ 35.190   ☒ 35.290   ☐ 35.390   ☐ 35.390 + generator experience in 35.290(c)(1)(ii)(G)

c.   For 35.590 only, provide documentation of training on use of the device.

Device	Type of Training	Location and Dates

d.   For 35.500 uses only, stop here. For 35.100 and 35.200 uses, skip to and complete Part II Preceptor Attestation.

PAGE 3

APPENDIX E

U.S. NUCLEAR REGULATORY COMMISSION

AUTHORIZED USER TRAINING AND EXPERIENCE AND PRECEPTOR ATTESTATION (continued)

3. <u>Training and Experience for Proposed Authorized User</u> (continued)

   b. Supervised Work Experience. (continued)

Description of Experience Must Include:	Location of Experience/License or Permit Number of Facility	Confirm	Dates of Experience*
Calculating, measuring, and safely preparing patient or human research subject dosages		☐ Yes ☒ No	
Using administrative controls to prevent a medical event involving the use of unsealed byproduct material		☐ Yes ☒ No	
Using procedures to contain spilled byproduct material safely and using proper decontamination procedures		☐ Yes ☒ No	
Administering dosages of radioactive drugs to patients or human research subjects		☐ Yes ☒ No	
Eluting generator systems appropriate for the preparation of radioactive drugs for imaging and localization studies, measuring and testing the eluate for radionuclidic purity, and processing the eluate with reagent kits to prepare labeled radioactive drugs	**Sample Medical Institution Limited** **1234 Main Street** **Anytown, WV 02120**	☒ Yes ☐ No	**August 2006** **to** **March 2007**

Supervising Individual **Jane Jones, MD**	License/Permit Number listing supervising individual as an authorized user **99-02120-01**

Supervisor meets the requirements below, or equivalent Agreement State requirements *(check one)*.

   ☐ 35.190   ☒ 35.290   ☐ 35.390   ☐ 35.390 + generator experience in 35.290(c)(1)(ii)(G)

   c. For 35.590 only, provide documentation of training on use of the device.

Device	Types of Training	Location and Dates

   d. For 35.500 uses only, stop here. For 35.100 and 35.200 uses, skip to and complete Part II Preceptor Attestation.

NRC FORM 313A (AUD)
(3-2007)

U.S. NUCLEAR REGULATORY COMMISSION

AUTHORIZED USER TRAINING AND EXPERIENCE AND PRECEPTOR ATTESTATION (continued)

PART II – PRECEPTOR ATTESTATION

Note: This part must be completed by the individual's preceptor. The preceptor does not have to be the supervising individual as long as the preceptor provides, directs, or verifies training and experience required. If more than one preceptor is necessary to document experience, obtain separate preceptor statement from each. (Not required to meet training requirements in 35.590)

By checking the boxes below, the preceptor is attesting that the individual has knowledge to fulfill the duties of the position sought and not attesting to the individual's "general clinical competency."

First Section

Check one of the following for each use requested:

For 35.190

Board Certification

☐ I attest that _____ has satisfactorily completed the requirements in
Name of Proposed Authorized User

10 CFR 35.290(a)(1) and has achieved a level of competency sufficient to function independently as an authorized user for the medical uses authorized under 10 CFR 35.100.

OR

Training and Experience

☐ I attest that _____ has satisfactorily completed the 60 hours of training and
Name of Proposed Authorized User

experience, including a minimum of 8 hours of classroom and laboratory training, required by 10 CFR 35.190(c)(1), and has achieved a level of competency sufficient to function independently as an authorized user for the medical uses authorized under 10 CFR 35.100.

For 35.290

Board Certification

☐ I attest that _____ has satisfactorily completed the requirements in
Name of Proposed Authorized User

10 CFR 35.290(a)(1) and has achieved a level of competency sufficient to function independently as an authorized user for the medical uses authorized under 10 CFR 35.100 and 35.200.

OR

Training and Experience

☒ I attest that **Noe Directive, MD** has satisfactorily completed the 700 hours of training
Name of Proposed Authorized User

and experience, including a minimum of 80 hours of classroom and laboratory training, required by 10 CFR 35.290(c)(1), and has achieved a level of competency sufficient to function independently as an authorized user for the medical uses authorized under 10 CFR 35.100 and 35.200.

- -

Second Section

Complete the following for preceptor attestation and signature:

☒ I meet the requirement below, or equivalent Agreement State requirements, as an authorized user for:

☐ 35.190   ☒ 35.290   ☐ 35.390   ☐ 35.390 + generator experience

Name of Preceptor	Signature	Telephone Number	Date
Jane Jones, MD	*Jane Jones*	(123) 456-7890	4-11-07

License/Permit Number/Facility Name

99-02120-01   Sample Medical Institution Limited

Noe Directive, M.D. Attachment 1 of 3

Table E.1 Sample Submission: Table C.2 Completed

(If using this checklist, check applicable rows and fill in details, and attach copy of checklist to the application.)

☒ Yes ☐ No	This response includes security-related sensitive information (see Section 5.2) which is included in Attachment 3 and marked "Security-related information – withhold under 10 CFR 2.390"			
Yes	**Radionuclide**	**Form or Manufacturer/ Model No.**	**Maximum Quantity**	**Purpose of Use**
☒	Any byproduct material permitted by 10 CFR 35.100	Any	As needed	Any uptake, dilution, and excretion study permitted by 10 CFR 35.100.
☒	Any byproduct material permitted by 10 CFR 35.200	Any	As needed	Any imaging and localization study permitted by 10 CFR 35.200.
	F-18	Any	_____ curies	Production of PET radioactive drugs under 10 CFR 30.32(j).
	O-15	Any	_____ curies	Production of PET radioactive drugs under 10 CFR 30.32(j).
	C-11	Any	_____ curies	Production of PET radioactive drugs under 10 CFR 30.32(j).
	Any byproduct material permitted by 10 CFR 35.300	Any	_____ millicuries	Any radiopharmaceutical therapy procedure permitted by 10 CFR 35.300.
	Iodine-131	Any	___ millicuries	Administration of I-131 sodium iodide.
	Byproduct material permitted by 10 CFR 35.400 (Radionuclide _____)	Sealed source or device (Manufacturer _____, Model No._____)	___ millicuries	Any brachytherapy procedure permitted by 10 CFR 35.400.
	Byproduct material permitted by 10 CFR 35.400 (Radionuclide _____)	Sealed source or device (Manufacturer _____, Model No._____)	___ millicuries	Any brachytherapy procedure permitted by 10 CFR 35.400.
	Byproduct material permitted by 10 CFR 35.400 (Radionuclide _____)	Sealed source or device (Manufacturer _____, Model No._____)	___ millicuries	Any brachytherapy procedure permitted by 10 CFR 35.400.
	Byproduct material permitted by 10 CFR 35.400 (Radionuclide _____)	Sealed source or device (Manufacturer _____, Model No._____)	___ millicuries	Any brachytherapy procedure permitted by 10 CFR 35.400.

Noe Directive, M.D. Attachment 2 of 3

Table E.2   Sample Submission: Table C.3 Completed

(Check all applicable rows and fill in details and attach a copy of the checklist to the application or provide information separately.)

Item Number and Title	Suggested Response	Check box to indicate material included in application
Item 7: Radiation Safety Officer Name: Patrick Physicist, Ph.D., RSO on NRC License 11-2222-33	*For an individual previously identified as an RSO on an NRC or Agreement State license or permit:*	
	Previous license number (if issued by the NRC), or a copy of a license (if issued by an Agreement State), or a copy of a permit (if issued by an NRC master materials licensee) on which the individual was specifically named as the RSO.	☒
	For an individual qualifying under 10 CFR 35.57(a)(3):	
	Documentation that the individual was: • the RSO for only the medical uses of accelerator-produced radioactive material or discrete sources of Ra-226 included in the definition of byproduct material as a result of the EPAct; • the RSO for the medical uses of these materials before or during the effective period of NRC's waiver of August 31, 2005.	☐
	For an individual qualifying under 10 CFR 35.50(a):	
	Copy of certification by a specialty board whose certification process has been recognized[15] by NRC or an Agreement State under 10 CFR 35.50(a). AND	☐
	Description of the training and experience specified in 10 CFR 35.50(e) demonstrating that the proposed RSO is qualified by training in radiation safety, regulatory issues, and emergency procedures as applicable to the types of use for which the applicant seeks approval of an individual to serve as RSO. AND	☐
	Written attestation, signed by a preceptor RSO, that the individual has satisfactorily completed training in and experience required for certification, as well as training in radiation safety, regulatory issues, and emergency procedures for the types of use for which the licensee seeks approval, and has achieved a level of radiation safety knowledge sufficient to function independently as an RSO. AND	☐
	If applicable, description of recent related continuing education and experience as required by 10 CFR 35.59.	☐

---

[15]The names of board certifications that have been recognized by the NRC or an Agreement State are posted on the NRC's Web site http://www.nrc.gov/materials/miau/med-use-toolkit.html.

Table E.2	Sample Submission: Table C.3 Completed	
(Check all applicable rows and fill in details and attach a copy of the checklist to the application or provide information separately.)		
Item Number and Title	**Suggested Response**	**Check box to indicate material included in application**
Item 7: Authorized Users for medical uses: Name(s), (including license number authorizing practice of medicine, podiatry, or dentistry if not provided previously or in attachment); Requested uses for each individual <u>Noe Directive, MD</u> 35.100, 35.200	*For an individual previously identified as an AU on an NRC or Agreement State license or permit:*	
	Previous license number (if issued by the NRC), or a copy of the license (if issued by an Agreement State), or a copy of a permit issued by an NRC master materials licensee, or a copy of a permit issued by an NRC or Agreement State broad-scope licensee, or a copy of a permit issued by an NRC Master Materials License broad-scope permittee on which the physician, dentist, or podiatrist was specifically named as an AU for the uses requested.	❏
	For an AU requesting authorization for an additional medical use:	
	Description of the additional training and experience to demonstrate the AU is also qualified for the new medical uses requested (e.g., training and experience needed to meet the requirements in 10 CFR 35.290 (b), 35.396, 35.390(b)(1)(ii)(G), or 35.690(c)). **AND**	❏
	A preceptor attestation, if required (e.g., attestation is required to meet the requirements in 10 CFR 35.396, 35.390(b)(1)(ii)(G), or 35.690(c)).	
	For an individual qualifying under 10 CFR 35.57(b)(3):	
	Documentation that the physician, podiatrist, or dentist: • used only accelerator-produced radioactive materials, or discrete sources of Ra-226, or both, for medical uses before or during the effective period of NRC's waiver of August 31, 2005; and • used these materials for the same medical uses requested.	❏
	For an individual qualifying under 10 CFR Part 35, Subparts D, E, F, G, and/or H, who is board-certified:	
	Copy of the certification(s) by a specialty board(s) whose certification process has been recognized[16] by the NRC under 10 CFR Part 35, Subpart D, E, F, G, or H, as applicable to the use requested. **AND**	❏

---

[16]The names of board certifications that have been recognized by the NRC or an Agreement State are posted on the NRC's Web site http://www.nrc.gov/materials/miau/med-use-toolkit.html.

Item Number and Title	Suggested Response	Check box to indicate material included in application
	For an individual qualifying under 10 CFR Part 35, Subparts D, E, F, G, and/or H, who is not board-certified:	
	A description of the training and experience identified in 10 CFR Part 35, Subparts D, E, F, G, and H, demonstrating that the proposed AU is qualified by training and experience for the use(s) requested. **AND**	☒
	For an individual seeking authorization under 10 CFR Part 35, Subpart H, description of the training specified in 10 CFR 35.690 (c) demonstrating that the proposed AU is qualified for the type(s) of use for which authorization is sought. **AND**	☐
	Written attestation, signed by a preceptor physician AU, that the above training and experience have been satisfactorily completed and that a level of competency sufficient to function independently as an AU for the medical uses authorized has been achieved. **AND**	☒
	If applicable, description of recent related continuing education and experience as required by 10 CFR 35.59.	☐
Item 7: Authorized Nuclear Pharmacists Name(s) and license to practice pharmacy:	*For an individual previously identified as an ANP on an NRC or Agreement State license or permit:*	
	Previous license number (if issued by the NRC), or a copy of the license (if issued by an Agreement State), or a copy of a permit issued by an NRC master materials licensee, or a copy of a permit issued by an NRC or Agreement State broad-scope licensee, or a copy of a permit issued by an NRC Master Materials License broad-scope permittee on which the individual was specifically named ANP.	☐
	For an individual qualifying under 10 CFR 35.57(a)(3):	
	Documentation that the nuclear pharmacist: • used only accelerator-produced radioactive materials or discrete sources of Ra-226, or both, in the practice of nuclear pharmacy before or during the effective period of NRC's waiver of August 31, 2005; and • used these materials for the same uses requested.	☐

Table E.2   Sample Submission: Table C.3 Completed

(Check all applicable rows and fill in details and attach a copy of the checklist to the application or provide information separately.)

Table E.2   Sample Submission: Table C.3 Completed

(Check all applicable rows and fill in details and attach a copy of the checklist to the application or provide information separately.)

Item Number and Title	Suggested Response	Check box to indicate material included in application
	Description of the training and experience specified in 10 CFR 35.51(c) demonstrating that the proposed AMP is qualified by training in the types of use for which he or she is requesting AMP status, including hands-on device operation, safety procedures, clinical use, and operation of a treatment planning system. **AND**	☐
	Written attestation, signed by a preceptor AMP, that the required training and experience have been satisfactorily completed and that a level of competency sufficient to function independently as an AMP has been achieved. **AND**	☐
	If applicable, description of recent related continuing education and experience as required by 10 CFR 35.59.	☐
Item 7: Authorized User for nonmedical uses	*Note:* For purposes of this section of the table, the term "authorized user" is used to mean individuals authorized for the nonmedical uses described. See Sections 8.11 and 8.12.	
	For an individual previously authorized for nonmedical use on an NRC or Agreement State license or permit:	
Name(s): Requested types, quantities, and nonmedical uses for each individual	Previous license number (if issued by the NRC), or a copy of the license (if issued by an Agreement State), or a copy of a permit issued by an NRC master materials licensee, or a copy of a permit issued by an NRC or Agreement State broad-scope licensee, or a copy of a permit issued by an NRC Master Materials License broad-scope permittee on which the individual was specifically named an AU for the types, quantities, and uses requested.	☐
	For individuals qualifying under 10 CFR 30.33(a)(3):	
	Documentation of the individual's training and experience demonstrating that the individual is qualified to use the types and quantities of licensed materials for the requested uses.	☐
Item 9: Facility Diagram	A diagram is enclosed that describes the facilities and identifies activities conducted in all contiguous areas surrounding the area(s) of use. The following information is included:	☒
	• Guidance in Section 5.2 was reviewed and security-related sensitive information provided is marked accordingly.	☒
	• Drawings should be to scale, indicating the scale used.	☒

Table E.2   Sample Submission: Table C.3 Completed

(Check all applicable rows and fill in details and attach a copy of the checklist to the application or provide information separately.)

Item Number and Title	Suggested Response	Check box to indicate material included in application
	• Location, room numbers, and principal use of each room or area where byproduct material is prepared, used or stored, location of direct transfer delivery tubes from a PET radionuclide/radioactive drug production facility or production area of PET radioactive drugs under 10 CFR 30.32(j), and areas where higher energy gamma- emitting radionuclides (e.g., PET radionuclides) are used;	☒
	• Location, room numbers, and principal use of each adjacent room (e.g., office, file, toilet, closet, hallway), including areas above, beside, and below therapy treatment rooms, indicating whether the room is a restricted or unrestricted area as defined in 10 CFR 20.1003; and	☒
	• Provide shielding calculations and include information about the type, thickness, and density of any necessary shielding to enable independent verification of shielding calculations, including a description of any portable shields used (e.g., shielding of proposed patient rooms used for implant therapy, including the dimensions of any portable shield, if one is used; source storage safe).	☐
	In addition to the above, for teletherapy and GSR facilities, applicants should provide the directions of primary beam usage for teletherapy units and, in the case of an isocentric unit, the plane of beam rotation.	☐
Item 9: Radiation Monitoring Instruments	A statement that: "Radiation monitoring instruments will be calibrated by a person qualified to perform survey meter calibrations." **AND/OR**	☒
	A statement that: "We have developed and will implement and maintain written survey meter calibration procedures in accordance with the requirements in 10 CFR 20.1501 and that meet the requirements of 10 CFR 35.61." **AND**	☐
	A description of the instrumentation (e.g., gamma counter, solid state detector, portable or stationary count rate meter, portable or stationary dose rate or exposure rate meter, single or multichannel analyzer, liquid scintillation counter, proportional counter) that will be used to perform required surveys. **AND**	☐
	A statement that: "We reserve the right to upgrade our survey instruments as necessary as long as they are adequate to measure the type and level of radiation for which they are used."	☐
Item 9: Dose Calibrator and Other Dosage Measuring Equipment N/A	A statement that: "Equipment used to measure dosages will be calibrated in accordance with nationally recognized standards or the manufacturer's instructions."	☐

Table E.2   Sample Submission: Table C.3 Completed

(Check all applicable rows and fill in details and attach a copy of the checklist to the application or provide information separately.)

Item Number and Title	Suggested Response	Check box to indicate material included in application
	When administering dosages of alpha-emitting unsealed byproduct material in other than unit dosages made by a manufacturer or preparer licensed under 10 CFR 32.72 or 10 CFR 30.32(j), • A statement that: "Dosages will be determined by relying on the provider's dose label for measurement of the radioactivity and a combination of volumetric measurement and mathematical calculation." **OR**	☐
	• We are providing a description of the dosage measurement equipment, the nationally recognized calibration standard (or manufacturer's calibration instructions), and dosage measurement procedures.	☐
Item 9: Therapy Unit - Calibration and Use N/A	We are providing the procedures required by 10 CFR 35.642, 10 CFR 35.643, and 10 CFR 35.645, if applicable to the license application.	☐
Item 9: Other Equipment and Facilities N/A	Guidance in Section 5.2 was reviewed and security-related information provided is marked accordingly.	☐
	Attached is a description, identified as Attachment 9.4, of additional facilities and equipment.	☐
	For manual brachytherapy facilities, we are providing a description of the emergency response equipment.	☐
	For PET radionuclide use, PET radioactive drug production, and radiopharmaceutical therapy programs, we are providing a description of the additional facilities and equipment for these uses.	☐
	For teletherapy, GSR, and remote afterloader facilities, we are providing a description of the following: • Warning systems and restricted area controls (e.g., locks, signs, warning lights and alarms, interlock systems) for each therapy treatment room;	☐
	• Area radiation monitoring equipment;	☐
	• Viewing and intercom systems (except for LDR units);	☐
	• Steps that will be taken to ensure that no two units can be operated simultaneously, if other radiation-producing equipment (e.g., linear accelerator, X-ray machine) is in the treatment room;	☐
	• Methods to ensure that whenever the device is not in use or is unattended, the console keys will be inaccessible to unauthorized persons; and	☐
	• Emergency response equipment.	☐

Table E.2   Sample Submission: Table C.3 Completed

(Check all applicable rows and fill in details and attach a copy of the checklist to the application or provide information separately.)

Item Number and Title	Suggested Response	Check box to indicate material included in application
Item 10: Safety Procedures and Instructions N/A	Attached are procedures required by 10 CFR 35.610.	❐
	Guidance in Section 5.2 was reviewed and security-related sensitive information provided is marked accordingly.	❐
Item 10: Occupational Dose	A statement that: "Either we will perform a prospective evaluation demonstrating that unmonitored individuals are not likely to receive, in 1 year, a radiation dose in excess of 10% of the allowable limits in 10 CFR Part 20 or we will provide dosimetry that meets the requirements listed under 'Criteria' in NUREG-1556, Vol. 9, Rev. 1, 'Consolidated Guidance About Materials Licenses: Program-Specific Guidance About Medical Use Licenses.' " **OR**	☒
	A description of an alternative method for demonstrating compliance with the referenced regulations.	❐
Item 10: Area Surveys	A statement that: "We have developed and will implement and maintain written procedures for area surveys in accordance with 10 CFR 20.1101 that meet the requirements of 10 CFR 20.1501 and 10 CFR 35.70."	☒
Item 10: Safe Use of Unsealed Licensed Material	A statement that: "We have developed and will implement and maintain procedures for safe use of unsealed byproduct material that meet the requirements of 10 CFR 20.1101 and 10 CFR 20.1301."	☒
Item 10: Spill/Contamination Procedures	A statement that: "We have developed and will implement and maintain written procedures for safe response to spills of licensed material in accordance with 10 CFR 20.1101."	☒
Item 10: Installation, Maintenance, Adjustment, Repair, and Inspection of Therapy Devices Containing Sealed Sources	Name of the proposed employee and types of activities requested: _____ **AND**	❐
	Description of the training and experience demonstrating that the proposed employee is qualified by training and experience for the use requested. **AND**	❐
	Copy of the manufacturer's training certification and an outline of the training in procedures to be followed.	❐
Item 10: Minimization of Contamination N/A	A response is not required under the following condition: the NRC will consider that the above criteria have been met if the information provided in applicant's responses satisfy the criteria in Sections 8.15, 8.16, 8.21, 8.25, 8.27, and 8.29, on the topics: facilities and equipment, facility diagram, Radiation Protection Program, safety program, and waste management.	N/A

Table E.2   Sample Submission: Table C.3 Completed

(Check all applicable rows and fill in details and attach a copy of the checklist to the application or provide information separately.)

Item Number and Title	Suggested Response	Check box to indicate material included in application
Item 11: Waste Management	A statement that: "We have developed and will implement and maintain written waste disposal procedures for licensed material in accordance with 10 CFR 20.1101, that also meet the requirements of the applicable section of 10 CFR Part 20, Subpart K, and of 10 CFR 35.92."	☒
	Attached is a description of the radioactive waste incinerator facility and related portions of the Radiation Safety Program (10 CFR 20.2004).	☐
	Attached is a request to receive potentially contaminated radiation transport shields from consortium members receiving PET radioactive drugs noncommercially transferred under 10 CFR 30.32(j) authorization.	☐

Noe Directive, M.D. Attachment 3 of 3

Dr. Noe Directive

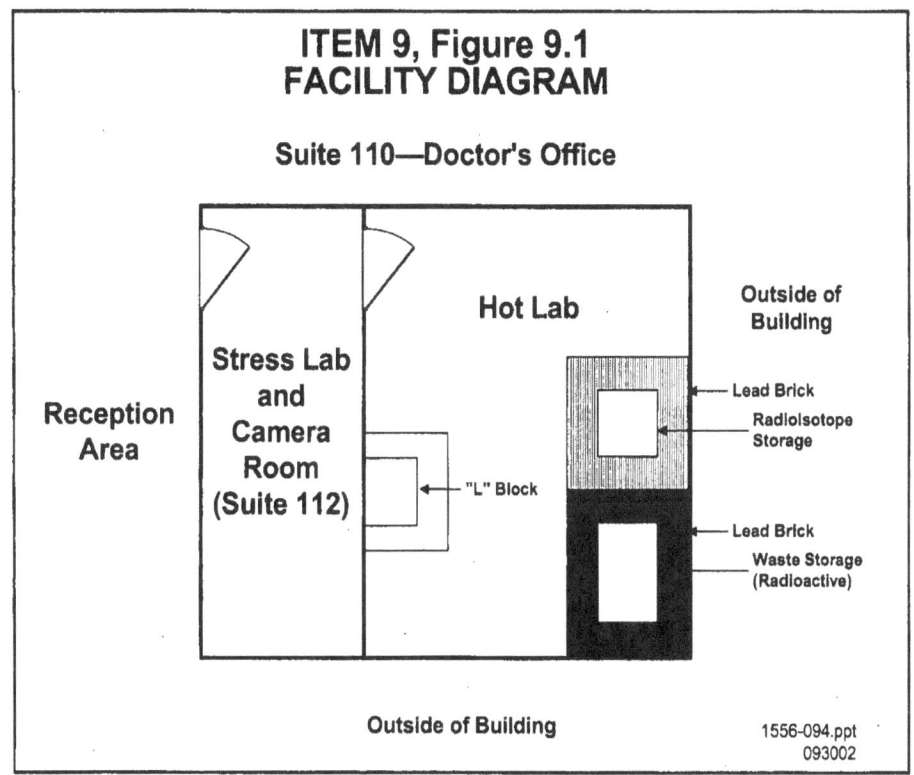

Figure E.1   Sample License Application: Facility Diagram

Notes:

1)   *Radioactive material delivered to hot lab.*

2)   *Counter surfaces are stainless steel and floors are seamless vinyl to facilitate cleanup and minimize permanent contamination.*

3)   *Unoccupied basement located underneath facility and Suite 212 (a doctor's office) located above facility.*

4)   *Description of Instrumentation:*

      Ludlum Model 14C GM Survey meter
      Ludlum Model 3 GM Survey meter
      Capintec Caprac - R600 well/wipe test counter

*For the purposes of this NUREG, the facility diagram is marked appropriately for an application. This particular diagram does not contain real security-related information.

APPENDIX F

Sample Licenses

Sample Licenses

The license conditions listed in the sample licenses come from the standard conditions in NUREG-1556, Volume 20, "Consolidated Guidance About Materials Licenses: Guidance About Administrative Licensing Procedures," with some modifications to reflect provisions of 10 CFR Part 35. The modified conditions are as follows:

- Standard tie-down condition (standard condition 38) modified to reflect 10 CFR 35.26,

- Decay-in-storage condition (standard condition 140) modified to reflect 10 CFR 35.92, and

- Sealed sources leak test condition (standard condition 165) modified to reflect 10 CFR 35.67.

When preparing licenses, refer to the latest revision of NUREG-1556, Volume 20, for the most current versions of the license conditions.

Broad-Scope License

In accordance with 10 CFR 35.12(e), an applicant that satisfies the requirements specified in 10 CFR 33.13 may apply for a Type A specific license of broad scope. Because NRC grants significant decision-making authority to broad-scope licensees through the license, a broad-scope license is not normally issued to a new licensee. An applicant for a broad-scope license typically has several years of experience operating under a limited-scope license and a good regulatory performance history. As opposed to limited-scope licenses, which typically identify specific isotopes that may be possessed, the broad-scope license generally authorizes the possession and use of a wide range of byproduct radioactive materials. Volume 11 of NUREG-1556, "Consolidated Guidance About Materials Licenses: Program-Specific Guidance About Broad-Scope Licenses," provides additional guidance to assist the experienced limited-scope licensees in preparing an application for a broad-scope license.

Sealed Sources and Devices For Broad-Scope Licensees

Under 10 CFR 35.15(g) broad-scope licensees are exempt from the provisions of 10 CFR 35.49(a).

Section 10 CFR 35.49(a) requires that, for medical use, a licensee may only use sealed sources or devices manufactured and distributed in accordance with a license issued under 10 CFR Part 30 and 10 CFR 32.74 or equivalent requirements of an Agreement State. 10 CFR 32.74 requires manufacturers and distributors of sources or devices containing byproduct material for medical use to submit for NRC review, information used for registration of the sealed source or device. This exemption, therefore, grants broad-scope licensees the authority to use sealed sources and/or devices that they have fabricated or obtained from vendors without prior NRC or Agreement State review and registration. However, these licensees have the responsibility for conducting the necessary evaluations and using such devices safely. Pursuant to 10 CFR 33.13(c)(3)(iii), the licensee's Radiation Safety Committee is required to assure that radiation safety evaluations commensurate with the intended use of the sources and/or devices have been performed. If the source and/or device is presently listed in NRC's Registry of Sealed Sources and Devices as approved for the licensee's intended use, no radiation safety evaluation by the licensee is required. If the source and/or device has not been registered, or the source and/or device has not been approved for the licensee's intended use, then the licensee must perform a safety evaluation as required by 10 CFR 33.13(c)(3)(ii).

Sample SR 90 Eye Applicator Materials License*

1.	Norma L. Vision, M.D.	3. License number
2.	Suite 201	4. Expiration date
	1234 Bright Sun Drive	5. Docket No.
	Sun City, Puerto Rico 02210	Reference No.

6. Byproduct, source, and/or special nuclear material	7. Chemical and/or physical form	8. Maximum amount that licensee may possess at any one time under this license
A. Strontium-90 permitted by 10 CFR 35.400	A. Sealed Source (DuPont Merck Pharmaceutical Co. Model NB-1)	A. 120 millicuries

9. Authorized use:

   A. Strontium-90 for ophthalmic radiotherapy permitted by 10 CFR 35.400.

CONDITIONS

10. Licensed material may be used or stored only at the licensee's facilities located at Suite 201, 1234 Bright Sun Drive, Sun City, Puerto Rico.

11. The Radiation Safety Officer for this license is Cecil Source, Ph.D.

12. Licensed material is only authorized for use by, or under the supervision of:

    A. Individuals permitted to work as authorized users and/or authorized medical physicists in accordance with 10 CFR 35.13 and 10 CFR 35.14.

    B. Authorized user and use: Norma L. Vision, M.D. - Strontium-90 for ophthalmic radiotherapy.

    C. Authorized medical physicist: Cecil Source, Ph.D.

13. The licensee is authorized to transport licensed material in accordance with the provisions of 10 CFR Part 71, "Packaging and Transportation of Radioactive Material."

14. Except as specifically provided otherwise in this license, the licensee shall conduct its program in accordance with the statements, representations, and procedures contained in the documents, including any enclosures, listed below. This license condition applies only to those procedures that are required to be submitted in accordance with the regulations. Additionally, this license condition does not limit the licensee's ability to make changes to the Radiation Protection Program as provided for in 10 CFR 35.26. The U.S. Nuclear Regulatory Commission's regulations shall govern unless the statements, representations, and procedures in the licensee's application and correspondence are more restrictive than the regulations.

    A. Application dated March 15, 2005.

    U.S. Nuclear Regulatory Commission

*Note: Certain information about quantities and locations of radioactive materials is no longer released to the public. See Section 5.2.

Sample Medical Institution Limited Materials License*

1. Sample Medical Institution Limited	3. License number	
2. 1234 Main Street Anytown, Missouri 02120	4. Expiration date	
	5. Docket No. Reference No.	

6. Byproduct, source, and/or special nuclear material	7. Chemical and/or physical form	8. Maximum amount that licensee may possess at any one time under this license
A. Any byproduct material permitted by 10 CFR 35.100	A. Any	A. As needed
B. Any byproduct material permitted by 10 CFR 35.200	B. Any	B. As needed
C. Any byproduct material permitted by 10 CFR 35.300	C. Any	C. 900 millicuries
D. Any PET radionuclide	D. Any	D. 20 curies
E. Any byproduct material permitted by 10 CFR 35.400	E. Sealed Sources (US Atomic Models Ir-192L, Cs-137V, and I-125M) Pd 103 PR	E. 2 curies
F. Any byproduct material permitted by 10 CFR 35.500	F. Sealed Sources (US Atomic Model I-125P and GD-153A)	F. 0.3 curie per source and 2 curies total
G. Any byproduct material permitted by 10 CFR 31.11	G. Prepackaged Kits	G. 5 millicuries
H. Strontium-90 permitted by 10 CFR 35.1000	H. Sealed Sources (BEBIG Model Sr0.S03 or AEAT SICW.2 series)	H. 5 millicuries per source and 800 millicuries total
I. Iodine-125 permitted by 10 CFR 35.1000	I. Liquid brachytherapy source Proxima I-125 Iotrex	I. 2 curies
J. Yttrium-90 permitted by 10 CFR 35.1000	J. Sealed sources MDS Nordion Therasphere microspheres	J. 2.5 curies
K. Iridium-192 permitted by 10 CFR 35.600	K. Sealed Sources (US Atomic Model IR-192HDR2)	K. 10 curies per source and 20 curies total
L. Cesium-137	L. Sealed Source (US Atomic Model CS-137C)	L. 200 millicuries
M. Depleted Uranium	M. Metal	M. 999 kilograms

9. Authorized use:

   A. Any uptake, dilution and excretion study permitted by 10 CFR 35.100.

   B. Any imaging and localization study permitted by 10 CFR 35.200.

   C. Any use permitted by 10 CFR 35.300.

*Note: Certain information about quantities and locations of radioactive materials is no longer released to the public. See Section 5.2.

Sample Medical Institution Limited Materials License (Cont.)

D. Production and noncommercial transfer under 10 CFR 30.32(j) of PET radioactive drugs to medical use consortium members and potential contamination on returned "empty" radiation transport shields.

E. Any manual brachytherapy use permitted by 10 CFR 35.400.

F. Diagnostic medical use of sealed sources permitted by 10 CFR 35.500 in compatible devices registered pursuant to 10 CFR 30.32(g).

G. *In vitro* studies.

H. One source assembly for medical use in each Novoste A1000 series model for intravascular brachytherapy permitted by 10 CFR 35.1000.

I. For <u>temporary</u> manual brachytherapy in Proxima Therapeutics Gliasite RTS system permitted by 10 CFR 35.1000.

J. For <u>permanent</u> manual brachytherapy using MDS Nordion Therasphere Y-90 microspheres and delivery system permitted by 10 CFR 35.1000.

K. One source for medical use described in 10 CFR 35.600, in a US Atomic Model IR-192THER remote afterloader unit. The source activity may not exceed 10 curies at the time of medical use. One source in its shipping container as necessary for replacement of the source in the remote afterloader unit.

L. For use in a US Atomic Model CS-137SC for calibrations and checking of licensee's survey instruments.

M. For shielding in a linear accelerator.

CONDITIONS

10. Licensed material may be used or stored only at the licensee's facilities located at 1234 Main Street, Anytown, Missouri.

11. The Radiation Safety Officer for this license is Melba Physicist, M.S.

12. Licensed material is only authorized for use by, or under the supervision of:

    A. Individuals permitted to work as authorized users, authorized nuclear pharmacists, and/or authorized medical physicists in accordance with 10 CFR 35.13 and 35.14.

    B. The following individuals are authorized users for the material and medical uses indicated:

	Material and Use
Jane Jones, M.D.	35.100; 35.200; 35.300; 35.500; *In vitro* studies
Thomas Group, D.O.	35.100; 35.200; 35.300 except iodine-131
Gilbert Lawrence, M.D.	35.100; 35.200; 35.300 sodium iodide I-131 in quantities less than or equal to 33 millicuries only for oral administration for imaging and localization studies[‡]; 35.500

[‡]The example provided in the condition of use for Dr. Lawrence in this sample license illustrates the authorization of a physician who is permitted, under 10 CFR 35.57, to continue use of I-131 for uses for which he was previously authorized but for which he would not now qualify because of new requirements for training and experience (in 10 CFR 35.390) for authorized medical use of byproduct material for which a written directive is now required.

See the discussion in Section 8 of this guide under "8.10 ITEM 7: INDIVIDUAL(S) RESPONSIBLE FOR RADIATION SAFETY PROGRAM AND THEIR TRAINING AND EXPERIENCE," and in "8.12 ITEM 7: AUTHORIZED USERS (AUs)."

Sample Medical Institution Limited Materials License (Cont.)

John Therapy, M.D.   35.400; 35.600 only iridium-192 for use in a High
                              Dose-Rate Remote Afterloader Unit; 35.1000 only
                              for Strontium-90 for intravascular brachytherapy;
                              Depleted Uranium

Mary Innovative, MD   35.1000 only Yttrium-90 microspheres

Newton Technology, MD   35.1000 only Iodine-125 Gliasite RTS system

C.   The following individuals are authorized users for nonmedical uses:

Material and Use

James Pathology   *In vitro* studies

Cecil Source, Ph.D.   Cesium-137 for calibration of instruments

Doug Producer   Production of PET radioactive drugs under
                              10 CFR 30.32(j)

D.   The following individual is an authorized medical physicist:

Material and Use

Melba Physicist, M.S.   Iridium-192 for use in a High Dose-Rate Remote
                              Afterloader Unit

E.   Intravascular brachytherapy procedures shall be conducted under the supervision of the authorized user, who will consult with the interventional cardiologist/physician and authorized medical physicist prior to initiating treatment. The procedures shall be conducted in the physical presence of the authorized user or the authorized medical physicist.

13. In addition to the possession limits in Item 8, the licensee shall further restrict the possession of licensed material to quantities below the minimum limit specified in 10 CFR 30.35(d) for establishing decommissioning financial assurance.

14. The intravascular brachytherapy afterloader device shall be inspected and serviced at intervals recommended by the manufacturer, and maintenance and repair shall be performed only by the manufacturer or persons specifically licensed by NRC or an Agreement State to perform such services.

15. For sealed sources not associated with 10 CFR Part 35 use, the following conditions apply:

A.   Sealed sources shall be tested for leakage and/or contamination at intervals not to exceed the intervals specified in the certificate of registration issued by the U.S. Nuclear Regulatory Commission under 10 CFR 32.210 or under equivalent regulations of an Agreement State.

B.   Notwithstanding Paragraph A of this Condition, sealed sources designed primarily to emit alpha particles shall be tested for leakage and/or contamination at intervals not to exceed 3 months.

C.   In the absence of a certificate from a transferor indicating that a leak test has been made within the intervals specified in the certificate of registration issued by the U.S. Nuclear Regulatory Commission under 10 CFR 32.210 or under equivalent regulations of an Agreement State, prior to the transfer, a sealed source received from another person shall not be put into use until tested and the test results received.

D.   Sealed sources need not be tested if they contain only hydrogen-3, or they contain only a radioactive gas, or the half-life of the isotope is 30 days or less, or they contain not more than 100 microcuries of beta- and/or gamma-emitting material or not more than 10 microcuries of alpha-emitting material.

E.   Sealed sources need not be tested if they are in storage and are not being used; however, when they are removed from storage for use or transferred to another person and have not

Sample Medical Institution Limited Materials License (Cont.)

been tested within the required leak test interval, they shall be tested before use or transfer. No sealed source shall be stored for a period of more than 10 years without being tested for leakage and/or contamination.

F.   The leak test shall be capable of detecting the presence of 0.005 microcurie (185 becquerels) of radioactive material on the test sample. If the test reveals the presence of 0.005 microcurie (185 becquerels) or more of removable contamination, a report shall be filed with the U.S. Nuclear Regulatory Commission in accordance with 10 CFR 30.50(c)(2), and the source shall be removed immediately from service and decontaminated, repaired, or disposed of in accordance with Commission regulations.

G.   Tests for leakage and/or contamination, including leak test sample collection and analysis, shall be performed by the licensee or by other persons specifically licensed by the U.S. Nuclear Regulatory Commission or an Agreement State to perform such services.

H.   Records of leak test results shall be kept in units of microcuries and shall be maintained for 5 years.

16.   The licensee shall conduct a physical inventory every 6 months, or at other intervals approved by the U.S. Nuclear Regulatory Commission, to account for all sources and/or devices received and possessed under the license. Records of inventories shall be maintained for 5 years from the date of each inventory and shall include the radionuclides, quantities, manufacturer's name and model numbers, and the date of the inventory.

17.   Sealed sources or detector cells containing licensed material shall not be opened or sources removed from source holders by the licensee.

18.   The licensee is authorized to transport licensed material in accordance with the provisions of 10 CFR Part 71, "Packaging and Transportation of Radioactive Material."

19.   Except as specifically provided otherwise in this license, the licensee shall conduct its program in accordance with the statements, representations, and procedures contained in the documents, including any enclosures, listed below. This license condition applies only to those procedures that are required to be submitted in accordance with the regulations. Additionally, this license condition does not limit the licensee's ability to make changes to the Radiation Protection Program as provided for in 10 CFR 35.26. The U.S. Nuclear Regulatory Commission's regulations shall govern unless the statements, representations, and procedures in the licensee's application and correspondence are more restrictive than the regulations.

A.   Application dated June 10, 2002.

B.   Letter dated September 30, 2002.

C.   Letter dated February 3, 2008.

U.S. Nuclear Regulatory Commission

Sample I-131 Medical Materials License*

1.	Thomas I. Royed, M.D.	3.	License number
2.	Suite 301	4.	Expiration date
	2 Physician Circle Parkway	5.	Docket No.
	Anytown, West Virginia 02200		Reference No.

6. Byproduct, source, and/or special nuclear material	7. Chemical and/or physical form	8. Maximum amount that licensee may possess at any one time under this license
A. Iodine-131 permitted by 10 CFR 35.300	A. Any	A. 500 millicuries

9. Authorized use:

   A. Any iodine-131 procedure permitted by 10 CFR 35.300 for which the patient can be released under the provisions of 10 CFR 35.75.

CONDITIONS

10. Licensed material may be used or stored only at the licensee's facilities located at Suite 301, 2 Physician Circle Parkway, Anytown, West Virginia.

11. The Radiation Safety Officer for this license is Roger O. Blation, M.D.

12. Licensed material is only authorized for use by, or under the supervision of:

    A. Individuals permitted to work as authorized users in accordance with 10 CFR 35.13 and 35.14.

    B. The following individuals are authorized users for the materials and medical use indicated:

	Material and Use
Roger O. Blation, M.D.	Oral administration of sodium iodide I-131
Thomas I. Royed, M.D.	Oral administration of sodium iodide I-131 in quantities less than or equal to 33 millicuries

13. The licensee is authorized to transport licensed material in accordance with the provisions of 10 CFR Part 71, "Packaging and Transportation of Radioactive Material."

14. Except as specifically provided otherwise in this license, the licensee shall conduct its program in accordance with the statements, representations, and procedures contained in the documents, including any enclosures, listed below. This license condition applies only to those procedures that are required to be submitted in accordance with the regulations. Additionally, this license condition does not limit the licensee's ability to make changes to the Radiation Protection Program as provided for in 10 CFR 35.26. The U.S. Nuclear Regulatory Commission's regulations shall govern unless the statements, representations, and procedures in the licensee's application and correspondence are more restrictive than the regulations.

    A. Application dated October 30, 2002.

    U.S. Nuclear Regulatory Commission

*Note: Certain information about quantities and locations of radioactive materials is no longer released to the public. See Section 5.2.

Sample Manual Brachytherapy Medical Materials License*

1.	Manuel U. Seeds, M.D.	3.	License number
2.	Suite 106	4.	Expiration date
	3 Physician Circle Parkway	5.	Docket No.
	Anytown, Idaho 02200		Reference No.

6.	Byproduct, source, and/or special nuclear material	7.	Chemical and/or physical form	8.	Maximum amount that licensee may possess at any one time under this license
A.	Any byproduct material permitted by 10 CFR 35.400	A.	Sealed Sources (US Atomic Models US-I-125-10L and Pd-103P)	A.	500 millicuries

9. Authorized use:

   A. Any manual brachytherapy use permitted by 10 CFR 35.400 for which the patient can be released under the provisions of 10 CFR 35.75.

CONDITIONS

10. Licensed material may be used or stored only at the licensee's facilities located at Suite 106, 3 Physician Circle Parkway, Anytown, Idaho.

11. The Radiation Safety Officer for this license is Manuel U. Seeds, M.D.

12. Licensed material is only authorized for use by, or under the supervision of:

   A. Individuals permitted to work as authorized users in accordance with 10 CFR 35.13 and 35.14.

   B. The following individual is an authorized user for the material and medical uses indicated:

<u>Material and Use</u>

Manuel U. Seeds, M.D.   35.400

13. The licensee is authorized to transport licensed material in accordance with the provisions of 10 CFR Part 71, "Packaging and Transportation of Radioactive Material."

14. Except as specifically provided otherwise in this license, the licensee shall conduct its program in accordance with the statements, representations, and procedures contained in the documents, including any enclosures, listed below. This license condition applies only to those procedures that are required to be submitted in accordance with the regulations. Additionally, this license condition does not limit the licensee's ability to make changes to the Radiation Protection Program as provided for in 10 CFR 35.26. The U.S. Nuclear Regulatory Commission's regulations shall govern unless the statements, representations, and procedures in the licensee's application and correspondence are more restrictive than the regulations.

   A. Application dated July 20, 2008.

   U.S. Nuclear Regulatory Commission

---

*Note: Certain information about quantities and locations of radioactive materials is no longer released to the public. See Section 5.2.

Sample No Written Directive Medical Materials License*

1.	Noe Directive, M.D.	3.	License number
2.	Suite 112 ·	4.	Expiration date
	2 Physician Circle Parkway	5.	Docket No.
	Anytown, West Virginia 02201		Reference No.

6. Byproduct, source, and/or special nuclear material	7. Chemical and/or physical form	8. Maximum amount that licensee may possess at any one time under this license
A. Any byproduct material permitted by 10 CFR 35.100	A. Any	A. As needed
B. Any byproduct material permitted by 10 CFR 35.200	B. Any, except generators	B. As needed

9. Authorized use:

   A. Any uptake, dilution and excretion study permitted by 10 CFR 35.100.

   B. Any imaging and localization study permitted by 10 CFR 35.200.

CONDITIONS

10. Licensed material may be used or stored only at the licensee's facilities located at Suite 112, 2 Physician Circle Parkway, Anytown, West Virginia.

11. The Radiation Safety Officer for this license is Patrick Physicist, Ph.D.

12. Licensed material is only authorized for use by, or under the supervision of:

    A. Individuals permitted to work as an authorized user in accordance with 10 CFR 35.13 and 35.14.

    B. The following individual is an authorized user for the material and medical uses indicated:

    <u>Material and Use</u>

    Noe Directive, M.D.   35.100; 35.200

13. In addition to the possession limits in Item 8, the licensee shall further restrict the possession of licensed material to quantities below the minimum limit specified in 10 CFR 30.35(d) for establishing decommissioning financial assurance.

14. The licensee is authorized to transport licensed material in accordance with the provisions of 10 CFR Part 71, "Packaging and Transportation of Radioactive Material."

15. Except as specifically provided otherwise in this license, the licensee shall conduct its program in accordance with the statements, representations, and procedures contained in the documents, including any enclosures, listed below. This license condition applies only to those procedures that are required to be submitted in accordance with the regulations. Additionally, this license condition does not limit the licensee's ability to make changes to the Radiation Protection Program as provided for in 10 CFR 35.26. The U.S. Nuclear Regulatory Commission's regulations shall govern unless the statements, representations, and procedures in the licensee's application and correspondence are more restrictive than the regulations.

    A. Application dated April 11, 2007.

    U.S. Nuclear Regulatory Commission

---

*Note: Certain information about quantities and locations of radioactive materials is no longer released to the public. See Section 5.2.

Sample Mobile Medical Materials License*

1.	Sample Mobile Nuclear Medicine	3.	License number
2.	Suite 214	4.	Expiration date
	2 Physician Circle Parkway	5.	Docket No.
	Anytown, Missouri 02220		Reference No.

6.	Byproduct, source, and/or special nuclear material	7.	Chemical and/or physical form	8.	Maximum amount that licensee may possess at any one time under this license
A.	Any byproduct material permitted by 10 CFR 35.100	A.	Any	A.	As needed
B.	Any byproduct material permitted by 10 CFR 35.200	B.	Any, except generators	B.	As needed

9. Authorized use:

   A. Any uptake, dilution, and excretion study permitted by 10 CFR 35.100.

   B. Any imaging and localization study permitted by 10 CFR 35.200.

CONDITIONS

10. Licensed material may be used or stored at the licensee's facilities located at Suite 214, 2 Physician Circle Parkway, Anytown, Missouri, and may be used at temporary job sites of the licensee anywhere in the United States where the U.S. Nuclear Regulatory Commission maintains jurisdiction for regulating the use of licensed material, including areas of exclusive Federal jurisdiction within Agreement States.

    If the jurisdiction status of a Federal facility within an Agreement State is unknown, the licensee should contact the Federal agency controlling the job site in question to determine whether the proposed job site is an area of exclusive Federal jurisdiction. Authorization for use of radioactive materials at job sites in Agreement States not under exclusive Federal jurisdiction shall be obtained from the appropriate State regulatory agency.

11. Licensed material is only authorized for use by, or under the supervision of:

    A. Individuals permitted to work as an authorized user in accordance with 10 CFR 35.13 and 35.14.

    B. The following individual is an authorized user for the material and medical uses indicated:

	Material and Use
Thomas Group, D.O.	35.100; 35.200

12. The Radiation Safety Officer for this license is Thomas Group, D.O.

13. In addition to the possession limits in Item 8, the licensee shall further restrict the possession of licensed material to quantities below the minimum limit specified in 10 CFR 30.35(d) for establishing decommissioning financial assurance.

14. The licensee is authorized to transport licensed material in accordance with the provisions of 10 CFR Part 71, "Packaging and Transportation of Radioactive Material."

15. Except as specifically provided otherwise in this license, the licensee shall conduct its program in accordance with the statements, representations, and procedures contained in the documents, including any enclosures, listed below. This license condition applies only to those procedures that are required to be submitted in accordance with the regulations. Additionally, this license condition does not limit the licensee's ability to make changes to the Radiation Protection Program as provided for in 10 CFR 35.26. The U.S. Nuclear Regulatory Commission's regulations shall govern unless the statements, representations, and procedures in the licensee's application and correspondence are more restrictive than the regulations.

    A. Application dated November 15, 2002.

    U.S. Nuclear Regulatory Commission

*Note: Certain information about quantities and locations of radioactive materials is no longer released to the public. See Section 5.2.

Sample Teletherapy Medical Materials License*

1.	Sample Teletherapy	3.	License number
2.	200 Cobalt Street	4.	Expiration date
	Anytown, Missouri 02300	5.	Docket No.
			Reference No.

6. Byproduct, source, and/or special nuclear material	7. Chemical and/or physical form	8. Maximum amount that licensee may possess at any one time under this license
A. Cobalt-60 permitted by 10 CFR 35.600	A. Sealed Sources (US Atomic Model US-CO-60TELE)	A. 5,500 curies per source and 11,000 curies total
B. Depleted Uranium	B. Metal	B. 999 kilograms

9. Authorized use:

    A.   One source for medical use permitted by 10 CFR 35.600, in a US Atomic Model TELE teletherapy unit. One source in its shipping container as necessary for replacement of the source in the teletherapy unit.

    B.   Shielding in a teletherapy unit.

CONDITIONS

10.   Licensed material may be used or stored only at the licensee's facilities located at 200 Cobalt Street, Anytown, Missouri.

11.   The Radiation Safety Officer for this license is Sarah Smith, M.S.

12.   Licensed material is only authorized for use by, or under the supervision of:

    A.   Individuals permitted to work as authorized users, and/or authorized medical physicists in accordance with 10 CFR 35.13 and 35.14.

    B.   The following individual is an authorized user for the material and medical uses indicated:

	Material and Use
David Jones, M.D.	Cobalt-60 for medical uses in a Teletherapy Unit; Depleted Uranium

    C.   The following individual is an authorized medical physicist:

	Material and Use
Sarah Smith, M.S.	Cobalt-60 in a Teletherapy Unit for calibrations, spot-checks, and training

13.   The licensee is exempt from decommissioning financial assurance requirements for possession of licensed material in sealed sources in quantities greater than the limits in 10 CFR 30.35(d) for the purpose of source changes only. This exemption is granted for no more than 30 days for any one source change.

14.   The licensee is authorized to transport licensed material in accordance with the provisions of 10 CFR Part 71, "Packaging and Transportation of Radioactive Material."

*Note: Certain information about quantities and locations of radioactive materials is no longer released to the public. See Section 5.2.

Sample Teletherapy Medical Materials License (Cont.)

15.   Except as specifically provided otherwise in this license, the licensee shall conduct its program in accordance with the statements, representations, and procedures contained in the documents, including any enclosures, listed below. This license condition applies only to those procedures that are required to be submitted in accordance with the regulations. Additionally, this license condition does not limit the licensee's ability to make changes to the Radiation Protection Program as provided for in 10 CFR 35.26. The U.S. Nuclear Regulatory Commission's regulations shall govern unless the statements, representations, and procedures in the licensee's application and correspondence are more restrictive than the regulations.

    A.   Application dated March 19, 2003.

    U.S. Nuclear Regulatory Commission

Sample Gamma Stereotactic Materials License*

1. Sample Gamma Stereotactic
2. 100 Main Street
   Anytown, Indiana 02310
3. License number
4. Expiration date
5. Docket No.
   Reference No.

6. Byproduct, source, and/or special nuclear material	7. Chemical and/or physical form	8. Maximum amount that licensee may possess at any one time under this license
A. Cobalt-60 permitted by 10 CFR 35.600	A. Sealed Sources (US Atomic Model US-CO-60STER)	A. 33 curies per source and 10,000 curies total

9. Authorized use:

   A. For medical use permitted by 10 CFR 35.600, in a US Atomic Model STEREO gamma stereotactic radiosurgery unit. Sources in the shipping container as necessary for replacement of the sources in the gamma stereotactic radiosurgery unit.

CONDITIONS

10. Licensed material may be used or stored only at the licensee's facilities located at 100 Main Street, Anytown, Indiana.

11. The Radiation Safety Officer for this license is Kimberly Therapy, Ph.D.

12. Licensed material is only authorized for use by, or under the supervision of:

    A. Individuals permitted to work as authorized users, and/or authorized medical physicists in accordance with 10 CFR 35.13 and 35.14.

    B. The following individuals are authorized users for the material and medical uses indicated:

	Material and Use
John Smith, M.D.	35.600 only Cobalt-60 for medical use in a Gamma Stereotactic Radiosurgery Unit
Jessica Water, M.D.	35.600 only Cobalt-60 for medical use in a Gamma Stereotactic Radiosurgery Unit

    C. The following individuals are authorized medical physicists for the material and uses indicated:

	Material and Use
Kimberly Therapy, Ph.D.	Cobalt-60 for use in a Gamma Stereotactic Radiosurgery Unit
Ronald Stereo, M.S.	Cobalt-60 for use in a Gamma Stereotactic Radiosurgery Unit

13. The licensee is authorized to transport licensed material in accordance with the provisions of 10 CFR Part 71, "Packaging and Transportation of Radioactive Material."

*Note: Certain information about quantities and locations of radioactive materials is no longer released to the public. See Section 5.2.

Sample Gamma Stereotactic Materials License (Cont.)

14. Except as specifically provided otherwise in this license, the licensee shall conduct its program in accordance with the statements, representations, and procedures contained in the documents, including any enclosures, listed below. This license condition applies only to those procedures that are required to be submitted in accordance with the regulations. Additionally, this license condition does not limit the licensee's ability to make changes to the Radiation Protection Program as provided for in 10 CFR 35.26. The U.S. Nuclear Regulatory Commission's regulations shall govern unless the statements, representations, and procedures in the licensee's application and correspondence are more restrictive than the regulations.

   A. Application dated December 15, 2002.

   B. Letter dated March 4, 2003.

   U.S. Nuclear Regulatory Commission

Sample Pacemaker Medical Materials License*

1.	Sample Pacemaker License	3.	License number
2.	100 Medical Center Drive	4.	Expiration date
	Anytown, West Virginia 22160	5.	Docket No.
			Reference No.

6.	Byproduct, source, and/or special nuclear material	7.	Chemical and/or physical form	8.	Maximum amount that licensee may possess at any one time under this license
A.	Plutonium (principal radionuclide Pu-238)	A.	Sealed Sources (US Atomic Model US-PU-238)	A.	5 curies per source and 50 curies total

9. Authorized use:

    A.   As a component of US Atomic Model PACE nuclear-powered pacemakers for clinical evaluation in accordance with manufacturer's protocol dated March 25, 1974. This authorization includes: follow-up, explantation, recovery, and disposal, but not implantation.

CONDITIONS

10. Licensed material may be used or stored only at the licensee's facilities located at 100 Medical Center Drive, Anytown, West Virginia.

11. The Radiation Safety Officer for this license is Chief Radiologist, M.D.

12. The physicians responsible for follow-up, explantation, and return of nuclear-powered pacemakers to the manufacturer for proper disposal are Chief Cardiosurgeon, M.D.

13. The specified possession limit for nuclear-powered pacemakers includes all licensed material possessed by the licensee under this license whether in storage, implanted in patients, or otherwise in use.

14. The licensee shall continue patient follow-up and replacement procedures for the nuclear-powered pacemaker during the life of the patient. Procedures for recovery and authorized disposal of the nuclear-powered pacemaker by return to the manufacturer shall be followed upon the death of the patient.

15. The licensee shall report to the U.S. Nuclear Regulatory Commission's Regional Office referenced in Appendix D of 10 CFR Part 20, within 10 days after discovery of loss of contact with a nuclear-powered pacemaker patient.

16. The licensee shall report to the U.S. Nuclear Regulatory Commission's Regional Office referenced in Appendix D of 10 CFR Part 20, within 24 hours of occurrence, the death of any nuclear pacemaker patient, and any adverse reaction and/or malfunction involving a pacemaker system, including the leads. A written report giving details of the adverse reaction and/or malfunction shall be submitted within 30 days.

17. Sealed sources or detector cells containing licensed material shall not be opened or sources removed from source holders by the licensee.

18. The licensee is authorized to transport licensed material in accordance with the provisions of 10 CFR Part 71, "Packaging and Transportation of Radioactive Material."

19. Except as specifically provided otherwise in this license, the licensee shall conduct its program in accordance with the statements, representations, and procedures contained in the documents, including any enclosures, listed below. The U.S. Nuclear Regulatory Commission's regulations shall govern unless the statements, representations, and procedures in the licensee's application and correspondence are more restrictive than the regulations.

    A.   Application dated September 30, 2002.

    B.   Letter dated October 15, 2002.

    U.S. Nuclear Regulatory Commission

*Note: Certain information about quantities and locations of radioactive materials is no longer released to the public. See Section 5.2.

Sample Medical Broad-Scope Materials License*

1. Sample Medical Broad Scope
2. 300 Main Street
   Anytown, Missouri 02110

3. License number
4. Expiration date
5. Docket No.
   Reference No.

6. Byproduct, source, and/or special nuclear material	7. Chemical and/or physical form	8. Maximum amount that licensee may possess at any one time under this license
A. Any byproduct material with atomic numbers 1 through 83	A. Any	A. 200 millicuries per radionuclide and 15 curies total
B. Any byproduct material with atomic numbers 3 through 83	B. Sealed Sources	B. 1.5 curies per radionuclide and 15 curies total
C. Hydrogen-3	C. Any	C. 2 curies
D. Carbon-14	D. Any	D. 1 curie
E. Phosphorus-32	E. Any	E. 2 curies
F. Sulfur-35	F. Any	F. 2 curies
G. Chromium-51	G. Any	G. 500 millicuries
H. Molybdenum-99	H. Any	H. 10 curies
I. Technetium-99m	I. Any	I. 10 curies
J. Any PET radionuclide	J. Any	J. 30 curies
K. Iridium-192	K. Sealed Sources (US Atomic Model IR-192HDR)	K. 12 curies per square and 24 curies total
L. Cobalt-60	L. Sealed Sources (US Atomic Model US CO-60 STER)	L. 33 curies per source and 10,000 curies total

9. Authorized use:

   A. - I. Medical diagnosis, therapy, and research in humans. Research and development as defined in 10 CFR 30.4, including animal studies; instrument calibration; student instruction; and in-vitro studies.

   J. Production and noncommercial transfer under 10 CFR 30.32(j) of PET radioactive drugs to medical use consortium members and potential contamination on returned "empty" radiation transport shields.

   K. One source in a US Atomic Model IR-192THER remote afterloader unit for medical therapy and research in humans. The source activity may not exceed 10 curies at the time of use. One source in its shipping container as necessary for replacement of the source in the remote afterloader unit.

   L. Sources in a US Atomic Model STEREO gamma stereotactic radiosurgery unit for medical therapy and research in humans. Sources in the shipping container as necessary for replacement of the sources in the gamma stereotactic radiosurgery unit.

*Note: Certain information about quantities and locations of radioactive materials is no longer released to the public. See Section 5.2.

Sample Medical Broad-Scope Materials License (Cont.)

CONDITIONS

10. Licensed material may be used or stored only at the licensee's facilities located at 300 Main Street, Anytown, Missouri.

11. A. The Radiation Safety Officer for this license is Patty Melt, Ph.D.

   B. The use of licensed material in or on humans shall be by an authorized user as defined in 10 CFR 35.2.

   C. Individuals designated to work as authorized users, authorized nuclear pharmacists, or authorized medical physicists as defined in 10 CFR 35.2, shall meet the training, experience, and recentness of training criteria established in 10 CFR Part 35, and shall be designated, in writing, by the licensee's Radiation Safety Committee.

   D. Licensed material for other than human use shall be used by, or under the supervision of, individuals designated by the Radiation Safety Committee.

12. The licensee shall not use licensed material in field applications where it is released except as provided otherwise by a specific condition of this license.

13. Experimental animals, or the products from experimental animals, that have been administered licensed materials shall not be used for human consumption.

14. Each sealed source fabricated by the licensee shall be inspected and tested for construction defects, leakage, and contamination prior to any use or transfer as a sealed source.

15. For sealed sources not associated with 10 CFR Part 35 use, the following conditions apply:

   A. Sealed sources shall be tested for leakage and/or contamination at intervals not to exceed the intervals specified in the certificate of registration issued by the U.S. Nuclear Regulatory Commission under 10 CFR 32.210 or under equivalent regulations of an Agreement State.

   B. Notwithstanding Paragraph A of this Condition, sealed sources designed primarily to emit alpha particles shall be tested for leakage and/or contamination at intervals not to exceed 3 months.

   C. In the absence of a certificate from a transferor indicating that a leak test has been made within the intervals specified in the certificate of registration issued by the U.S. Nuclear Regulatory Commission under 10 CFR 32.210 or under equivalent regulations of an Agreement State, prior to the transfer, a sealed source received from another person shall not be put into use until tested and the test results received.

   D. Sealed sources need not be tested if they contain only hydrogen-3, or they contain only a radioactive gas, or the half-life of the isotope is 30 days or less, or they contain not more than 100 microcuries of beta- and/or gamma-emitting material or not more than 10 microcuries of alpha-emitting material.

   E. Sealed sources need not be tested if they are in storage and are not being used; however, when they are removed from storage for use or transferred to another person and have not been tested within the required leak test interval, they shall be tested before use or transfer. No sealed source shall be stored for a period of more than 10 years without being tested for leakage and/or contamination.

   F. The leak test shall be capable of detecting the presence of 0.005 microcuries (185 becquerels) of radioactive material on the test sample. If the test reveals the presence of 0.005 microcuries (185 becquerels) or more of removable contamination, a report shall be filed with the U.S. Nuclear Regulatory Commission in accordance with 10 CFR 30.50(c)(2), and the source shall be removed immediately from service and decontaminated, repaired, or disposed of in accordance with Commission regulations.

   G. Tests for leakage and/or contamination, including leak test sample collection and analysis, shall be performed by the licensee or by other persons specifically licensed by the U.S. Nuclear Regulatory Commission or an Agreement State to perform such services.

Sample Medical Broad-Scope Materials License (Cont.)

16. Sealed sources or detector cells containing licensed material shall not be opened or sources removed from source holders by the licensee.

17. The licensee shall conduct a physical inventory every 6 months, or at other intervals approved by the U.S. Nuclear Regulatory Commission, to account for all sources and/or devices received and possessed under the license.

18. A. Detector cells containing a titanium tritide foil or a scandium tritide foil shall only be used in conjunction with a properly operating temperature control mechanism that prevents the foil temperature from exceeding that specified in the certificate of registration issued by NRC pursuant to 10 CFR 32.210 or the equivalent regulations from an Agreement State.

    B. When in use, detector cells containing a titanium tritide foil or a scandium tritide foil shall be vented to the outside.

19. For radioactive material held for decay-in-storage other than that held in accordance with 10 CFR 35.92, the licensee is authorized to hold radioactive material with a physical half-life of less than or equal to 120 days for decay-in-storage before disposal in ordinary trash, provided the licensee:

    A. Monitors byproduct material at the surface before disposal and determines that its radioactivity cannot be distinguished from the background radiation level with an appropriate radiation detection survey meter set on its most sensitive scale and with no interposed shielding;

    B. Removes or obliterates all radiation labels, except for radiation labels on materials that are within containers and that will be managed as biomedical waste after they have been released from the licensee; and

    C. Maintains records of the disposal of licensed materials for 3 years. The record must include the date of the disposal, the survey instrument used, the background radiation level, the radiation level measured at the surface of each waste container, and the name of the individual who performed the disposal.

20. The licensee is authorized to transport licensed material in accordance with the provisions of 10 CFR Part 71, "Packaging and Transportation of Radioactive Material."

21. Except as specifically provided otherwise in this license, the licensee shall conduct its program in accordance with the statements, representations, and procedures contained in the documents, including any enclosures, listed below. This license condition applies only to those procedures that are required to be submitted in accordance with the regulations. Additionally, this license condition does not limit the licensee's ability to make changes to the Radiation Protection Program as provided for in 10 CFR 35.26. The U.S. Nuclear Regulatory Commission's regulations shall govern unless the statements, representations, and procedures in the licensee's application and correspondence are more restrictive than the regulations.

    A. Application dated December 20, 2002.

    B. Letter dated February 15, 2003.

    U.S. Nuclear Regulatory Commission

APPENDIX G

Information Needed for Transfer of Control

Information Needed for Transfer of Control

The following information is taken from NUREG-1556, Volume 15, "Consolidated Guidance About Materials Licenses: Program-Specific Guidance About Changes of Control and About Bankruptcy Involving Byproduct, Source, or Special Nuclear Material Licenses."

Definitions

Control: Control of a license is in the hands of the person or persons who are empowered to decide when and how that license will be used. That control is to be found in the person or persons who, because of ownership or authority explicitly delegated by the owners, possess the power to determine corporate policy and thus the direction of the activities under the license.

Transferee: A transferee is an entity that proposes to purchase or otherwise gain control of an NRC-licensed operation.

Transferor: A transferor is an NRC licensee selling or otherwise giving up control of a licensed operation.

Licensees must provide full information and obtain NRC's *prior written consent* before transferring control of the license. Provide the following information concerning changes of control by the applicant (transferor and/or transferee, as appropriate). If any items are not applicable, so state.

1. Provide a complete description of the transaction (transfer of stocks or assets, or merger). Indicate whether the name has changed and include the new name. Include the name and telephone number of a licensee contact whom NRC may contact if more information is needed.

2. Describe any changes in personnel or duties that relate to the licensed program. Include training and experience for new personnel.

3. Describe any changes in the organization, location, facilities, equipment, or procedures that relate to the licensed program.

4. Describe the status of the surveillance program (surveys, wipe tests, quality control) at the present time and the expected status at the time that control is to be transferred.

5. Confirm that all records concerning the safe and effective decommissioning of the facility will be transferred to the transferee or to NRC, as appropriate. These records include documentation of surveys of ambient radiation levels and fixed and/or removable contamination, including methods and sensitivity.

6. Confirm that the transferee will abide by all constraints, conditions, requirements, and commitments of the transferor or that the transferee will submit a complete description of the proposed licensed program.

APPENDIX H

NRC Form 314
"Certificate of Disposition of Materials"

NRC FORM 314 (9-2007) 10 CFR 30.36(j)(1); 40.42(j)(1); 70.38(j)(1); and 72.54(k)(5)(1)(1) **CERTIFICATE OF DISPOSITION OF MATERIALS**	U.S. NUCLEAR REGULATORY COMMISSION	APPROVED BY OMB: NO. 3150-0028 EXPIRES: 08/31/2010 Estimated burden per response to comply with this mandatory collection request: 30 minutes. This submittal is used by NRC as part of the basis for its determination that the facility is released for unrestricted use. Send comments regarding burden estimate to the Records and FOIA/Privacy Services Branch (T-5 F52), U.S. Nuclear Regulatory Commission, Washington, DC 20555-0001, or by internet e-mail to infocollects@nrc.gov, and to the Desk Officer, Office of Information and Regulatory Affairs, NEOB-10202, (3150-0028), Office of Management and Budget, Washington, DC 20503. If a means used to impose an information collection does not display a currently valid OMB control number, the NRC may not conduct or sponsor, and a person is not required to respond to, the information collection.

LICENSEE NAME AND ADDRESS	LICENSE NUMBER	DOCKET NUMBER
	LICENSE EXPIRATION DATE	

A. LICENSE STATUS *(Check the appropriate box)*

☐ This license has expired.   ☐ This license has not yet expired; please terminate it.

B. DISPOSAL OF RADIOACTIVE MATERIAL
(Check the appropriate boxes and complete as necessary. If additional space is needed, provide attachments)

The licensee, or any individual executing this certificate on behalf of the licensee, certifies that:

☐ 1. No radioactive materials have ever been procured or possessed by the licensee under this license.

☐ 2. All activities authorized by this license have ceased, and all radioactive materials procured and/or possessed by the licensee under this license number cited above have been disposed of in the following manner.

    ☐ a. Transfer of radioactive materials to the licensee listed below:

    ☐ b. Disposal of radioactive materials:

       ☐ 1. Directly by the licensee:

       ☐ 2. By licensed disposal site:

       ☐ 3. By waste contractor:

    ☐ c. All radioactive materials have been removed such that any remaining residual radioactivity is within the limits of 10 CFR Part 20, Subpart E, and is ALARA.

C. SURVEYS PERFORMED AND REPORTED

☐ 1. A radiation survey was conducted by the licensee. The survey confirms:

    ☐ a. the absence of licensed radioactive materials

    ☐ b. that any remaining residual radioactivity is within the limits of 10 CFR 20, Subpart E, and is ALARA.

☐ 2. A copy of the radiation survey results:

    ☐ a. is attached; or ☐ b. is not attached (Provide explanation); or ☐ c. was forwarded to NRC on: _____
                                                                     Date

☐ 3. A radiation survey is not required as only sealed sources were ever possessed under this license, and

    ☐ a. The results of the latest leak test are attached; and/or   ☐ b. No leaking sources have ever been identified.

The person to be contacted regarding the information provided on this form:

NAME	TITLE	TELEPHONE *(Include Area Code)*	E-MAIL ADDRESS

Mail all future correspondence regarding this license to:

C. CERTIFYING OFFICIAL
I CERTIFY UNDER PENALTY OF PERJURY THAT THE FOREGOING IS TRUE AND CORRECT

PRINTED NAME AND TITLE	SIGNATURE	DATE

WARNING: FALSE STATEMENTS IN THIS CERTIFICATE MAY BE SUBJECT TO CIVIL AND/OR CRIMINAL PENALTIES. NRC REGULATIONS REQUIRE THAT SUBMISSIONS TO THE NRC BE COMPLETE AND ACCURATE IN ALL MATERIAL RESPECT. 18 U.S.C. SECTION 1001 MAKES IT A CRIMINAL OFFENSE TO MAKE A WILLFULLY FALSE STATEMENT OR REPRESENTATION TO ANY DEPARTMENT OR AGENCY OF THE UNITED STATES AS TO ANY MATTER WITHIN ITS JURISDICTION.

APPENDIX H

CERTIFICATE OF DISPOSITION OF MATERIALS

PLEASE READ THESE INSTRUCTIONS BEFORE COMPLETING NRC FORM 314.

Subpart E of 10 CFR Part 20 establishes the radiological criteria for license terminations/decommissioning of facilities licensed under 10 CFR Parts 30, 40, 50, 60, 61, 70, and 72, as well as other facilities subject to the Commission's jurisdiction under the Atomic Energy Act of 1954, as amended, and the Energy Reorganization Act of 1974, as amended.

INSTRUCTIONS

Section B, Item 2.
Licensees should describe the specific radioactive material transfer actions. If radioactive wastes were generated in terminating this license, the licensee should describe the disposal actions taken, including the disposition of low-level radioactive waste, mixed waste, greater-than-Class-C waste, and sealed sources.

Section B, Item 2.a.
The information provided concerning the transfer of radioactive material to another licensee should specify the date of the transfer, the name of the licensee recipient, an individual contact name and telephone number for the licensee recipient, and the recipient's NRC or Agreement State license number.

Section B, Item 2.b.
For disposal of radioactive materials, licensees should describe the specific disposal method or procedure (e.g., decay-in-storage). For those cases when radioactive materials are disposed of by a licensed disposal site or by a waste contractor, the licensee should specify the name, address, and telephone number of the licensed disposal site operator or waste contractor.

Section B, Item 2.c.
"Residual radioactivity," as defined in 10 CFR 20.1003, means radioactivity in 'areas' (structures, materials, soils, etc.) remaining as a result of activities (licensed and unlicensed) under the licensee's control from sources used by the licensee, excluding background radiation. ALARA is defined in 10 CFR 20.1003.

FILE CERTIFICATES AS FOLLOWS:

IF YOU ARE LOCATED IN:

ALABAMA, CONNECTICUT, DELAWARE, DISTRICT OF COLUMBIA, FLORIDA, GEORGIA, KENTUCKY, MAINE, MARYLAND, MASSACHUSETTS, NEW HAMPSHIRE, NEW JERSEY, NEW YORK, NORTH CAROLINA, PENNSYLVANIA, PUERTO RICO, RHODE ISLAND, SOUTH CAROLINA, TENNESSEE, VERMONT, VIRGINIA, VIRGIN ISLANDS, OR WEST VIRGINIA, SEND CERTIFICATES TO:

LICENSING ASSISTANT SECTION
NUCLEAR MATERIALS SAFETY BRANCH
U.S. NUCLEAR REGULATORY COMMISSION, REGION I
475 ALLENDALE ROAD
KING OF PRUSSIA, PA 19406-1415

ILLINOIS, INDIANA, IOWA, MICHIGAN, MINNESOTA, MISSOURI, OHIO, OR WISCONSIN, SEND CERTIFICATES TO:

MATERIALS LICENSING SECTION
U.S. NUCLEAR REGULATORY COMMISSION, REGION III
2443 WARRENVILLE ROAD, SUITE 210
LISLE, IL 60532-4352

IF YOU ARE LOCATED IN:

ALASKA, ARIZONA, ARKANSAS, CALIFORNIA, COLORADO, HAWAII, IDAHO, KANSAS, LOUISIANA, MISSISSIPPI, MONTANA, NEBRASKA, NEVADA, NEW MEXICO, NORTH DAKOTA, OKLAHOMA, OREGON, PACIFIC TRUST TERRITORIES, SOUTH DAKOTA, TEXAS, UTAH, WASHINGTON, OR WYOMING, SEND CERTIFICATES TO:

MATERIAL RADIATION PROTECTION SECTION
U. S. NUCLEAR REGULATORY COMMISSION, REGION IV
611 RYAN PLAZA DRIVE, SUITE 400
ARLINGTON, TX 76011-8064

NUREG - 1556, Vol. 9, Rev. 2

H-2

APPENDICES I-W

MODEL PROCEDURES FOR
INFORMATION PURPOSES ONLY

The following model procedures provide one method of complying with the regulations and are not intended to be the only means for satisfying the requirements for licensees.

These model procedures were originally developed for medical uses only. With implementation of the EPAct and addition of NARM materials and nonmedical uses, such as a 10 CFR 30.32(j) authorization, to medical use licenses, the model procedures may have to be supplemented to address the new materials and nonmedical uses. When adopting one of the model procedures, the applicant needs to ensure that all appropriate aspects of its licensed program are addressed.

APPENDIX I

Typical Duties and Responsibilities of the Radiation Safety Officer and Sample Delegation of Authority

Typical Duties and Responsibilities of the Radiation Safety Officer and Sample Delegation of Authority

Model Radiation Safety Officer Duties and Responsibilities

The duties and responsibilities of the Radiation Safety Officer (RSO) include ensuring radiological safety and compliance with NRC and DOT regulations and the conditions of the license. Model procedures for describing the RSO's duties and responsibilities appear below. Applicants may either adopt these model procedures or develop alternative procedures to meet the requirements of 10 CFR 35.24. As a result of implementation of the EPAct, licensed material now includes accelerator-produced radioactive materials and discrete sources of Ra-226. Licensees authorized under 10 CFR 30.32(j) to produce and noncommercially transfer PET radioactive drugs to consortium members should review the model duties and responsibilities below, expanding on them as necessary to ensure radiation safety oversight of the production and transfer only to medical use consortium members.

Typically, these duties and responsibilities include ensuring the following:

- Unsafe activities involving licensed material are stopped;

- Radiation exposures are ALARA;

- Up-to-date radiation protection procedures in the daily operation of the licensee's byproduct material program are developed, distributed, and implemented;

- Possession, use, and storage of licensed material are consistent with the limitations in the license, the regulations, the SSDR certificate(s), and the manufacturer's recommendations and instructions;

- Individuals installing, relocating, maintaining, adjusting, or repairing devices containing sealed sources are trained and authorized by an NRC or Agreement State license;

- Personnel training is conducted and is commensurate with the individual's duties regarding licensed material;

- Documentation is maintained to demonstrate that individuals are not likely to receive, in 1 year, a radiation dose in excess of 10% of the allowable limits or that personnel monitoring devices are provided;

- When necessary, personnel monitoring devices are used and exchanged at the proper intervals, and records of the results of such monitoring are maintained;

- Licensed material is properly secured;

- Documentation is maintained to demonstrate, by measurement or calculation, that the total effective dose equivalent to the individual likely to receive the highest dose from the licensed operation does not exceed the annual limit for members of the public;

- Proper authorities are notified of incidents such as loss or theft of licensed material, damage to or malfunction of sealed sources, and fire;

- Medical events and precursor events are investigated and reported to NRC, cause(s) and appropriate corrective action(s) are identified, and timely corrective action(s) are taken;

- Audits of the Radiation Protection Program are performed at least annually and documented;

- If violations of regulations, license conditions, or program weaknesses are identified, effective corrective actions are developed, implemented, and documented;

- Licensed material is transported, or offered for transport, in accordance with all applicable DOT requirements;

- Licensed material is disposed of properly;

- Appropriate records are maintained; and

- An up-to-date license is maintained, and amendment and renewal requests are submitted in a timely manner.

Model Delegation of Authority

Memo To: Radiation Safety Officer

From: Chief Executive Officer

Subject: Delegation of Authority

You, _____, have been appointed Radiation Safety Officer and are responsible for ensuring the safe use of radiation. You are responsible for managing the Radiation Protection Program; identifying radiation protection problems; initiating, recommending, or providing corrective actions; verifying implementation of corrective actions; stopping unsafe activities; and ensuring compliance with regulations. You are hereby delegated the authority necessary to meet those responsibilities, including prohibiting the use of byproduct material by employees who do not meet the necessary requirements and shutting down operations where justified to maintain radiation safety. You are required to notify management if staff does not cooperate and does not address radiation safety issues. In addition, you are free to raise issues with the Nuclear Regulatory Commission at any time. It is estimated that you will spend _____ hours per week conducting radiation protection activities.

_____   _____
Signature of Management Representative   Date

I accept the above responsibilities,

_____   _____
Signature of Radiation Safety Officer   Date

cc: Affected department heads

APPENDIX J

Model Training Program

This Appendix was originally developed for medical uses only. With implementation of the EPAct and addition of NARM materials and nonmedical uses, such as authorizations under 10 CFR 30.32(j) to medical use licenses, the procedures may have to be supplemented in this Appendix to address the new materials and nonmedical uses.

Model Training Program

Model procedures for describing training programs appear below. These models provide examples of topics to be chosen for training, based on the experience, duties, and previous training of trainees. The topics chosen will depend on the purpose of the training, the audience, and the state of learning (background knowledge) of the audience. These models also may be useful to identify topics for annual refresher training. Refresher training should include topics with which the individual is not involved frequently and topics that require reaffirmation. Topics for refresher training need not include review of procedures or basic knowledge that the trainee routinely uses. Applicants may either adopt these model procedures or develop an alternative program to meet NRC requirements. Guidance on requirements for training and experience for AMPs and AUs for medical use who engage in certain specialized practices is also included.

Note: With the implementation of the EPAct, the NRC now has regulatory authority for accelerator-produced radioactive material and discrete sources of Ra-226. Personnel should be provided new training on the application of the NRC requirements and license conditions to these materials when NRC's waiver of August 31, 2005, is terminated for the medical use facility. The waiver was terminated on November 30, 2007, for Government agencies, Federally recognized Indian tribes, Delaware, the District of Columbia, Puerto Rico, the U.S. Virgin Islands, Indiana, Wyoming, and Montana. The appropriate NRC Regional Office should be contacted to confirm the waiver termination date for other medical use facilities.

Model Training Program for Medical and Nonmedical Uses of Radionuclides, Sealed Sources, and Medical Devices Containing Sealed Sources

Personnel will receive instruction before assuming duties with, or in the vicinity of, radioactive materials during annual refresher training, and whenever there is a significant change in duties, regulations, terms of the license, or type of radioactive material or therapy device used. Records of worker training will be maintained for at least 3 years. The training records will include the date of the instruction or training and the name(s) of the attendee(s) and instructor(s).

Training for Individuals Involved in the Medical Usage of Byproduct Material

Training for professional staff (e.g., AU, AMP, ANP, RSO, nurse, dosimetrist, technologist, therapist) may contain the following elements for those who provide or are involved in the care of patients during diagnostic or therapeutic procedures, *commensurate with their duties*:

- Basic radiation biology (e.g., interaction of ionizing radiation with cells and tissues);

- Basic radiation protection to include concepts of time, distance, and shielding;

- Concept of maintaining exposure ALARA (10 CFR 20.1101);

- Risk estimates, including comparison with other health risks;

- Posting requirements (10 CFR 20.1902);

- Proper use of personnel dosimetry (when applicable);

- Access control procedures (10 CFR 20.1601, 10 CFR 20.1802);

- Proper use of radiation shielding, if used;

- Patient release procedures (10 CFR 35.75);

- Instruction in procedures for notification of the RSO and AU, when responding to patient emergencies or death, to ensure that radiation protection issues are identified and addressed in a timely manner. The intent of these procedures should in no way interfere with or be in lieu of appropriate patient care (10 CFR 19.12, 10 CFR 35.310, 10 CFR 35.410, 10 CFR 35.610);

- Occupational dose limits and their significance (10 CFR 20.1201);

- Dose limits to the embryo/fetus, including instruction on declaration of pregnancy (10 CFR 20.1208);

- Worker's right to be informed of occupational radiation exposure (10 CFR 19.13);

- Each individual's obligation to report unsafe conditions to the RSO (10 CFR 19.12);

- Applicable regulations, license conditions, information notices, bulletins, etc. (10 CFR 19.12);

- Where copies of the applicable regulations, the NRC license, and its application are posted or made available for examination (10 CFR 19.11);

- Proper recordkeeping required by NRC regulations (10 CFR 19.12);

- Appropriate surveys to be conducted (10 CFR 20.1501);

- Proper calibration of required survey instruments (10 CFR 20.1501);

- Emergency procedures;

- Decontamination and release of facilities and equipment (10 CFR 20.1406, 10 CFR 30.36);

- Dose to individual members of the public (10 CFR 20.1301); and

- Licensee's operating procedures (e.g., survey requirements, instrument calibration, waste management, sealed-source leak testing) (10 CFR 35.27, 10 CFR 30.32(a)(3)).

Training for Individuals Involved in Nonmedical Usage of Byproduct Material

Training for staff working with byproduct material for nonmedical uses or animals containing byproduct material may include, as appropriate, the elements that are listed above for medical uses. Licensees authorized under 10 CFR 30.32(j) to produce PET radioactive drugs for noncommercial transfer to other medical use licensees in the consortium should also provide training on the production of PET radioactive drugs and special requirements in 10 CFR 30.32(j) and 10 CFR 30.34(j) for this activity. All training should be commensurate with the individual's duties.

Training for the Staff Directly Involved in Administration to or Care of Patients Administered Byproduct Material for which a Written Directive Is Required (Including Greater-than-30 microcuries of I-131), or Therapeutic Treatment Planning

Note: Byproduct material now includes accelerator-produced radionuclides and discrete sources of Ra-226.

In addition to the topics identified above, the following topics may be included in instruction for staff involved in the therapy treatment of patients (e.g., nursing, RSO, AMP, AU, and dosimetrist), *commensurate with their duties*:

- Leak testing of sealed sources (10 CFR 35.67);

- Emergency procedures (including emergency response drills) (10 CFR 35.310, 10 CFR 35.410, 10 CFR 35.610);

- Operating instructions (10 CFR 35.27, 10 CFR 35.610);

- Computerized treatment planning system (10 CFR 35.657);

- Dosimetry protocol (10 CFR 35.630);

- Detailed pretreatment quality assurance checks (10 CFR 35.27, 10 CFR 35.610);

- Safe handling (when applicable) of the patient's dishes, linens, excretions (saliva, urine, feces), and surgical dressings that are potentially contaminated or that may contain radioactive sources (10 CFR 35.310, 10 CFR 35.410);

- Patient control procedures (10 CFR 35.310, 10 CFR 35.410, 10 CFR 35.610);

- Visitor control procedures, such as visitors' stay times and safe lines in radiation control areas (patient's room) (10 CFR 35.310, 10 CFR 35.410, 10 CFR 35.610);

- Licensee's WD Procedures, to ensure that each administration is in accordance with the WD, patient identity is verified, and where applicable, attention is paid to correct positioning of sources and applicators to ensure that treatment is to the correct site (or, for GSR, correct positioning of the helmet) (10 CFR 35.41);

- Proper use of safety devices and shielding to include safe handling and shielding of dislodged sources (or, in the case of remote afterloaders, disconnected sources) (10 CFR 35.410, 10 CFR 35.610);

- Size and appearance of different types of sources and applicators (10 CFR 35.410, 10 CFR 35.610);

- Previous incidents, events, and/or accidents; and

- For remote afterloaders, teletherapy units, and GSR units, initial training provided by the device manufacturer or by individuals certified by the device manufacturer that is device model-specific and includes:

  — Design, use, and function of the device, including safety systems and interpretation of various error codes and conditions, displays, indicators, and alarms;

  — Hands-on training in actual operation of the device under the direct supervision of an experienced user, including "dry runs" (using dummy sources) of routine patient set-up and treatment and implementation of the licensee's emergency procedures;

  — A method, such as practical examinations, to determine each trainee's competency to use the device for each type of proposed use.

Additional Training for Authorized Medical Physicists

Applicants for licenses to include AMPs who plan to engage in certain tasks requiring special training should ensure that the AMP is trained in the activities specific to the different types of uses listed in 10 CFR 35.51(b)(1). Note, for example, that additional training is necessary for AMP planning tasks such as remote afterloader therapy, teletherapy, GSR therapy, the use of the treatment planning system that applicants contemplate using, as well as the calculation of activity of Sr-90 sources used for opthalmic treatments (10 CFR 35.433). Medical physicists must also have training for the type(s) of use for which authorization is sought that includes hands-on device operation, safety procedures, clinical use, and the operation of a treatment planning system, as required in 10 CFR 35.51(c).

Additional Training for Authorized Users for Medical Uses of Byproduct Materials for Which a Written Directive Is Required

Applicants for licenses should carefully consider the type of radiation therapy that is contemplated. In addition to the training and experience requirements of 10 CFR 35.390, 10 CFR 35.394, 10 CFR 35.396, 10 CFR 35.490, 10 CFR 35.491, and 10 CFR 35.690, attention should be focused on the additional training and experience necessary for treatment planning and quality control systems, and clinical procedures. Refer to the training and experience requirements associated with specialized uses discussed in Sections 35.390, 35.490, 35.491, and 35.690 of 10 CFR Part 35.

Training for Ancillary Staff

For the purposes of this section, ancillary staff includes personnel engaged in janitorial and/housekeeping duties, dietary, laboratory, security, and life-safety services. The training program for ancillary staff performing duties that are likely to result in a dose in excess of 1 mSv (100 mrem) will include instruction commensurate with potential radiological health protection problems present in the work place. Alternatively, prohibitions on entry into controlled or restricted areas may be applied to ancillary personnel unless escorted by trained personnel. Topics of instruction may include the following:

- Storage, transfer, or use of radiation and/or radioactive material (10 CFR 19.12);

- Potential biological effects associated with exposure to radiation and/or radioactive material, precautions or procedures to minimize exposure, and the purposes and functions of protective devices (e.g., basic radiation protection concepts of time, distance, and shielding) (10 CFR 19.12);

- The applicable provisions of NRC regulations and licenses for the protection of personnel from exposure to radiation and/or radioactive material (e.g., posting and labeling of radioactive material) (10 CFR 19.12);

- Responsibility to report promptly to the licensee any condition that may lead to or cause a violation of NRC regulations and licenses or unnecessary exposure to radiation and/or radioactive material (e.g., notification of the RSO regarding radiation protection issues) (10 CFR 19.12);

- Appropriate response to warnings made in the event of any unusual occurrence or malfunction that may involve exposure to radiation and/or radioactive material (10 CFR 19.12);

- Radiation exposure reports that workers may request, as per 10 CFR 19.13 (10 CFR 19.12).

APPENDIX K

General Radiation Monitoring Instrument Specifications and Model Survey Instrument Calibration Program

General Radiation Monitoring Instrument Specifications and Model Survey Instrument Calibration Program

Model procedures for describing the specifications for monitoring instruments and a program for calibration of survey instruments appear below. Applicants may either adopt these model procedures or adopt alternative procedures.

Facilities and Equipment

- To reduce doses received by individuals not calibrating instruments, calibrations should be conducted in an isolated area of the facility or at times when no one else is present.

- Individuals conducting calibrations will wear assigned dosimetry, if required.

Equipment Selection

- Low-energy beta emitters, such as carbon-14 and sulfur-35, are difficult to detect with Geiger-Mueller (GM) probes. The detection efficiency generally is about 2% for low-energy beta emitters. The proper surveying method (e.g., speed and height above surface) is important to perform adequate surveys. Additionally, wipes should be taken and counted on a liquid scintillation counter to verify potential contamination.

- Medium- to high-energy beta emitters, such as P-32 and Ca-45, can be detected with a pancake GM. The efficiency ranges from 15% to 40%, depending on the beta energy.

- Low-energy gamma emitters, such as I-125, can be detected with a sodium iodide (NaI) probe or a thin window GM probe (pancake or thin end-window). If the sodium iodide probe possesses a thin window and thin crystal, the detection efficiency is approximately 20%. If a pancake or thin end-window GM probe is used, the detection efficiency is significantly lower and care should be taken to ensure that the GM probe is capable of detecting the trigger levels.

- Medium- to high-energy gamma emitters, such as I-131, can be detected with either GM or sodium iodide probes, depending on the required sensitivity. In general, the sensitivity of GM probes is much lower than for sodium iodide probes.

- The following table (Table K.1) (except for items marked with an asterisk (*)), extracted from "The Health Physics & Radiological Health Handbook," Revised Edition, 1992, may be helpful in selecting instruments:

Table K.1 Typical Survey Instruments			
Portable Instruments Used for Contamination and Ambient Radiation Surveys			
Detectors	**Radiation**	**Energy Range**	**Efficiency**
Exposure Rate Meters	Gamma, X-ray	mR-R	N/A
Count Rate Meters			
GM	Alpha	All energies (dependent on window thickness)	Moderate
	Beta	All energies (dependent on window thickness)	Moderate
	Gamma	All energies	< 1%
NaI Scintillator	Gamma	All energies (dependent on crystal thickness)	Moderate
Plastic Scintillator	Beta	C-14 or higher (dependent on window thickness)	Moderate
Stationary Instruments Used to Measure Wipe, Bioassay, and Effluent Samples			
Detectors	**Radiation**	**Energy Range**	**Efficiency**
Liquid Scintillation Counter*	Alpha	All energies	High
	Beta	All energies	High
	Gamma		Moderate
Gamma Counter (NaI)*	Gamma	All energies	High
Gas Proportional	Alpha	All energies	High
	Beta	All energies	Moderate
	Gamma	All energies	< 1%

*Not extracted from source handbook: "The Health Physics and Radiological Handbook," Revised Edition, 1992.

Model Procedure for Calibrating Survey Instruments

This model provides acceptable procedures for survey instrument calibrations. Licensees may either adopt these model procedures or develop their own procedures to meet the requirements of 10 CFR Part 20 and 10 CFR 35.61. (Detailed information about survey instrument calibration may be obtained by referring to ANSI N323A-1997, "Radiation Protection Instrumentation Test and Calibration, Portable Survey Instruments." Copies may be obtained from the American National Standards Institute at 25 West 43rd Street, 4th Floor, New York, NY 10036, or by ordering electronically from http://www.ansi.org.)

Procedures for calibration of survey instruments:

- Radiation survey instruments will be calibrated with a radioactive source in accordance with 10 CFR 35.61. Electronic calibrations alone are not acceptable. Survey meters must be calibrated at least annually, before first use, and after servicing or repairs that affect calibration. (Battery changes are not considered "servicing.") Instruments used to monitor higher energies are most easily calibrated in known radiation fields produced by sources of gamma rays of approximately the same energies as those to be measured. An ideal calibration source would emit the applicable radiation (e.g., alpha, beta, or gamma) with an energy spectrum similar to that to be measured and have a suitably long half-life.

- Use a radioactive sealed source(s) that:

  — Approximates a point source;

  — Is a certified, NIST-traceable, standard source that has an activity or exposure rate accurate to within 5%; if the activity or exposure rate is determined by measurement, document the method used to make the determination and traceability to NIST;

  — Emits the type of radiation measured;

  — Approximates the same energy (e.g., Cs-137, Co-60) as the environment in which the calibrated device will be employed; and

  — Provides a radiation dose rate sufficient to reach the full scale (<1000 mR/hr) of the instrument calibrated.

- Use the inverse square and radioactive decay laws, as appropriate, to correct for changes in exposure rate due to changes in distance or source decay.

- A record must be made of each survey meter calibration and retained for 3 years after each record is made (10 CFR 20.2103(a) and 10 CFR 35.2061).

- Before use, perform a daily check (with a dedicated check source) and battery checks.

- Instrument readings should be within ± 10% of known radiation values at calibration points; however, readings within ± 20% are acceptable if a calibration chart or graph is prepared and made available with the instrument.

- The kinds of scales frequently used on radiation survey meters should be calibrated as follows:

  — Calibrate Linear-Readout Instruments at no fewer than two points on each scale. Calibration will be checked near the ends of each scale (at approximately 20% and 80%).

  — Calibrate Logarithmic-Readout Instruments at two points on each decade.

  — Calibrate Digital-Readout Instruments with either manual or automatic scale switching for indicating exposure rates at no fewer than two points on each scale. Check calibrations near the ends of each scale (at approximately 20% and 80% of each scale).

  — Calibrate Digital-Readout Instruments without scale switching for indicating exposure rates at two points on each decade.

—　Calibrate Integrating Instruments at two dose rates (at approximately 20% and 80% of the dose rate range).

- Readings above 1000 mR/hr (250 microcoulomb/kilogram of air per hour) need not be calibrated; however, such scales may be checked for operation and approximately correct response.

- Include in survey meter calibration records the procedure used and the data obtained. Record the following:

—　A description of the instrument, including the manufacturer's name, model number, serial number, and type of detector;

—　A description of the NIST-traceable calibration source, including the calibration procedure, exposure rate, distance at which it was measured, and date of measurement;

—　For each calibration point, the calculated exposure rate, the indicated exposure rate, the calculated correction factor (the calculated exposure rate divided by the indicated exposure rate), and the scale selected on the instrument;

—　The exposure reading indicated with the instrument in the "battery check" mode (if available on the instrument);

—　For instruments with external detectors, the angle between the radiation flux field and the detector (i.e., parallel or perpendicular);

—　For instruments with internal detectors, the angle between the radiation flux field and a specified surface of the instrument;

—　For detectors with removable shielding, an indication of whether the shielding was in place or removed during the calibration procedure;

—　The exposure rate from a check source, if used;

—　The name of the person who performed the calibration and the date it was performed.

- The following information should be attached to the instrument as a calibration sticker or tag:

—　The source that was used to calibrate the instrument;

—　The proper deflection in the battery check mode (unless this is clearly indicated on the instrument);

—　Special use conditions (e.g., an indication that a scale or decade was checked only for function but not calibrated);

—　The date of calibration and the next calibration due date;

—　The apparent exposure rate from the check source, if used.

Sensitivity of Counting System

Follow the procedures in Appendix Q to determine minimum detectable activity (MDA) if there is a question concerning the ability to measure small quantities of radioactivity.

Determining the Efficiency of NaI(Tl) Uptake Probes

Sodium iodide (thallium doped) [NaI(Tl)] uptake probes are commonly used for bioassays of personnel administering I-131 radionuclides in the form of sodium iodide. Refer to 10 CFR Part 20, Appendix B, for the Annual Limits on Intake (ALIs) and Derived Air Concentrations (DACs) for occupational exposure to radionuclides. Convert count rates (e.g., in cpm) to units of activity (dpm, μCi) when performing bioassays to determine thyroid burdens of radioiodines. Use the following procedure to calibrate the probe for uptake measurements:

• Frequency: perform calibrations annually, before first use, and after repairs that affect calibrations;

• Check the instrument's counting efficiency using either a standard source of the same radionuclide as the source being tested or one with similar energy characteristics. Accuracy of standards will be within ± 5% of the stated value and traceable to a primary radiation standard such as those maintained by NIST.

• Calculate the efficiency of the instrument.

For example:

$$Eff_a = \frac{[(cpm\ from\ std) - (cpm\ from\ bkg)]}{(activity\ of\ std\ in\ microcuries)}$$

where:

Eff_a = efficiency[1],
cpm = counts per minute,
std = standard, and
bkg = background.

Operational and calibration checks, using a dedicated check source, should be conducted on each day the instrument is used.

The date of the efficiency test should be attached to the instrument as a calibration sticker or tag and the following information should be included:

• The due date of the next efficiency test, and

• Results of efficiency calculation(s).

---

[1] The absolute efficiency is dependent on the counting geometry. Applicants may elect to use the intrinsic efficiency, which no longer includes the solid angle subtended by the detector and is much less dependent on the counting geometry.

Calculating the Gamma Well Efficiency of Counting Equipment

Gamma well counting equipment is often used for assaying the wipe testing of packages, sealed sources, and areas where unsealed byproduct material is prepared, administered, or stored. Converting cpm to dpm using smear wipes is required when dealing with radiation surveys of sealed and unsealed radioactive materials. Calculate the efficiency of all instruments used for assaying wipe tests on an annual basis, before first use, and/or after repair, using the following procedure:

- Check the instrument's counting efficiency, using either a standard source of the same radionuclide as the source being tested or one with similar energy characteristics. Accuracy of standards will be within ± 5% of the stated value and traceable to a primary radiation standard such as those maintained by NIST.

- Calculate the efficiency of the instrument.

    For example,
    $$Eff = \frac{[(cpm \ from \ std) - (cpm \ from \ bkg)]}{(activity \ of \ std \ in \ microcuries)}$$

    where:

    Eff = efficiency, in cpm/microcurie,
    cpm = counts per minute,
    std = standard, and
    bkg = background.

Operational and calibration checks, using a dedicated check source, should be conducted on each day the instrument is used.

The date of the efficiency test should be attached to the instrument as a calibration sticker or tag and the following information should be included:

- The due date of the next efficiency test, and;

- Results of efficiency calculation(s).

Reference: Draft RG FC 413-4, "Guide for the Preparation of Applications for Licenses for the Use of Radioactive Materials in Calibrating Radiation Survey and Monitoring Instruments," dated June 1985.

APPENDIX L

Model Medical Licensee Audit

This Appendix was originally developed for medical uses only. With implementation of the EPAct and addition of NARM materials and nonmedical uses, such as authorizations under 10 CFR 30.32(j) to medical use licenses, the procedures may have to be supplemented in this Appendix to address the new materials and nonmedical uses.

Model Medical Licensee Audit

Annual Radiation Protection Medical Licensee Audit

Note: All areas indicated in audit notes may not be applicable to every license and may not need to be addressed during each audit. For example, licensees do not need to address areas that do not apply to the licensee's activities, and activities that have not occurred since the last audit need not be reviewed at the next audit. Also, the audit notes may not be complete for nonmedical uses authorized on the license. Licensees should review audit lists in other volumes of the NUREG 1556 series and information provided in Appendix AA of this volume, as appropriate, when completing the audit list that is specific to their nonmedical uses (e.g., production of PET radioactive drugs under 10 CFR 30.32(j)).

With the implementation of the EPAct, the NRC now has regulatory authority over accelerator-produced radioactive materials and discrete sources of radium-226. Therefore, all audits must include these materials after NRC's waiver of August 31, 2005, is terminated for medical use facilities. The waiver was terminated on November 30, 2007, for Government agencies, Federally recognized Indian tribes, Delaware, the District of Columbia, Puerto Rico, the U.S. Virgin Islands, Indiana, Wyoming, and Montana. The NRC Regional Offices should be contacted to confirm the waiver termination date for other medical use facilities.

Date of This Audit: _____   Date of Last Audit: _____

Next Audit Date: _____

Auditor: _____   Date: _____
                 (Signature)

Management Review: _____   Date: _____
                          (Signature)

All references are to 10 CFR Parts unless noted otherwise.

Audit History

A.   Were previous audits conducted annually [20.1101]?

B.   Were records of previous audits maintained [20.2102]?

C.   Were any deficiencies identified during previous audit?

D.   Were corrective actions taken? (Look for repeated deficiencies.)

Organization and Scope of Program

A.   Radiation Safety Officer:

    1.   If the RSO was changed, was the license amended [35.13]?

    2.   Does the new RSO meet NRC training requirements [35.50, 35.57, 35.59]?

3. If the scope of the program expands, does the RSO have training in radiation safety, regulatory issues, and emergency procedures for the new uses [35.50(e)]?

4. Is the RSO fulfilling all duties [35.24]?

5. Is the written agreement in place for a new RSO [35.24(b)]?

B. Multiple places of use? If yes, list locations.

C. Are all locations listed on license? Includes locations of accelerator-produced radioactive materials and discrete sources of radium-226?

D. Were annual audits performed at each location? If no, explain.

E. Describe the scope of the program (staff size, number of procedures performed, etc.).

F. Licensed Material:

1. Isotope, chemical form, quantity, and use as authorized? Includes accelerator-produced radioactive materials and discrete sources of radium-226?

2. Does the total amount of radioactive material possessed require financial assurance [30.35(a)]? If so, is the financial assurance adequate?

3. Calibration, transmission, and reference sources [35.65]?

a. Sealed sources manufactured and distributed by a person licensed pursuant to 10 CFR 32.74, equivalent Agreement State regulations, or redistributed by a licensee authorized to redistribute sealed sources, and sources do not exceed 30 millicuries each [35.65(a) and (b)]?

b. Any byproduct material with a half-life not longer than 120 days in individual amounts not to exceed 15 millicuries [35.65(c)]?

c. Any byproduct material with a half-life longer than 120 days in individual amounts not to exceed the smaller of 200 microcuries or 1000 times the quantities in Appendix B of Part 30 [35.65(d)]?

d. Technetium-99m in individual amounts as needed [35.65(e)]?

4. Unsealed materials used under 10 CFR 35.100, 35.200, and 35.300 are:

a. Obtained from a manufacturer or preparer licensed under 10 CFR 32.72?

OR

b. Obtained from a producer of PET radioactive drugs under 10 CFR 30.32(j)?

OR

c. Prepared by a physician AU, an ANP, or an individual under the supervision of an ANP or physician AU?

OR

d. Obtained and prepared for research in accordance with 10 CFR 35.100, 10 CFR 35.200, and 10 CFR 35.300, as applicable?

5. Production of PET radioactive drugs

- Authorized under 10 CFR 30.32(j)?

- For internal use from licensee's PET radionuclide production facility as authorized in 10 CFR 35.100(b), 35.200(b), or 35.300(b)?

G. Are the sealed sources possessed and used as described in the Sealed Source and Device Registry (SSDR) certificate in 10 CFR 32.210, 35.400, 35.500, 35.600? Are copies of (or access to) SSDR certificates possessed? Are manufacturers' manuals for operation and maintenance of medical devices possessed?

H. Are there sealed sources containing accelerator-produced radioactive materials or discrete sources of radium-226 that do not have an SSDR certificate? If the sealed source is not generally licensed or exempt from licensing, seek a license amendment providing information under 10 CFR 32(g)(2) or (3).

I. Are the actual uses of medical devices consistent with the authorized uses listed on the license?

J. If places of use changed, was the license amended [35.13(e)]?

K. If control of the license was transferred or bankruptcy filed, was NRC's prior consent obtained or notification made [30.34(b) and 30.34(h) respectively]?

Radiation Safety Program

A. Minor changes to program [10 CFR 35.26 or license condition for 10 CFR 35.1000 medical uses]?

B. Records of changes maintained for 5 years [35.2026]?

C. Content and implementation reviewed annually by the licensee [20.1101(c)]?

D. Records of reviews maintained [20.2102]?

E. Changes include addition of accelerator-produced radioactive materials or discrete sources of radium-226 to NRC-regulated Radiation Safety Program?

F. Changes include authorization to produce PET radioactive drugs for noncommercial distribution to other medical use licensees in the consortium [10 CFR 30.32(j)]?

Use by Authorized Individuals

Compliance is established by meeting at least one criterion under each category.

A. Authorized Nuclear Pharmacist [35.55, 35.57, 35.59] (*Note:* Does not apply to facilities that are registered with FDA as the owner or operator of a drug establishment that engages in the manufacture, preparation, propagation, compounding, or processing of a drug under 21 CFR 207.20(a) or registered with the State as a drug manufacturer or PET drug production facility with distribution regulated under 10 CFR 32.72):

_____ 1. Certified by specialty board?

_____ 2. Identified on NRC or Agreement State license?

_____ 3. Identified on permit issued by broad-scope or master materials licensee?

_____ 4. Identified on permit issued by master materials permittee of broad scope?

_____ 5. Identified as an ANP by a commercial nuclear pharmacy that has been authorized to identify ANPs?

_____ 6. Designated as an ANP in accordance with 10 CFR 32.72(b)(4)?

_____ 7. Meets requirements in 35.57(a)(3)?

_____ 8. Listed on facility license?

B. Authorized User [35.57, 35.59, and 35.190, 35.290, 35.390, 35.392, 35.394, 35.396, 35.490, 35.491, 35.590, 35.690]:

_____ 1. Certified by specialty board?

_____ 2. Identified on NRC or Agreement State license?

_____ 3. Identified on permit issued by broad-scope or master materials licensee?

_____ 4. Identified on permit issued by master materials permittee of broad scope?

_____ 5. Meets requirements in 35.57(b)(3)?

_____ 6. Listed on facility license?

C. Authorized Medical Physicist [35.51, 35.57, 35.59]:

_____ 1. Certified by specialty board?

_____ 2. Identified on NRC or Agreement State license?

_____ 3. Identified on permit issued by broad-scope or master materials licensee?

_____ 4. Identified on permit issued by master materials permittee of broad scope?

_____ 5. Meets requirements in 35.57(a)(3)?

_____ 6. Listed on facility license?

D. Nonmedical use authorized users [30.33(a)(3)]:

_____ Listed on facility license for same materials and uses?

Mobile Medical Service

A. Operates services per 35.80, 35.647?

B. Compliance with 20.1301 evaluated and met?

C. Letter signed by management of each client [35.80(a)]?

D. Licensed material not delivered to client's address (unless client was authorized) [35.80(b)]?

E. Dosage measuring instruments checked for proper function before use at each address of use or on each day of use, if more frequent [35.80(a)]?

F. Survey instruments checked for proper operation before use at each address of use [35.80(a)]?

G. Survey of all areas of use prior to leaving each client address [35.80(a)]?

H. Additional technical requirements for mobile remote afterloaders per 35.647?

Amendments Since Last Audit [35.13]

A. Any Amendments since last audit [35.13]?

B. Security-related sensitive information was properly marked?

Notifications Since Last Audit [35.14]

A. Any Notifications since last audit [35.14]?

B. Appropriate documentation provided to NRC, for ANP, AMP, or AU, no later than 30 days after the individual starts work [35.14(a), 30.34(j)(4)]?

C. NRC notified within 30 days after: AU, ANP, AMP, or RSO stops work or changes name; licensee's mailing address changes; licensee's name changes without a transfer of control of the license; or licensee has added to or changed an area of use for 10 CFR 35.100 or 35.200 use, if the change does not include addition or relocation of either an area where PET radionuclides are produced or a radionuclide delivery line from a PET radionuclide production area [35.14(b)]?

Training, Retraining, and Instructions to Workers

A. Have workers been provided with required instructions [19.12, 35.27, 35.310, 35.410, 35.610]?

B. Have workers been informed of NRC's regulatory authority for accelerator-produced radioactive materials and discrete sources of radium-226?

C. Is the individual's understanding of current procedures and regulations adequate?

D. Is the training program implemented?

    1. Operating procedures [35.27, 35.310, 35.410, 35.610]?

    2. Emergency procedures [35.27, 35.310, 35.410, 35.610]?

    3. Periodic training required and implemented [35.310, 35.410, 35.610]?

    4. Were all workers who are likely to exceed 1 mSv (100 mrem) in a year instructed and was refresher training provided, as needed [19.12]?

    5. Was each supervised user instructed in the licensee's written radiation protection procedures and administration of written directives, as appropriate [35.27]?

    6. Are initial and periodic training records maintained for each individual [35.2310]?

7. Briefly describe training program.

E. Do additional therapy device instructions/training include:

1. Unit operation, inspection, associated equipment, survey instruments?

2. License conditions applicable to the use of the unit?

3. Emergency drills [35.610]?

F. 10 CFR Part 20 – Are workers cognizant of requirements for:

1. Radiation Safety Program [35.24, 35.26, 20.1101]?

2. Annual dose limits [20.1201, 20.1301, 20.1302]?

3. NRC Forms 4 and 5?

4. 10% monitoring threshold [20.1502]?

5. Dose limits to embryo/fetus and declared pregnant worker [20.1208]?

6. "Grave Danger" Posting [20.1902(c)]?

7. Procedures for opening packages [20.1906]?

G. Is supervision of individuals by AU and/or ANP in accordance with 10 CFR 35.27?

Training for Manual Brachytherapy and Use of Unsealed Byproduct Material for Which a Written Directive Is Required

A. Does safety instruction to personnel include [35.310, 35.410]:

1. Control of patient and visitors?

2. Routine visitation to patients in accordance with 10 CFR 20.1301?

3. Contamination control and size/appearance of sources?

4. Safe handling and shielding instructions?

5. Waste control?

6. RSO and AU notification if patient had a medical emergency or died?

7. Records retained [35.2310]?

Facilities

A. Facilities as described in license application?

B. Therapy device facilities provided with electrical interlock system, viewing and intercom systems, radiation monitor, source retraction mechanism, and source indicator lights?

C. Emergency source recovery equipment available [35.415, 35.615]?

D. Storage areas:

1. Materials secured from unauthorized removal or access [20.1801]?

    2.   Licensee controls and maintains constant surveillance of licensed material not in storage [20.1802]?

E.   Therapy unit operation:

    1.   Unit, console, console keys, and treatment room controlled adequately [20.1801, 35.610(a)(1)]?

    2.   Restricted to certain source orientations and/or gantry angles?

    3.   Ceases to operate in restricted orientation(s)?

    4.   Only one radiation device can be placed in operation at a time within the treatment room [35.610(a)(3)]?

Dose or Dosage Measuring Equipment

A.   Possession, use, and calibration of instruments to measure activities of unsealed radionuclides [35.60] or PET radioactive drugs produced by licensee [30.34(j)]:

    1.   Types of equipment listed?

    2.   Approved procedures for use of instrumentation followed?

    3.   Constancy, accuracy, linearity, and geometry dependence tests performed in accordance with nationally recognized standards or the manufacturer's instructions?

    4.   Instrument repaired or replaced or dosages mathematically corrected, as required, when tests do not meet the performance objectives provided in the nationally recognized standard or manufacturer's instructions (e.g., ±10%)?

    5.   Records maintained and include required information [35.2060]?

B.   Determination of dosages of unsealed byproduct material [35.63, 30.34(j)]?

    1.   Each dosage determined and recorded prior to medical use [35.63(a)]? Or transfer [30.34(j)]?

    2.   Measurement of unit dosages of photon- or beta-emitting radionuclides made either by direct measurement or by decay correction [35.63(b), 30.34(j)(2)(ii)]?

    3.   Measurement of unit dosage of alpha-emitting radionuclide by decay correction of the activity provided by the producer licensed in accordance with 10 CFR 32.72 or 30.32(j)?

    4.   For other than unit dosages of photon- or beta-emitting radionuclides, measurement made by direct measurement of radioactivity or by combination of radioactivity or volumetric measurement and calculation [35.63(c), 30.34(j)(2)(ii)]?

    5.   For other than unit dosages of alpha-emitting radionuclide, measurement made by combination using the activity provided by the producer licensed in accordance with 10 CFR 32.72, or 30.32(j) volumetric measurement, and calculation [35.63(c)]?

C.   Licensee uses generators?

    1.   First eluate after receipt tested for Mo-99 breakthrough [35.204(b)]?

2. No radiopharmaceuticals administered with Mo-99 concentrations over 0.15 μCi per mCi of Tc-99m [35.204(a)(1)]?

3. First eluate after receipt tested for strontium-82 and strontium-85 when eluting rubidium-82 [35.204(c)]?

4. No radiopharmaceuticals administered with strontium-82 concentrations over 0.02 μCi per mCi of rubidium-82 or strontium-85 concentrations over 0.2 μCi per mCi of rubidium-82 [35.204(a)(2)]?

5. Records maintained [35.2204]?

D. Dosimetry Equipment [35.630]:

1. Calibrated system available for use [35.630(a)]?

2. Calibrated by NIST or an AAPM-accredited lab within previous 2 years and after servicing [35.630(a)(1)] OR calibrated by intercomparison per 10 CFR 35.630(a)(2)?

3. Calibrated within the previous 4 years [35.630(a)(2)]?

4. Licensee has available for use a dosimetry system for spot-check measurements [35.630(b)]?

5. Record of each calibration, intercomparison, and comparison maintained [35.2630]?

Radiation Protection And Control of Radioactive Material (this now includes accelerator-produced radioactive materials and discrete sources of radium-226)

A. Use of radiopharmaceuticals and production of PET radioactive drugs:

1. Protective clothing worn?

2. Personnel routinely monitor their hands?

3. No eating/drinking in use/storage areas?

4. No food, drink, or personal effects kept in use/storage areas?

5. Proper dosimetry worn?

6. Radioactive waste disposed of in proper receptacles?

7. Syringe shields and vial shields used?

8. Proper use of remote handling tools and radiation shields?

B. Leak tests and inventories:

1. Leak test performed on sealed sources and brachytherapy sources [35.67(b)(1) or leak test license condition]?

2. Inventory of sealed sources and brachytherapy sources performed semiannually [35.67(g)]?

3. Records maintained [35.2067]?

Radiation Survey Instruments

A.   Survey instruments used to show compliance with 10 CFR Part 20 and
10 CFR 30.33(a)(2):

    1.   Appropriate operable survey instruments possessed or available [10 CFR Part 20]?

    2.   Calibrations [35.61(a) and (b)]:

        a.   Before first use, annually, and after repairs?

        b.   Within 20% on each scale or decade of interest?

    3.   Records maintained [35.2061]?

B.   Radiation surveys performed in accordance with the licensee's procedures and the
regulatory requirements [20.1501, 35.70]? If producing PET radioactive drugs under
10 CFR 30.32(j) or 35.100(b), 35.200(b), or 35.300(b), the survey frequencies described
below should be reviewed and adjusted as necessary.

    1.   Daily in all areas where radiopharmaceuticals requiring a written directive are
prepared or administered (except patient rooms) [35.70]?

    2.   Weekly in all areas where radiopharmaceuticals or waste are stored?

    3.   Weekly for wipes in all areas where radiopharmaceuticals are routinely prepared,
administered, or stored?

    4.   Trigger levels established?

    5.   Corrective action taken and documented if trigger level exceeded?

    6.   Techniques can detect 0.1 mR/hr, 2000dpm?

    7.   Surveys made to assure that the maximum radiation levels and average radiation
levels from the surface of the main source safe with the sources(s) in the shielded
position do not exceed the levels stated in the Sealed Source and Device Registry
[35.652(a)] and records maintained [35.2652]?

        a.   After new source installation?

        b.   Following repairs to the source(s) shielding, the source(s) driving unit, or other
electronic or mechanical mechanism that could expose the source, reduce the
shielding around the source(s), or compromise the radiation safety of the unit or
the source(s)?

Public Dose (this now includes dose from accelerator-produced radioactive materials and discrete sources of radium-226)

A.   Is licensed material used in a manner to keep doses below 1mSv (100 mrem) in a year
[20.1301(a)(1)]?

B.   Has a survey or evaluation been performed per 20.1501(a)?

C.   Have there been any additions or changes to the storage, security, or use of surrounding
areas that would necessitate a new survey or evaluation?

D.   Do unrestricted area radiation levels exceed 0.02 mSv (2 mrem) in any 1 hour [20.1301(a)(2)]?

E.   Is licensed material used or stored in a manner that would prevent unauthorized access or removal [20.1801 and 20.1802]?

F.   Are records maintained [20.2103, 20.2107]?

Patient Release

A.   Individuals released when TEDE is less than 0.5 rem [35.75(a)]?

B.   Instructions to the released individual, including breast-feeding women, include required information [35.75(b)]?

C.   Release records maintained [35.2075(a)]?

D.   Records of instructions given to breast-feeding women maintained, if required [35.2075(b)]?

Unsealed Byproduct Material for Which a Written Directive Is Required (this now includes written directives for accelerator-produced radioactive materials and discrete sources of radium-226)

A.   Safety precautions implemented to include patient facilities, posting, stay times, patient safety guidance, release, and contamination controls [35.315(a)]?

B.   RSO and AU promptly notified if patient had a medical emergency or died [35.315(b)]?

Brachytherapy or Brachytherapy Source Use

A.   Safety precautions implemented to include patient facilities, posting, stay times, and emergency response equipment [35.415]?

B.   Survey immediately after implant [35.404(a)]?

C.   Patients surveyed immediately after removing the last temporary implant source [35.404(b)]?

D.   RSO and AU promptly notified if patient had a medical emergency or died [35.415(c)]?

E.   Records maintained [35.2404]?

Radioactive Waste (this now includes waste containing accelerator-produced radioactive materials and discrete sources of radium-226)

A.   Disposal:

   1.   Decay-in-storage [35.92]?

   2.   Procedures followed?

3. Labels removed or defaced [20.1904, 35.92]?

B. Special procedures performed as required?

C. Authorized disposals [20.2001]?

D. Records maintained [20.2103(a), 20.2108, 35.2092]?

E. Effluents:

1. Release to sanitary sewer [20.2003]?

   a. Material is readily soluble or readily dispersible [20.2003(a)(1)]?

   b. Monthly average release concentrations do not exceed 10 CFR Part 20, Appendix B, Table 2 values?

   c. No more than 5 Ci of H-3, 1 Ci of C-14, and 1 Ci of all other radionuclides combined, released in a year [20.2003]?

   d. Procedures to ensure representative sampling and analysis implemented [20.1501]?

2. Release to septic tanks [20.2003]? Within unrestricted limits [10 CFR Part 20, Appendix B, Table 2]?

3. Waste incinerated?

   a. License authorizes [20.2004(a)(3)]?

   b. Exhaust directly monitored?

   c. Airborne releases evaluated and controlled [20.1302, 20.1501]?

4. Air effluents and ashes controlled [20.1101, 20.1201, 20.1301, 20.1501, 20.2001]? (See also IP 87102, RG 8.37.) If applicable, includes air effluent releases from production of PET radioactive drugs?

   a. Air effluent less than 10 mrem constraint limit [20.1101]?

      i. If no, reported appropriate information to NRC?

      ii. If no, corrective actions implemented and on schedule?

   b. Description of effluent program:

      i. Monitoring system hardware adequate?

      ii. Equipment calibrated, as appropriate?

      iii. Air samples/sampling technique (e.g., charcoal, HEPA) analyzed with appropriate instrumentation?

F. Waste storage:

1. Protection from elements and fire?

2. Control of waste maintained [20.1801]?

3. Containers properly labeled and area properly posted [20.1902, 20.1904]?

4. Package integrity adequately maintained?

G. Waste disposal:

    1. Sources transferred to authorized individuals [20.2006, 20.2001, 30.41]?

    2. Name of organization: _____ .

H. Records of surveys and material accountability maintained [20.2103, 20.2108, 35.2092]?

Receipt and Transfer of Radioactive Material (this now includes receipt and transfer of accelerator-produced radioactive materials and discrete sources of radium-226)

A. Description of how packages are received and by whom?

B. Written package-opening procedures established and followed [20.1906(e)]?

C. All incoming packages with a DOT label monitored for radioactive contamination, unless exempted (gases and special form) [20.1906(b)(1)]?

D. Incoming packages surveyed [20.1906(b)(2)]? When authorized for return, includes "empty" transport radiation shields from other consortium members receiving PET radioactive drugs under 10 CFR 30.32(j) authorization?

E. Monitoring in (C) and (D) performed within time specified [20.1906(c)]?

F. Transfer(s) performed per [30.41]?

G. If authorized under 10 CFR 30.32(j) for production and noncommercial transfer of PET radioactive drugs, all transfers of these drugs for medical use are to medical use licensees within the consortium?

H. All sources surveyed before shipment and transfer [20.1501(a)]?

I. Records of surveys and receipt/transfer maintained [20.2103(a), 30.51]?

J. Package receipt/distribution activities evaluated for compliance with 20.1301?

Transportation (10 CFR 71.5(a) and 49 CFR 171-189)

A. Shipments, including shipments of accelerator-produced radioactive materials and discrete sources of radium-226, and PET radioactive drugs produced for noncommercial transfer to other medical use licensees in the consortium, are:

    1. Delivered to common carriers?

    2. Transported in own private vehicle?

    3. Both?

    4. No shipments since last audit?

B. Return radiopharmacy doses to drug manufacture or commercial nuclear pharmacy or sealed sources to source or device manufacturer? *Note:* Licensees authorized under 10 CFR 30.32(j) for production and noncommercial transfer of PET radioactive drugs are not authorized to receive unused dosages or empty syringes and vials back from consortium members.

1. Licensee assumes shipping responsibility?

2. If "NO," describe arrangements made between licensee and radiopharmacy for shipping responsibilities.

C. Packages:

1. Authorized packages used?

2. Performance test records on file?

   a. DOT-7A packages

   b. Special form sources

3. Two labels (White-I, Yellow-II, Yellow-III) with Transport Index (TI), Nuclide, Activity, and Hazard Class?

4. Properly marked (Shipping Name, UN Number, Package Type, Reportable Quantity, "This End Up" (liquids), Name and Address of consignee)?

5. Closed and sealed during transport?

D. Shipping Papers:

1. Prepared and used?

2. Contain proper entries (Shipping Name; Hazard Class; Identification Number (UN Number); Total Quantity; Package Type; Nuclide; Reportable Quantity; Physical and Chemical Form; Activity; Category of Label; TI; Shipper's Name, Certification and Signature; Emergency Response Telephone Number; "Limited Quantity" (if applicable); "Cargo Aircraft Only" (if applicable))?

3. Readily accessible during transport?

Teletherapy and Gamma Stereotactic Radiosurgery Servicing

A. Inspection and servicing performed following source replacement or at intervals not to exceed 5 years [35.655(a)]?

B. Needed service arranged for as identified during the inspection?

C. Service performed by persons specifically authorized to do so [35.655(b)]?

Full Calibration-Therapeutic Medical Devices

A. Proper protocol(s) used (e.g., TG-21, AAPM 54, TG-56, TG-40)?

B. Performed prior to first patient use [35.632(a)(1), 35.633(a)(1), 35.635(a)(1)]?

C. At intervals not to exceed 1 year for teletherapy, gamma stereotactic, and LDR remote afterloader; at intervals not exceeding 1 quarter for HDR, MDR, and PDR remote afterloaders [35.632(a)(3)], 35.633(a)(3) and (4), 35.635(a)(3)]?

D. Whenever spot-checks indicate output differs from expected by ±5% [35.632(a)(2)(i), 35.635(a)(2)(i)]?

E. After source exchange, relocation, and major repair or modification [35.632(a)(2), 35.633(a)(2), 35.635(a)(2)]?

F. Performed with properly calibrated instrument [35.632(c), 35.633(c), 35.635(c)]?

G. Includes:

   1. For teletherapy:

      a. Output measured within ±3% of expected for the range of field sizes, range of distances [35.632(b)(1)]?

      b. Coincidence of radiation field and field light localizer [35.632(b)(2)]?

      c. Uniformity of radiation field and beam angle dependence [35.632(b)(3)]?

      d. Timer accuracy and linearity over the range of use [35.632(b)(4)]?

      e. On-off error [35.632(b)(5)]?

      f. Accuracy of all measuring and localization devices [35.632(b)(6)]?

   2. For remote afterloaders:

      a. Output measured within ±5% of expected [35.633(b)(1)]?

      b. Source positioning accuracy within ±1 millimeter [35.633(b)(2)]?

      c. Source retraction with backup battery upon power failure [35.633(b)(3)]?

      d. Length of source transfer tubes [35.633(b)(4)]?

      e. Timer accuracy and linearity over the typical range of use [35.633(b)(5)]?

      f. Length of the applicators [35.633(b)(6)]?

      g. Function of source transfer tubes, applicators, and transfer tube-applicator interfaces [35.633(b)(7)]?

      h. Autoradiograph quarterly of the LDR source(s) to verify source(s) arrangement and inventory [35.633(e)]?

   3. For gamma stereotactic radiosurgery:

      a. Output measured within ±3% of expected [35.635(b)(1)]?

      b. Helmet factors [35.635(b)(2)]?

      c. Isocenter coincidence [35.635(b)(3)]?

      d. Timer accuracy and linearity over the range of use [35.635(b)(4)]?

      e. On-off error [35.635(b)(5)]?

f. Trunnion centricity [35.635(b)(6)]?

g. Treatment table retraction mechanism, using backup battery power or hydraulic backups with the unit off [35.635(b)(7)]?

h. Helmet microswitches [35.635(b)(8)]?

i. Emergency timing circuit [35.635(b)(9)]?

j. Stereotactic frames and localizing devices (trunnions) [35.635(b)(10)]?

H. Output corrected mathematically for decay [35.632(e), 35.633(g), 35.635(e)]?

I. Records maintained [35.2632]?

Periodic Spot Checks For Therapeutic Devices

A. Performed at required frequency [35.642(a), 35.643(a), 35.645(a)]?

B. Procedures established by AMP [35.642(b), 35.643(b), 35.645(b)]?

C. Procedures followed?

D. Medical physicist reviews results within 15 days [35.642(c), 35.643(c), 35.645(b)]?

E. Performed with properly calibrated instrument [35.642(a)(5), 35.645(c)(2)(i)]?

F. Output and safety spot checks include:

1. For teletherapy:

a. Timer accuracy and linearity over the range of use [35.642(a)(1)]?

b. On-off error [35.642(a)(2)]?

c. Coincidence of radiation field and field light localizer [35.642(a)(3)]?

d. Accuracy of all measuring and localization devices [35.642(a)(4)]?

e. The output for one typical set of operating conditions [35.642(a)(5)]?

f. Difference between measured and expected output [35.642(a)(6)]?

g. Interlock systems [35.642(d)(1)]?

h. Beam stops [35.642(d)(2)]?

i. Source exposure indicator lights [35.642(d)(3)]?

j. Viewing and intercom systems [35.642(d)(4)]?

k. Treatment room doors, inside and out [35.642(d)(5)]?

l. Electrical treatment doors with power shut off [35.642(d)(6)]?

2. For remote afterloaders:

a. Interlock systems [35.643(d)(1)]?

b. Source exposure indicator lights [35.643(d)(2)]?

    c. Viewing and intercom systems, except for low dose-rate (LDR) [35.643(d)(3)]?

    d. Emergency response equipment [35.643(d)(4)]?

    e. Radiation monitors used to indicate source position [35.643(d)(5)]?

    f. Timer accuracy [35.643(d)(6)]?

    g. Clock (date and time) in the unit's computer [35.643(d)(7)]?

    h. Decayed source(s) activity in the unit's computer [35.643(d)(8)]?

3. For gamma stereotactic radiosurgery:

    a. Treatment table retraction mechanism [35.645(c)(1)(i)]?

    b. Helmet microswitches [35.645(c)(1)(ii)]?

    c. Emergency timing circuits [35.645(c)(1)(iii)]?

    d. Stereotactic frames and localizing devices [35.645(c)(1)(iv)]?

    e. The output for one typical set of operating conditions [35.645(c)(2)(i)]?

    f. Difference between measured and expected output [35.645(c)(2)(ii)]?

    g. Source output compared against computer calculation of output [35.645(c)(2)(iii)]?

    h. Timer accuracy and linearity over the range of use [35.645(c)(2)(iv)]?

    i. On-off error [35.645(c)(2)(v)]?

    j. Trunnion centricity [35.645(c)(2)(vi)]?

    k. Interlock systems [35.645(d)(1)]?

    l. Source exposure indicator lights [35.645(d)(2)]?

    m. Viewing and intercom systems [35.645(d)(3)]?

    n. Timer termination [35.645(d)(4)]?

    o. Radiation monitors used to indicate room exposures [35.645(d)(5)]?

    p. Emergency off buttons [35.645(d)(6)]?

G. Licensee promptly repaired items found to be not operating properly and did not use unit until repaired, if required [35.642(e), 35.643(e), 35.645(f)]?

H. Records maintained [35.2642, 35.2643, 35.2645]?

Installation, Maintenance, and Repair of Therapy Devices

A. Only authorized individuals perform installation, maintenance, adjustment, repair, and inspection [35.605, 35.655]? Name of organization/individual.

B. Records maintained [35.2605, 35.2655]?

Operating Procedures For Therapy Devices

A. Instructions on location of emergency procedures and emergency response telephone numbers posted at the device console [35.610(c)]?

B. Copy of the entire procedures physically located at the device console [35.610(b)]?

C. Procedures include:

    1. Instructions for responding to equipment failures and the names of the individuals responsible for implementing corrective actions [35.610(a)(4)]?

    2. The process for restricting access to and posting of the treatment area to minimize the risk of inadvertent exposure [35.610(a)(4)]?

    3. The names and telephone numbers of the AUs, the AMP, and the RSO to be contacted if the unit or console operates abnormally [35.610(a)(4)]?

D. Radiation survey of patient is performed to ensure source is returned to shielded position [35.604(a)]?

E. Records of radiation surveys maintained for 3 years [35.2404]?

F. AMP and AU:

    1. Physically present during initiation of patient treatment with remote afterloaders? (*Note:* for MDR and PDR, an appropriately trained physician under the supervision of the AU may be physically present instead of the AU) [35.615(f)(1) and (2)].

    2. Physically present throughout all patient treatments with a gamma stereotactic radiosurgery device [35.615(f)(3)]?

Personnel Radiation Protection (this now includes exposures from accelerator-produced radioactive materials and discrete sources of radium-226)

A. Exposure evaluation performed [20.1501]? Includes evaluation for uses of accelerator-produced radioactive materials and discrete sources of radium-226?

B. ALARA program implemented [20.1101(b)]?

C. External Dosimetry:

    1. Monitors workers per [20.1502(a)]? Includes workers using or working near accelerator-produced radioactive materials and discrete sources of radium-226?

    2. External exposures account for contributions from airborne activity [20.1203]?

    3. Supplier _____ Frequency _____

    4. Supplier is NVLAP-approved [20.1501(c)]?

    5. Dosimeters exchanged at required frequency?

D.   Internal Dosimetry:

    1.   Monitors workers per 20.1502? Includes workers using or working near accelerator-produced radioactive materials and discrete sources of radium-226?

    2.   Program for monitoring and controlling internal exposures [20.1701, 20.1702] briefly described?

    3.   Monitoring/controlling program implemented (includes bioassays)?

    4.   Respiratory protection equipment [20.1703]?

E.   Review of Records and Reports:

    1.   Reviewed by _____ Frequency _____

    2.   Auditor reviewed personnel monitoring records for period _____ to _____

    3.   Prior dose determined for individuals likely to receive doses [20.2104]?

    4.   Maximum exposures TEDE _____ Other _____

    5.   Maximum CDEs _____ Organs _____

    6.   Maximum CEDE

    7.   Internal and external summed [20.1202]?

    8.   Occupational limits met [20.1201]?

    9.   NRC forms or equivalent [20.2104(d), 20.2106(c)]?

        a.   NRC-4 Complete:

        b.   NRC-5 Complete:

    10.   If a worker declared her pregnancy during the audit period, was the dose in compliance [20.1208] and were the records maintained [20.2106(e)]?

F.   Any planned special exposures (number of people involved and doses received) [20.1206, 20.2104, 20.2105, 20.2204]?

G.   Records of exposures, surveys, monitoring, and evaluations maintained [20.2102, 20.2103, 20.2106]?

Confirmatory Measurements

Detail location and results of confirmatory measurements.

Medical Events

If medical events, including those with accelerator-produced radioactive materials and discrete sources of radium-226, [criteria in 35.3045] have occurred since the last audit, evaluate the incident(s) and procedures for implementing and administering written directives using the existing guidance.

1.   Event date _____ Information Source

2.   Notifications:

      NRC Ops Center   NRC Region

      Referring Physician   Patient

      In writing/By telephone

      If notification did not occur, why not?

3.   Written Reports [35.3045]: Submitted to Region within 15 days?

Notification and Reports (this now includes notifications and reports for accelerator-produced radioactive materials and discrete sources of radium-226)

A.   In compliance with 10 CFR 19.13, and 10 CFR 30.50 (reports to individuals, public and occupational, monitored to show compliance with Part 20)?

B.   In compliance with 10 CFR 20.2201, and 10 CFR 30.50 (theft or loss)?

C.   In compliance with 10 CFR 20.2202, and 10 CFR 30.50 (incidents)?

D.   In compliance with 10 CFR 20.2203, and 10 CFR 30.50 (overexposure and high radiation levels)?

E.   Aware of NRC Operations Center telephone number?

F.   In compliance with 10 CFR 20.2203 (constraint on air emissions)?

Posting and Labeling

A.   NRC Form 3, "Notice to Workers" is posted [19.11]?

B.   10 CFR Parts 19, 20, 21, Section 206 of Energy Reorganization Act, procedures adopted pursuant to 10 CFR Part 21, and license documents are posted, or a notice indicating where documents can be examined is posted [19.11, 21.6]?

C.   Other posting and labeling per 10 CFR 20.1902, 20.1904, and not exempted by 20.1903, 20.1905?

Recordkeeping for Decommissioning (this now includes records for accelerator-produced radioactive materials and discrete sources of radium-226 produced before, on, or after the August 8, 2005 EPAct).

A.   Records of information important to the safe and effective decommissioning of the facility maintained in an independent and identifiable location until license termination [30.35(g)]?

B.   Records include all information outlined in 10 CFR 30.35(g)?

Bulletins and Information Notices

A. Bulletins, Information Notices, NMSS Newsletters, etc., received?

B. Appropriate action in response to Bulletins, Generic Letters, etc.?

Special License Conditions or Issues

A. Special license conditions or issues to be reviewed:

1. If authorized for the production and noncommercial distribution of PET radioactive drugs under 10 CFR 30.32(j), review the program for conformance with the requirements in 10 CFR 30.34(j).

2. If authorized for 10 CFR 35.1000 medical uses, review the program for conformance with license application commitments, license conditions, and regulations.

3. Other

B. Evaluation.

Audits and Findings

A. Summary of findings.

B. Corrective and preventive actions.

APPENDIX M

Model Procedures for an Occupational Dose Program

Model Procedures for an Occupational Dose Program

With the implementation of the EPAct, the NRC now has regulatory authority over accelerator-produced radioactive materials and discrete sources of radium-226. Therefore, after NRC's waiver of August 31, 2005, is terminated for medical use facilities, occupational dose programs must also include occupational doses to workers if they are only exposed to the radiation from these materials. Previously, the dose from these materials was only included for NRC purposes if the worker was exposed to radiation from NRC-regulated materials which excluded NARM materials. The NRC waiver that applied to Government agencies, Federally recognized Indian tribes, Delaware, the District of Columbia, Puerto Rico, the U.S. Virgin Islands, Indiana, Wyoming, and Montana was terminated on November 30, 2007. The NRC Regional Offices should be contacted to confirm the waiver termination date for other medical use facilities.

This model provides acceptable procedures for an external occupational dose program and references for developing an internal occupational dose program. Applicants may either adopt these model procedures for an external occupational dose program or develop alternative procedures to meet the requirements of 10 CFR 20.1101 and Subparts C and F of 10 CFR Part 20. The model includes guidance as well as a discussion of regulatory requirements that are to be reflected in the elements of an occupational dose program.

"Dosimetry" is a broad term commonly applied to the use of monitoring devices, bioassay, and other methods to measure or otherwise quantify radiation doses to individuals. The licensee must control occupational doses and provide individuals with monitoring devices in accordance with the requirements of 10 CFR 20.1502(a). The occupational dose limits for adults are provided in 10 CFR 20.1201, while 10 CFR 20.1502 provides, in part, that adults likely to receive in 1 year a dose in excess of 10 percent of those dose limits must be provided with dosimetry. Definitions of relevant terms such as Total Effective Dose Equivalent (TEDE), deep-dose equivalent (DDE), and committed effective dose equivalent (CEDE) can be found in 10 CFR 20.1003, "Definitions." In addition, if monitoring is required pursuant to 10 CFR 20.1502, each licensee shall maintain records of doses received (see 10 CFR 20.2106), and individuals must be informed of their doses on at least an annual basis (see 10 CFR 19.13(b)).

If an individual, including an individual only exposed to accelerator-produced radioactive materials or discrete sources of radium-226 or radiation from these materials, is likely to receive more than 10% of the annual dose limits, the NRC requires the licensee to monitor the dose, to maintain records of the dose, and, on at least an annual basis, to inform the worker of his/her dose.

The As-Low-As-Reasonably-Achievable "ALARA" Program

Section 10 CFR 20.1101 states that "each licensee shall develop, document, and implement a Radiation Protection Program commensurate with the scope and extent of licensed activities..." and "the licensee shall use, to the extent practical, procedures and engineering controls based upon sound radiation protection principles to achieve occupational doses and doses to members of the public that are as low as is reasonably achievable (ALARA)." Additionally,

10 CFR 20.1101 requires that licensees periodically review the content of the Radiation Protection Program and its implementation.

External Exposure

It is necessary to assess doses to radiation workers to demonstrate compliance with regulatory limits on radiation dose and to help demonstrate that doses are maintained at ALARA levels.

Providing for the safe use of radioactive materials and radiation is a management responsibility. It is important that management recognize the importance of radiation monitoring in the overall requirements for radiation protection.

There are three dose limits included in 10 CFR 20.1201 that apply to external exposure: deep dose to the whole body (5 rem or 0.05 Sv), shallow dose to the skin or extremities (50 rem or 0.5 Sv), and dose to the lens of the eye (15 rem or 0.15 Sv). According to the definitions in 10 CFR 20.1003, the DDE to the whole body is considered to be at a tissue depth of 1 cm (1000 mg/cm^2), shallow-dose equivalent to the skin or extremities at 0.007 cm (7 mg/cm^2), and eye dose equivalent at 0.3 cm (300 mg/cm^2). In evaluating the eye dose equivalent, it is acceptable to take credit for the shielding provided by protective lenses.

Under 10 CFR 20.1502(a), the use of individual monitoring devices is required for the following:

- Adults likely to receive, in 1 year, from sources external to the body, a dose in excess of 10% of the occupational dose limits in 10 CFR 20.1201(a). Monitoring devices are accordingly required for adults with an annual dose in excess of:
  — 0.5 rem (0.005 Sv) DDE,
  — 1.5 rem (0.015 Sv) eye dose equivalent,
  — 5 rem (0.05 Sv) shallow-dose equivalent to the skin,
  — 5 rem (0.05 Sv) shallow-dose equivalent to any extremity.

- Minors who are likely to receive an annual dose in excess of:
  — 0.1 rem (1.0 mSv) DDE,
  — 0.15 rem (1.5 mSv) eye dose equivalent,
  — 0.5 rem (5 mSv) shallow-dose equivalent to the skin, or
  — 0.5 rem (5 mSv) shallow-dose equivalent to any extremity.

- Declared pregnant women likely to receive an annual dose in excess of 0.1 rem (1.0 mSv) DDE during the entire pregnancy.

- Individuals entering a high- or a very-high-radiation area.

To demonstrate that monitoring of occupational exposure is not necessary for a group of radiation workers, including those working with or near accelerator-produced radioactive materials or discrete sources of radium-226, it must be demonstrated that doses will not exceed

10% of the applicable limits. In these cases, the NRC does not require licensees to monitor radiation doses for this class of worker.

The following methods may be used to demonstrate that doses are expected to be within 10% of regulatory limits:

- Prior Experience: Reviews of radiation dose histories for workers in a specific work area show that they are not likely to receive a dose in excess of 10% of the limits;

- Area Surveys: Demonstrate through the conduct of appropriate radiation level surveys (e.g., using a survey meter or area thermoluminescent dosimeters (TLDs)) in the work area, combined with estimates of occupancy rates and calculations, that doses to workers are not likely to exceed 10% of the limits (exposures associated with reasonable "accident" scenarios should also be evaluated);

- The licensee performs a reasonable calculation, based upon source strength, distance, shielding, and time spent in the work area, that shows that workers are not likely to receive a dose in excess of 10% of the limits.

External dose is determined by using individual monitoring devices, such as film badges, optically stimulated luminescence dosimeters, or TLDs. These devices must be evaluated by a processor that is National Voluntary Laboratory Accreditation Program (NVLAP)-approved, as required by 10 CFR 20.1501.

The device for monitoring the whole body dose, eye dose, skin dose, or extremity dose shall be placed near the location expected to receive the highest dose during the year (10 CFR 20.1201(c)). When the whole body is exposed fairly uniformly, the individual monitoring device is typically worn on the front of the upper torso.

If the radiation dose is highly nonuniform, causing a specific part of the whole body (head, trunk, arms above the elbow, or legs above the knees) to receive a substantially higher dose than the rest of the whole body, the individual monitoring device shall be placed near that part of the whole body expected to receive the highest dose. For example, if the dose rate to the head is expected to be higher than the dose rate to the trunk of the body, a monitoring device shall be located on or close to the head.

If, after the exposure is received, the licensee somehow learns that the maximum dose to a part of the whole body, eye, skin, or extremity was substantially higher than the dose measured by the individual monitoring device, an evaluation shall be conducted to estimate the actual maximum dose.

Under 10 CFR 20.2106, individual monitoring must be recorded on NRC Form 5 or equivalent. The Form 5 is used to record doses received for the calendar year. The monitoring year may be adjusted as necessary to permit a smooth transition from one monitoring year to another, as long as the year begins and ends in the month of January, the change is made at the beginning of the year, and no day is omitted or duplicated in consecutive years.

Because evaluation of dose is an important part of the Radiation Protection Program, it is important that users return dosimeters on time. Licensees should be vigorous in their effort to recover any missing dosimeters. Delays in processing a dosimeter can result in the loss of the stored information.

If an individual's dosimeter is lost, the licensee needs to perform and document an evaluation of the dose the individual received and to add it to the employee's dose record in order to demonstrate compliance with occupational dose limits in 10 CFR 20.1201. Sometimes the most reliable method for estimating an individual's dose is to use his/her recent dose history. In other cases, particularly if the individual does nonroutine types of work, it may be better to use doses of co-workers as the basis for the dose estimate. It also may be possible to estimate doses by modeling and calculation (i.e., reconstruction) of scenarios leading to dose.

Investigational Levels – External Dose Monitoring

The NRC has emphasized that the Investigational Levels in this program are not new dose limits but, as noted in ICRP Report 26, "Recommendations of the International Commission on Radiological Protection," Investigational Levels serve as check points above which the results are considered sufficiently important to justify investigation.

In cases where a worker's dose or the dose for a group of workers needs to exceed an Investigational Level, a new, higher Investigational Level may be established for that individual or group on the basis that it is consistent with good ALARA practices. Justification for new Investigational Levels should be documented.

When the cumulative annual exposure to a radiation worker exceeds Investigational Level I in Table M.1 (i.e., 10% of the annual limit for occupational exposure), the RSO or the RSO's designee should investigate the exposure and review the actions that might be taken to reduce the probability of recurrence. When the cumulative annual exposure exceeds Investigational Level II in Table M.1 (i.e., 30% of the annual limit for occupational exposure), the RSO or the RSO's designee will investigate the exposure and review actions to be taken to reduce the probability of recurrence, and management should review the report of the actions to be taken to reduce the probability of occurrence.

Table M.1 Investigational Levels		
Part of Body	Investigational Level I (mrem per year)	Investigational Level II (mrem per year)
whole body, head, trunk including male gonads, arms above the elbow, or legs above the knee	500 (5 mSv)	1500 (15 mSv)
hands, elbows, arms below the elbow, feet, knees, legs below the knee, or skin	5000 (50 mSv)	15,000 (150 mSv)
lens of the eye	1500 (15 mSv)	4500 (45 mSv)

Review and record on NRC Form 5, "Current Occupational External Radiation Exposures," or an equivalent form (e.g., dosimeter processor's report), results of personnel monitoring. Take the actions listed below when the investigation levels listed in Table M.1 are reached:

- Personnel dose less than Investigational Level I.

    Except when deemed appropriate by the RSO or the RSO's designee, no further action will be taken if an individual's dose is less than Table M.1 values for Investigational Level I.

- Personnel dose equal to or greater than Investigational Level I but less than Investigational Level II.

    When the dose of an individual equals or exceeds Investigational Level I, the RSO or the RSO's designee should conduct a timely investigation and review the actions that might be taken to reduce the probability of recurrence, following the period when the dose was recorded. If the dose does not equal or exceed Investigational Level II, no action related specifically to the exposure is required unless deemed appropriate by the RSO or the RSO's designee. Consider investigating the factors that led to the radiation exposure and the radiation doses and work habits of other individuals engaged in similar tasks to determine if improvements or additional safety measures are needed to reduce exposures. Evaluate, in the context of ALARA program quality, and record the results of investigations and evaluations.

- Personnel dose equal to or greater than Investigational Level II.

    The RSO should investigate in a timely manner the causes of all personnel doses equaling or exceeding Investigational Level II. The RSO should consider actions to reduce the probability of occurrence, and a report of the actions should be reviewed by the licensee's management at its first meeting following completion of the investigation.

- Reestablishment of Investigational Level II to a level above that listed in Table M.1.

Declared Pregnancy and Dose to Embryo/Fetus

Section 10 CFR 20.1208 states that the licensee shall ensure that the dose to an embryo/fetus during the entire pregnancy, due to occupational exposure of a declared pregnant woman, does not exceed 0.5 rem (5 mSv). This includes exposure to accelerator-produced radioactive materials or discrete sources of radium-226 or radiation from these materials. The licensee shall make efforts to avoid substantial variation above a uniform monthly exposure rate to a declared pregnant woman. If the pregnancy is declared in writing and includes the worker's estimated date of conception, the dose equivalent to an embryo/fetus shall be taken as the sum of:

- The deep-dose equivalent to the declared pregnant woman, and

- The dose equivalent to the embryo/fetus from radionuclides in the embryo/fetus and radionuclides in the declared pregnant woman.

References:

- Methods for calculating the radiation dose to the embryo/fetus can be found in Regulatory Guide 8.36, "Radiation Dose to the Embryo/Fetus."

- NUREG/CR-5631, PNL-7445, Rev. 2, "Contribution of Maternal Radionuclide Burdens to Prenatal Radiation Doses" (1996).

Internal Exposure

With respect to internal exposure, licensees are required to monitor occupational intake of radioactive material, including accelerator-produced radioactive materials or discrete sources of radium-226, and assess the resulting dose if it appears likely that personnel will receive greater than 10% of the annual limit on intake (ALI) from intakes in 1 year (10 CFR 20.1502). Terms for radionuclide intakes by means of inhalation and ingestion (i.e., derived air concentration (DAC) and ALI) are provided in 10 CFR Part 20.

The DAC for each class of radionuclide is the concentration of airborne radioactivity in μCi/ml that, if an occupational worker were to be continuously exposed to it for 2,000 hours (1 year), would result in either a CEDE of 5 rem (0.05 Sv) to the whole body or a committed dose equivalent of 50 rem (0.5 Sv) to any individual organ or tissue, with no consideration for the contribution of external dose. The ALI and DAC for each radionuclide in a specific chemical form are listed in Appendix B of 10 CFR Part 20.

For each class of each radionuclide, there are two ALIs, one for ingestion and one for inhalation. The ALI is the quantity of radioactive material that, if taken into the body of an adult worker by the corresponding route, would result in a committed effective dose equivalent of 5 rem (0.05 Sv) or a committed dose equivalent of 50 rem (0.5 Sv) to any individual organ or tissue; again, with no consideration for the contribution of external dose.

The total effective dose equivalent concept makes it possible to combine both the internal and external doses in assessing the overall risk to the health of an individual. The ALI and DAC numbers in 10 CFR Part 20 reflect the doses to all principal organs that are irradiated. The ALI and DAC were derived by multiplying a unit intake by the appropriate organ weighting factors (W_T), for the organs specifically targeted by the radionuclide compound, and then summing the organ-weighted doses to obtain a whole-body risk-weighted "effective dose." Per 10 CFR Part 20, Appendix B, when an ALI is defined by the stochastic dose limit, this value alone is given. When the ALI is determined by the nonstochastic dose limit to an organ, the organ or tissue to which the limit applies is shown, and the ALI for the stochastic limit is shown in parentheses.

The types and quantities of radioactive material manipulated at most medical facilities do not provide a reasonable possibility for an internal intake by workers. However, uses such as preparing radioiodine capsules from liquid solutions, and opening and dispensing radioiodine from vials containing millicurie quantities, require particular caution. To monitor internal exposures from such operations, a routine bioassay program to periodically monitor workers should be established.

If a licensee determines that a program for performing thyroid uptake bioassay measurements is necessary, a program should be established. The program should include:

- adequate equipment to perform bioassay measurements,

- procedures for calibrating the equipment, including factors necessary to convert counts per minute into becquerel or microcurie units,

- the technical problems commonly associated with performing thyroid bioassays (e.g., statistical accuracy, attenuation by neck tissue),

- the interval between bioassays,

- action levels, and

- the actions to be taken at those levels.

For guidance on developing bioassay programs and determination of internal occupational dose and summation of occupational dose, refer to Regulatory Guide 8.9, Revision 1, "Acceptable Concepts, Models, Equations and Assumptions for a Bioassay Program," dated July 1993; Regulatory Guide 8.34, "Monitoring Criteria and Methods to Calculate Occupational Radiation Doses," dated July 1992; and NUREG-1400, "Air Sampling in the Workplace," dated September 1993.

Recordkeeping

Records of measurement data, calculations of intakes, and methods for calculating dose must be maintained as required by 10 CFR 20.2106. For additional information on recordkeeping and reporting occupational exposure data, including intakes, refer to Revision 1 of Regulatory Guide 8.7, "Instructions for Recording and Reporting Occupational Radiation Exposure Data." Because these documents were developed before the EPAct, they may not include examples or values for all accelerator-produced radioactive materials or discrete sources of Ra-226 that NRC now regulates.

Summation of External and Internal Doses

Pursuant to 10 CFR 20.1202, the external and internal doses must be summed if required to monitor both under 10 CFR 20.1502.

Two documents that contain helpful information regarding occupational doses are:

- NRC Regulatory Issue Summary 2002-06, "Evaluating Occupational Dose for Individuals Exposed to NRC-Licensed Material and Medical X-Rays," and

- NRC Regulatory Issue Summary 2002-10, "Revision of Skin Dose Unit in 10 CFR Part 20."

Copies of Regulatory Issue Summaries are available on the NRC Web site in the Electronic Reading Room at http://www.nrc.gov/reading-rm/doc-collections/gen-comm/reg-issues/.

APPENDIX N

Model Emergency Procedures

This Appendix was originally developed for medical uses only. With implementation of the EPAct and addition of NARM materials and nonmedical uses, such as authorizations under 10 CFR 30.32(j) to medical use licenses, the procedures may have to be supplemented in this Appendix to address the new materials and nonmedical uses.

Model Emergency Procedures

Model Spill/Contamination, Emergency Surgery, and Autopsy Procedures

With the implementation of the EPAct, the NRC now has regulatory authority over accelerator-produced radioactive materials and discrete sources of radium-226. Therefore, after NRC's waiver of August 31, 2005, is terminated for medical use facilities, procedures for responding to spills, emergency surgery, and autopsies must also include responses when accelerator-produced radioactive materials and discrete sources of radium-226 are involved. The NRC waiver that applied to Government agencies, Federally recognized Indian tribes, Delaware, the District of Columbia, Puerto Rico, the U.S. Virgin Islands, Indiana, Wyoming, and Montana was terminated on November 30, 2007. The NRC Regional Offices should be contacted to confirm the waiver termination date for other medical use facilities.

Model Spill/Contamination Procedures – Low- and High-Dose Unsealed Sources (this now includes spills of and contamination from accelerator-produced radioactive materials or unsealed discrete sources of radium-226)

This model provides acceptable procedures for responding to medical use emergencies. This model does not address responding to emergencies associated with the production of PET radioactive drugs and their transfer. Applicants may either adopt this model or develop alternative procedures to meet the requirements of 10 CFR 20.1101. A medical use applicant that will produce PET radioactive drugs may need to supplement these model procedures to meet the requirements in 10 CFR 20.1101.

Minor Spills of Liquids and Solids (this now includes spills of and contamination from accelerator-produced radioactive materials or discrete sources of radium-226)

1. Notify persons in the area that a spill has occurred.

2. Prevent the spread of contamination by covering the spill with absorbent paper.

3. Wear gloves and protective clothing such as a lab coat and booties, and clean up the spill using absorbent paper. Carefully fold the absorbent paper with the clean side out and place in a bag labeled "caution radioactive material" for transfer to a radioactive waste container. Also put contaminated gloves and any other contaminated disposable material in the bag.

4. Survey the area with a low-range radiation detection survey instrument sufficiently sensitive to detect the radionuclide. Check for removable contamination to ensure contamination levels are below trigger levels. Check the area around the spill. Also check hands, clothing, and shoes for contamination.

5. Report the incident to the RSO.

Major Spills of Liquids and Solids (this now includes spills of or contamination from accelerator-produced radioactive materials or discrete sources of radium-226)

1. Clear the area. Notify all persons not involved in the spill to vacate the room.

2. Prevent the spread of contamination by covering the spill with absorbent paper labeled "caution radioactive material," but do not attempt to clean it up. To prevent the spread of contamination, clearly indicate the boundaries of the spill and limit the movement of all personnel who may be contaminated.

3. Shield the source if possible. Do this only if it can be done without further contamination or a significant increase in radiation exposure.

4. Close the room and lock or otherwise secure the area to prevent entry.

5. Notify the RSO immediately.

6. Decontaminate personnel by removing contaminated clothing and flushing contaminated skin with lukewarm water, then washing with mild soap. If contamination remains, the RSO may consider inducing perspiration. Then wash the affected area again to remove any contamination that was released by the perspiration.

The decision to implement a major spill/contamination procedure instead of a minor spill/contamination procedure depends on many incident-specific variables, such as the number of individuals affected, other hazards present, likelihood of contamination spread, types of surfaces contaminated, and radiotoxicity of the spilled material.

For some spills of radionuclides with half-lives shorter than 24 hours and in amounts less than five times the lowest ALI, an alternative spill/contamination procedure may be to restrict access pending complete decay.

Note: A report to NRC may be required pursuant to 10 CFR 30.50.

Use Table N.1 as general guidance to determine whether a major spill/contamination procedure or a minor spill/contamination procedure will be implemented. All spills/contaminations of radium-226 will be considered major spills.

Table N.1 Relative Hazards of Common Radionuclides			
Radionuclide	Millicurie	Radionuclide	Millicurie
P-32	1	Tc-99m	100
Cr-51	100	In-111	10
Co-57	10	I-123	10
Co-58	10	I-125	.1
Fe-59	1	I-131	1
Co-60	1	Sm-153	10
Ga-67	10	Yb-169	10
Se-75	1	Hg-197	10
Sr-85	10	Au-198	10
Sr-89	1	Tl-201	100

Estimate the amount of radioactivity spilled. Initiate a major or minor spill/contamination procedure, based on the following information. Spills above these mCi amounts are considered major, and below these levels are considered minor. Spills involving curie quantities of PET radionuclides should initially be considered major spills; either downgrade to a minor spill after decay or restrict access pending complete decay.

Spill/Contamination Kit

Assemble a spill/contamination kit that may contain the following items:

- Disposable gloves and housekeeping gloves,
- Disposable lab coats,
- Disposable head coverings,
- Disposable shoe covers,
- Roll of absorbent paper with plastic backing,
- Masking tape,
- Plastic trash bags with twist ties,
- "Radioactive Material" labeling tape,
- Marking pen,
- Pre-strung "Radioactive Material" labeling tags,
- Contamination wipes,
- Instructions for "Emergency Procedures,"
- Clipboard with copy of Radioactive Spill Report Form,
- Pencil, and
- Appropriate survey instruments, including batteries.

Emergency Surgery of Patients Who Have Received Therapeutic Amounts of Radionuclides (this now includes therapeutic amounts of accelerator-produced radioactive materials or any discrete sources of radium-226)

The following procedures should be followed:

1. If emergency surgery is performed within the first 24 hours following the administration of I-131 sodium iodide, fluids (e.g., blood, urine) will be carefully removed and contained in a closed system.

2. Protective eye wear will be worn by the surgeon and any personnel involved in the surgical procedure for protection of the eyes from possible splashing of radioactive material and exposure from beta radiation (if applicable).

3. The radiation safety staff will direct personnel in methods to keep doses ALARA during surgical procedures.

4. If an injury occurs during surgery that results in a cut or tear in the glove used, the individual involved will be monitored to determine if radioactive material was introduced into the wound. The RSO will be informed of any possible radiation hazard.

Autopsy of Patients Who Have Received Therapeutic Amounts of Radionuclides (this now includes therapeutic amounts of accelerator-produced radioactive materials or any discrete sources of radium-226)

The following procedures should be followed:

1. Immediately notify the AU in charge of the patient and the RSO upon death of a therapy patient.

2. An autopsy will be performed only after consultation and permission from the RSO. Radiation safety staff should evaluate the radiation hazard(s), direct personnel in safety and protection, and suggest suitable procedures in order to keep doses ALARA during the autopsy.

3. Protective eye wear should be worn by the pathologist and assisting staff for protection from possible splashing of radioactive material. Consider the need for protection against exposure from high-energy beta rays in cases involving therapy with P-32 and Y-90.

4. Remove tissues containing large activities early to help reduce exposure of autopsy personnel. Shield and dispose of contaminated tissues in accord with license conditions. In some cases, exposure reduction may be accomplished by removing tissues for dissection to a location where the exposure rate is lower.

5. If an injury occurs during the autopsy that results in a cut or tear in the glove, monitor the wound and decontaminate as appropriate to the situation; inform radiation safety staff.

References: NRCP Report No. 111, "Developing Radiation Emergency Plans for Academic, Medical, and Industrial Facilities," 1991, contains helpful information. It is available from the National Council on Radiation Protection and Measurements, 7910 Woodmont Avenue, Suite 400, Bethesda, Maryland 20814-3095. NCRP's telephone numbers are: (301) 657-2652 or 1-800-229-2652.

APPENDIX O

Model Procedures for
Ordering and Receiving Packages

This Appendix was originally developed for medical uses only. With the implementation of the EPAct and the addition of NARM materials and nonmedical uses, such as authorizations under 10 CFR 30.32(j) to medical use licenses, the procedures in this Appendix may have to be supplemented to address the new materials and nonmedical uses.

Model Procedures for Ordering and Receiving Packages

This model provides acceptable procedures for ordering and receiving packages containing licensed material. As a result of the EPAct, licensed materials now include accelerator-produced radioactive materials and discrete sources of radium-226. A medical use applicant that requests authorization for the production and noncommercial transfer of PET radioactive drugs may need to supplement these model procedures by developing procedures for filling orders for these drugs from other consortium members, to meet the requirements in 10 CFR 30.41, 30.32(j), and 30.34(j).

Applicants may either adopt this model or develop alternative procedures.

Model Guidance

- Authorize, through a designee (e.g., RSO), each order of radioactive materials, including orders of accelerator-produced radioactive materials and discrete sources of radium-226, and ensure that the requested materials and quantities are authorized by the license for use by the requesting AU and that possession limits are not exceeded.

- Establish and maintain a system for ordering and receiving radioactive material; include the following information:

  — Records that identify the AU or department, radionuclide, physical and/or chemical form, activity, and supplier;

  — Confirmation, through the above records, that material received was ordered through proper channels.

  — When ordering PET radioactive drugs produced under 10 CFR 30.32(j), confirm that the medical use licensee is a member of the consortium.

- For deliveries during normal working hours, tell carriers to deliver radioactive packages directly to a specified area.

- For deliveries during off-duty hours, tell security personnel or other designated persons to accept delivery of radioactive packages in accordance with procedures outlined in the sample memorandum for delivery of packages to the Nuclear Medicine Division, provided below. Develop a similar memorandum for delivery of packages to other divisions.

APPENDIX O

Sample Memorandum

MEMO TO: Chief of Security
FROM: Radiation Safety Officer
SUBJECT: Receipt of Packages Containing Radioactive Material

The security guard on duty will accept delivery of radioactive material that arrives outside normal working hours. Packages will be taken immediately to the Nuclear Medicine Division, Room ___. Unlock the door, place the package on top of the counter, and relock the door.

If the package appears to be damaged, immediately contact one of the individuals identified below. Ask the carrier to remain at the hospital until it can be determined that neither the driver nor the delivery vehicle is contaminated.

If you have any questions concerning this memorandum, please call our hospital Radiation Safety Officer, at extension _____.

	Name	**Home Telephone**
Radiation Safety Officer:		
Director of Nuclear Medicine:		
Nuclear Medicine Technologist Supervisor:		
Nuclear Medicine Technologist on call		
(call/page operator at extension _____)		
Nuclear Medicine Physician on call		
(call/page operator at extension _____)		

APPENDIX P

Model Procedure for Safely Opening Packages Containing Radioactive Material

Model Procedure for Safely Opening Packages Containing Radioactive Material

With the implementation of the EPAct, the NRC now has regulatory authority over accelerator-produced radioactive materials and discrete sources of radium-226. Therefore, after NRC's waiver of August 31, 2005, is terminated for medical use facilities, all procedures for the safe opening of packages containing radioactive materials must also be used for packages containing accelerator-produced radioactive materials and discrete sources of radium-226. The NRC waiver that applied to Government agencies, Federally recognized Indian tribes, Delaware, the District of Columbia, Puerto Rico, the U.S. Virgin Islands, Indiana, Wyoming, and Montana was terminated on November 30, 2007. The NRC Regional Offices should be contacted to confirm the waiver termination date for other medical use facilities.

This model provides acceptable procedures for opening packages containing radioactive material. Applicants may either adopt this model procedure or develop an alternative procedure to meet the requirements of 10 CFR 20.1906.

Special requirements must be followed for packages containing quantities of radioactive material in excess of the Type A quantity limits specified in Table A.1 of 10 CFR Part 71. Such packages must be received expeditiously when the carrier offers them for delivery or when the carrier notifies the licensee that the package has arrived at the carrier's terminal. For these and other packages for which monitoring is required, check for external radiation levels and surface contamination within 3 hours of receipt (if received during working hours) or no later than 3 hours from the beginning of the next working day (if received after working hours), in accordance with the requirements of 10 CFR 20.1906(c). The appropriate NRC Regional Office and the final delivery carrier must be notified if the following conditions apply:

- Removable radioactive surface contamination exceeds the limits of 10 CFR 71.87(i) and

- External radiation levels exceed the limits of 10 CFR 71.47.

Model Procedure

1. Put on gloves to prevent hand contamination.

2. Visually inspect the package for any sign of damage (e.g., wet or crushed). If damage is noted, stop the procedure and immediately notify the RSO (or the designee of the RSO if the RSO is not present).

3. Monitor the external surfaces of a labeled[1] package for radioactive contamination, unless the package contains only radioactive material in the form of a gas or in special form, as defined in 10 CFR 71.4.

4. Monitor the external surfaces of a labeled[1] package for radiation levels, unless the package contains quantities of radioactive material that are less than or equal to the Type A quantity, as defined in 10 CFR 71.4 and Table A to 10 CFR Part 71.

---

[1]Labeled with a Radioactive White I, Yellow II, or Yellow III label as specified in DOT regulations.

5. Monitor all packages known to contain radioactive material for radioactive contamination and radiation levels, if there is evidence of degradation of package integrity, such as packages that are crushed, wet, or damaged.

6. Remove the packing slip.

7. Open the outer package, following any instructions that may be provided by the supplier.

8. Open the inner package and verify that the contents agree with the packing slip.

9. Check the integrity of the final source container. Notify the RSO (or the RSO's designee) of any broken seals or vials, loss of liquid, condensation, or discoloration of the packing material.

10. If there is any reason to suspect contamination, wipe the external surface of the final source container and remove the wipe sample to a low-background area. Assay the wipe sample to determine if there is any removable radioactivity. An appropriate instrument with sufficient sensitivity will be used to assay the sample. For example, a NaI(T1) crystal and rate meter, a liquid scintillation counter, or a proportional flow counter may be used for these assays. The detection efficiency will be determined to convert wipe sample counts per minute to disintegrations per minute. *Note: a dose calibrator is not sufficiently sensitive for this measurement.* Take precautions against the potential spread of contamination.

11. Check the user request to ensure that the material received is the material that was ordered.

12. Monitor the packing material and the empty packages for contamination with a radiation detection survey meter before discarding. If contaminated, treat this material as radioactive waste. If not contaminated, remove or obliterate the radiation labels before discarding in in-house trash.

13. Make a record of the receipt.

For packages received under the general license in 10 CFR 31.11, implement the following procedure for opening each package:

1. Visually inspect the package for any sign of damage (e.g., wet or crushed). If damage is noted, stop the procedure and notify the RSO (or the RSO's designee) immediately.

2. Check to ensure that the material received is the material that was ordered.

For "empty" transport radiation shields being returned from consortium members, implement the following procedure for opening each package:

1. Monitor the package for radioactive contamination .

2. Visually inspect the contents to ensure that the transport radiation shield is empty. Notify the RSO if the transport radiation shield is not empty.

APPENDIX Q

Model Leak Test Program

Model Leak Test Program

With the implementation of the EPAct, the NRC now has regulatory authority over accelerator-produced radioactive materials and discrete sources of radium-226. Therefore, all leak test procedures must also be applied to these materials used after NRC's waiver of August 31, 2005, is terminated for medical use facilities. The NRC waiver that applied to Government agencies, Federally recognized Indian tribes, Delaware, the District of Columbia, Puerto Rico, the U.S. Virgin Islands, Indiana, Wyoming, and Montana was terminated on November 30, 2007. The NRC Regional Offices should be contacted to confirm the waiver termination date for other medical use facilities.

This model provides acceptable procedures for sealed source leak testing and analysis. Applicants may either adopt these model procedures or develop alternative procedures.

Facilities and Equipment

- To ensure achieving the required sensitivity of measurements, leak tests should be analyzed in a low-background area.

- Consider using a NaI(Tl) well counter system with a single or multichannel analyzer to analyze samples obtained from gamma-emitting sources (e.g., Cs-137).

- Consider using a liquid scintillation or gas-flow proportional counting system to analyze samples obtained from beta-emitting or alpha-emitting sources (e.g., Sr-90).

- Instrumentation used to analyze leak test samples must be capable of detecting 185 Bq (0.005 μCi) of radioactivity.

Model Procedure for Performing Gaseous Emanation Test for Individual Ra-226 Sealed Sources (ANSI/HPS N43.6-1997, "Sealed Radioactive Sources - Classification," Appendix A, Section A.2.1.5)

- For each source to be tested, list identifying information such as sealed source serial number, radionuclide, and activity.

- Number each container to correlate information for each source.

- Wear gloves.

- Put each Ra-226 sealed source into a separate small, gas-tight container with activated carbon or two cotton filters.

- Leave source in airtight container for 24 hours.

- Remove source.

- Close container.

- Measure immediately the activity of the Absorber. (See "Model Procedure for Analysis of Gaseous Emanation and Leak Test" below for: how to analyze the absorber, required records, leakage determination, and required response to a leaking source.)

• If activity corresponds to less than 1 nanocurie of radon or daughter products, the source is considered leak-free.

Model Procedure for Performing Leak Testing (on all sealed sources except individual Ra-226 sealed sources)

• For each source to be tested, list identifying information such as sealed source serial number, radionuclide, and activity.

• Use a separate wipe sample (e.g., cotton swab or filter paper) for each source.

• Number each wipe to correlate identifying information for each source.

• Wear gloves.

• Obtain samples at the most accessible area where contamination would accumulate if the sealed source were leaking.

Model Procedure for Analysis of Gaseous Emanation and Leak Test (for all sources)

• Measure the background count rate and record.

• Check the instrument's counting efficiency, using either a standard source of the same radionuclide as the source being tested or one with similar energy characteristics. Accuracy of standards should be within ± 5% of the stated value and traceable to a primary radiation standard, such as those maintained by NIST.

• If the sensitivity of the counting system is unknown, the minimum detectable activity (MDA) should be determined. The MDA may be determined using the following formula:

$$MDA = \frac{3 + 4.65(bkg/t)^{1/2}}{E}$$

where:

MDA	=	minimum detectable activity in disintegrations per minute (dpm)
bkg	=	background count rate in counts per minute (cpm)
t	=	background counting time in minutes
E	=	detector efficiency in counts per disintegration

For example:

where:

bkg	=	200 cpm
E	=	10%, or 0.1
t	=	2 minutes

$$MDA = \frac{3 + 4.65(200 \text{ cpm}/2 \text{ minutes})^{1/2}}{(0.1)}$$
$$= 495 \text{ dpm}$$

- Calculate efficiency of the instrument.

  For example,

$$Eff = \frac{[(cpm\ from\ std) - (cpm\ from\ bkg)]}{(activity\ of\ std\ in\ microcuries)}$$

  where: Eff = efficiency, in cpm/microcurie,
                cpm = counts per minute,
                std = standard, and
                bkg = background.

- Analyze each wipe (or absorber for a Ra-226 sealed source) sample to determine net count rate.

- For each sample, calculate the activity in microcuries and record.

- The activity on the wipe (or absorber) sample is given by:

$$\frac{[(cpm\ from\ wipe\ sample) - (cpm\ from\ bkg)]}{(Eff\ in\ cpm/microcuries)}$$

$$= activity\ on\ wipe\ sample\ in\ microcuries$$

- Leak test records (which include the gaseous emanation test) will be retained in accordance with 10 CFR 35.2067 or standard license condition for 3 years. Licensees should include the following in records:

  — The model number and serial number (if assigned) of each source tested;

  — The identity of each source radionuclide and its estimated activity;

  — The measured activity of each test sample expressed in microcuries;

  — A description of the method used to measure each test sample;

  — The date of the test; and

  — The name of the individual who performed the test.

- If the wipe test reveals 185 Bq (0.005μCi) [or 37 Bq (1 nano Ci) of radon] or greater:

  — Immediately withdraw the sealed source from use and store it, dispose of it, or cause it to be repaired in accordance with the requirements in 10 CFR Parts 20 and 30 [10 CFR 35.67 or standard license condition].

  — File a report within 5 days of the leak test in accordance with 10 CFR 35.3067 or standard license condition.

APPENDIX R

Model Procedure for Area Surveys

This Appendix was originally developed for medical uses only. With the implementation of the EPAct and the addition of NARM materials and nonmedical uses, such as authorizations under 10 CFR 30.32(j) to medical use licenses, the procedures in this Appendix may have to be supplemented to address the new materials and nonmedical uses.

Model Procedure for Area Surveys

With the implementation of the EPAct, the NRC now has regulatory authority over accelerator-produced radioactive materials and discrete sources of radium-226. Therefore, after NRC's waiver of August 31, 2005, is terminated for medical use facilities, all procedures for area surveys must also be used for surveying areas where accelerator-produced radioactive materials and discrete sources of radium-226 are (were) present. The NRC waiver that applied to Government agencies, Federally recognized Indian tribes, Delaware, the District of Columbia, Puerto Rico, the U.S. Virgin Islands, Indiana, Wyoming, and Montana was terminated on November 30, 2007. The NRC Regional Offices should be contacted to confirm the waiver termination date for other medical use facilities.

This model provides acceptable procedures for medical use area surveys. This model addresses some, but not all, area survey procedures associated with the production of PET radioactive drugs and their transfer or with other nonmedical uses. Applicants may either adopt these model procedures or develop alternative procedures to meet the requirements of 10 CFR 20.1101, 10 CFR 20.1501, and 10 CFR 35.70. A medical use applicant that will produce or transfer PET radioactive drugs or have other nonmedical uses may need to supplement the model procedures for those activities to meet the requirements of 10 CFR 20.1101 and 10 CFR 20.1501. Guidance for developing alternate trigger levels for contamination in restricted areas is included below.

Ambient Radiation Level Surveys (this now includes surveys for accelerator-produced radioactive materials or discrete sources of radium-226)

Procedures for ambient radiation level surveys (reference 10 CFR 20.1101, 10 CFR 20.1501, and 10 CFR 35.70):

- Perform surveys of dose rates in locations where:
  - Workers are exposed to radiation levels that might result in radiation doses in excess of 10% of the occupational dose limits; or
  - An individual is working in an environment with a dose rate of 2.5 mrem/hour or more (5 rem/year divided by 2,000 hour/year).

- Section 10 CFR 20.1301 requires that the TEDE to an individual member of the public from the licensed operation does not exceed 1 mSv (0.1 rem) in a year, and that the dose in any unrestricted area from external sources does not exceed 0.02 mSv (0.002 rem) in any 1 hour. Appropriate surveys will be conducted to ensure that the requirements of 10 CFR 20.1301 are met. *Note:* As a result of the EPAct, licensed operations now include the possession and use of accelerator-produced radioactive materials and discrete sources of radium-226.

- Perform radiation level surveys with a survey meter sufficiently sensitive to detect 0.1 milliroentgen (mR) per hour in the following areas, at the frequency specified:
  - Survey at the end of each day of use all radiopharmaceutical elution, preparation, assay and administration areas (except patient rooms, which will be surveyed at the end of the therapy instead of on the day of administration) when using

radiopharmaceuticals requiring a written directive (e.g., all therapy dosages and any iodine-131 dosage exceeding 30 μCi).

— Survey monthly all laboratory areas where only small quantities of gamma-emitting radioactive material are used (< 200 μCi at a time).

— Survey weekly all radionuclide use, storage, and waste storage areas. If diagnostic administrations are occasionally made in patients' rooms (e.g., bone scan injections, Tc-99m heart agents) and special care is taken to remove all paraphernalia, those rooms need not be surveyed.

— Survey quarterly all sealed-source and brachytherapy-source storage areas.

• If trigger levels are exceeded, follow internal procedures for responding and investigating what caused the trigger to be tripped. Examples of trigger levels for restricted and unrestricted areas are presented in Table R.1.

Table R.1 Ambient Dose Rate Trigger Levels		
Type of Survey	Area Surveyed	Trigger Level
Ambient Dose Rate	Unrestricted	0.1 mR/hr
Ambient Dose Rate	Restricted	5.0 mR/hr

Contamination Surveys (this now includes surveys for accelerator-produced radioactive materials or discrete sources of radium-226)

Facilities and equipment for contamination surveys:

To ensure achieving the required sensitivity of measurements, analyze survey samples in a low-background area. Table K-1, entitled "Stationary Instruments Used to Measure Wipe, Bioassay, and Effluent Samples," in Appendix K provides examples of appropriate instruments.

Perform contamination surveys using instruments suitable for removable and fixed contamination to identify areas of contamination that might result in doses to workers or to the public. Removable contamination can be detected and measured by conducting a wipe test of the surface, counted in an appropriate counting instrument, such as a liquid scintillation counter, a sodium iodide or germanium gamma counter, or a proportional alpha/beta counter.

Procedures for contamination surveys:

• Contamination surveys are performed in areas where unsealed forms of materials, including unsealed accelerator-produced radioactive materials or unsealed discrete sources of radium-226, are used:

— To evaluate radioactive contamination that could be present on surfaces of floors, walls, laboratory furniture, and equipment;

— After any spill or contamination event;

— When procedures or processes have changed;

—   To evaluate contamination of users and the immediate work area, at the end of the day, when licensed material is used;

—   In unrestricted areas at frequencies consistent with the types and quantities of materials in use, but not less frequently than monthly; and

—   In areas adjacent to restricted areas and in all areas through which licensed materials are transferred and temporarily stored before shipment.

• Use methods for conducting surveys for removable contamination that are sufficiently sensitive to detect contamination for those radionuclides in use and for which the most restrictive limits apply, as listed in Tables R.2 for restricted areas and R.3 for unrestricted areas (e.g., 200 dpm/100 cm^2 for isotopes of iodine-131 in unrestricted areas). Removable contamination survey samples should be measured in a low-background area. The following areas and frequencies should be followed:

—   Removable contamination surveys weekly for radiopharmaceutical elution, preparation, assay, and administration areas. If diagnostic administrations are occasionally made in patients' rooms (e.g., bone scan injections, Tc-99m heart agents), with special care taken to remove all paraphernalia, those rooms need not be surveyed.

—   Removable contamination surveys monthly of laboratory areas where only small quantities of photon-emitting radioactive material are used (<200 microcuries at a time).

—   Removable contamination surveys weekly for radionuclide storage and radionuclide waste storage areas.

• A radioactive source with a known amount of activity should be used to convert sample measurements (usually in cpm) to dpm.

• The area should be either decontaminated, shielded, or posted and restricted from use if it cannot be decontaminated.

• If trigger levels are exceeded, follow internal procedures for responding and investigating what caused the trigger to be tripped. Examples of trigger levels for restricted areas are presented in Table R.2. Contamination found in unrestricted areas and on personal clothing will be immediately decontaminated to background levels.

Table R.2	Surface Contamination Levels in Restricted Areas (dpm/100 cm^2)		
Area, clothing	alpha emitters	P-32, Co-58, Fe-59, Co-60, Se-75, Sr-85, Y-90, In-111, I-123, I-125, I-131, Sm-153, Yb-169, Lu-177, Au-198	Cr-51, Co-57, Ga-67, Tc-99m, Hg-197, Tl-201
Restricted areas, protective clothing used only in restricted areas	200	2000	20000

Table R.3 Surface Contamination Levels in Unrestricted Areas (dpm/100 cm²)			
Nuclide[1]	Average[2,3,6]	Maximum[2,4,6]	Removable[2,5,6]
I-125, I-126, I-131, I-133, Sr-90	1000	3000	200
Beta-gamma emitters (nuclides with decay modes other than alpha emission or spontaneous fission) except Sr-90 and others noted above.	5000	15000	1000
Ra-226	100	300	20

[1] Where surface contamination by multiple nuclides exists, the limits established for each nuclide should apply independently.

[2] As used in this table, dpm means the rate of emission by radioactive material, as determined by correcting the counts per minute observed by an appropriate detector for background, efficiency, and geometric factors associated with the instrumentation.

[3] Measurements of average contaminants should not be averaged over more than 1 square meter. For objects of less surface area, the average should be derived for each such object.

[4] The maximum contamination level applies to an area of not more than 100 cm².

[5] The amount of removable radioactive material per 100 cm² of surface area should be determined by wiping that area with filter or soft absorbent paper, applying moderate pressure, and assessing the amount of radioactive material on the wipe with an appropriate instrument of known efficiency. When removable contamination on objects of less surface area is determined, the pertinent levels should be reduced proportionally and the entire surface should be wiped.

[6] The average and maximum radiation levels associated with surface contamination resulting from beta-gamma emitters should not exceed 0.2 millirad/hour at 1 centimeter and 1.0 millirad/hour at 1 centimeter, respectively, measured through not more than 7 milligrams per square centimeter of total absorber.

Establishing Alternate Trigger Levels for Restricted Areas

The following guidance is provided for those applicants who plan to develop procedures for surveying and controlling contamination using action levels for controlling contamination that differ from those provided in Tables R.1 and R.2:

Alternate action levels for cleanup of contamination in restricted areas may be developed without prior NRC approval if:

- acceptable unrestricted area trigger levels are implemented (e.g., Tables R.1 and R.3);

- the action levels maintain occupational doses ALARA; and

- the action levels meet all other regulatory requirements (e.g., they should also be designed to minimize, to the extent practicable, contamination of the facility and the environment; facilitate eventual decommissioning; and minimize, to the extent practicable, the generation of radioactive waste).

Alternate Survey Frequency

A sample alternate survey frequency is described below using Tables R.4, R.5, and R.6. The objective is to determine how often to survey the laboratory. To do this, multiply the activity range for the appropriate group under LOW, MEDIUM, and HIGH survey frequency by the appropriate Modifying Factor to construct a new set of mCi ranges for LOW, MEDIUM, and HIGH survey frequency. For instance, if 30 millicuries of iodine-131 is used in the hot laboratory, the survey frequency for the hot laboratory would be daily; since the group for iodine-131 is Group 2, the survey frequency category for an activity of greater than 10 millicuries is high, and the modifying factor is 1.

Table R.4 Isotope Groups	
Group 1	Pb-210 Po-210 Ra-223 Ra-226 Ra-228 Ac-227 Th-230 Pa-231 Pu-238 Am-241 Am-243 Cm-242 Cm-243 Cm-244 Cm-245 Cm-246 Cf-249 Cf-250 Cf-252 Ra-226
Group 2	Na-22 Cl-36 Ca-45 Sc-46 Mn-54 Co-56 Co-60 Sr-89 Sr-90 Y-91 Zr-95 Ru-106 Ag-110m Cd-115m In-114m Sb-124 Sb-125 Te-127m Te-129m I-124 I-125 I-126 I-131 I-133 Cs-134 Cs-137 Ba-140 Ce-144 Eu-152 (13 y) Eu-154 Tb-160 Tm-170 Hf-181 Ta-182 Ir-192 Tl-204 Bi-207 Bi-210 At-211 Pb-212 Ra-224 Ac-228 Pa-230
Group 3	Be-7 C-14 F-18 Na-24 Cl-38 Si-31 P-32 S-35 Ar-41 K-42 K-43 Ca-47 Sc-47 Sc-48 V-48 Cr-51 Mn-52 Mn-56 Fe-52 Fe-55 Fe-59 Co-57 Co-58 Ni-63 Ni-65 Cu-64 Zn-65 Zn-69m Ga-72 As-73 As-74 As-76 As-77 Se-75 Br-82 Kr-85m Kr-87 Rb-86 Sr-85 Sr-91 Y-90 Y-92 Y-93 Zr-97 Nb-93m Nb-95 Mo-99 Tc-96 Tc-97m Tc-97 Tc-99 Ru-97 Ru-103 Ru-105 Rh-105 Pd-103 Pd-109 Ag-105 Ag-111 Cd-109 Cd-115 In-115m Sn-113 Sn-125 Sb-122 Te-125m Te-127 Te-129 Te-31m Te-132 I-130 I-132 I-134 I-135 Xe-135 Cs-131 Cs-136 Ba-31 La-140 Ce-141 Ce-143 Pr-142 Pr-143 Nd-147 Nd-149 Pm-147 Pm-149 Sm-151 Sm-153 Eu-152 Eu-155 Gd-153 Gd-159 Dy-165 Dy-166 Ho-166 Er-169 Er-171 (9.2 hr) Tm-171 Yb-175 Lu-177 W-181 W-185 W-187 Re-183 Re-186 Re-188 Os-185 Os-191 Os-193 Ir-190 Ir-194 Pt-191 Pt-193 Pt-197 Au-196 Au-198 Au-199 Hg-197 Hg-197m Hg-203 Tl-200 Tl-201 Tl-202 Pb-203 Bi-206 Bi-212 Rn-220 Rn-222
Group 4	H-3 O-15 Ar-37 Co-58m Ni-59 Zn-69 Ge-71 Kr-85 Sr-85m Rb-87 Y-91m Zr-93 Nb-97 Tc-96m Tc-99m Rh-103m In-113m I-129 Xe-131m Xe-133 Cs-134m Cs-135 Sm-147 Re-187 Os-191m Pt-193m Pt-197m

Table R.5   Classification of Laboratories for Alternate Survey Frequency

Group	\multicolumn{3}{c}{Survey Frequency Category}		
	Low	Medium	High
1	<0.1 mCi	0.1 mCi to 1 mCi	>1 mCi
2	<1 mCi	1 mCi to 10 mCi	>10 mCi
3	<100 mCi	100 mCi to 1 Ci	>1 Ci
4	<10 Ci	10 Ci to 100 Ci	>100 Ci

Survey Frequency:

- Low – Not less than once a month;
- Medium – Not less than once per week;
- High – Not less than once per normal working day.

Proportional fractions are to be used for more than one isotope.

Table R.6   Modifying Factors for Alternate Survey Frequency

Modifying Factors	Factors
Simple storage	x 100
Very simple wet operations (e.g., preparation of aliquots of stock solutions)	x 10
Normal chemical operations (e.g., analysis, simple chemical preparations)	x 1
Complex wet operations (e.g., multiple operations, or operations with complex glass apparatus)	x 0.1
Simple dry operations (e.g., manipulation of powders) and work with volatile radioactive compounds	x 0.1
Exposure of nonoccupational persons (including patients)	x 0.1
Dry and dusty operations (e.g., grinding)	x 0.01

Contents of Survey Records

- A diagram of the area surveyed,
- A list of items and equipment surveyed,
- Specific locations on the survey diagram where wipe tests were taken,
- Ambient radiation levels with appropriate units,
- Contamination levels with appropriate units,
- Make and model number of instruments used,
- Background levels, and
- Name of the person making the evaluation and recording the results and date.

Record contamination levels observed and procedures followed for incidents involving contamination of individuals. Include names of individuals involved, description of work activities, calculated dose, probable causes (including root causes), steps taken to reduce future incidents of contamination, times and dates, and the surveyor's signature.

APPENDIX S

Model Procedures for Developing, Maintaining, and Implementing Written Directives

Model Procedures for Developing, Maintaining, and Implementing Written Directives

With the implementation of the EPAct, the NRC now has regulatory authority over accelerator-produced radioactive materials and discrete sources of radium-226. Therefore, the requirements for written directives and procedures to assure that administrations are in accordance with these written directives also apply to the medical use of accelerator-produced radioactive materials and discrete sources of radium-226 after NRC's waiver of August 31, 2005, is terminated for medical use facilities. The NRC waiver that applied to Government agencies, Federally recognized Indian tribes, Delaware, the District of Columbia, Puerto Rico, the U.S. Virgin Islands, Indiana, Wyoming, and Montana was terminated on November 30, 2007. The NRC Regional Offices should be contacted to confirm the waiver termination date for other medical use facilities.

This model provides acceptable procedures for administrations that require written directives (WDs). Applicants may either adopt this model procedure or develop their own procedure to meet the requirements of 10 CFR 35.40 and 10 CFR 35.41.

Written Directive Procedures

This model provides guidance to licensees and applicants for developing, maintaining, and implementing procedures for administrations that require WDs. This model does not restrict the use of other guidance in developing, implementing, and maintaining written procedures for administrations requiring a WD. Such procedures are to provide high confidence that the objectives specified in 10 CFR 35.41 will be met.

The WD must be prepared for any administration of I-131 sodium iodide greater than 1.11 MBq (30 µCi), any therapeutic dosage of a radiopharmaceutical, and any therapeutic dose of radiation from byproduct material. The WD must contain the information described in 10 CFR 35.40 and be retained in accordance with 10 CFR 35.2040.

Discussion

The administration of radioactive materials can be a complex process for many types of diagnostic and therapeutic procedures in nuclear medicine or radiation oncology departments. A number of individuals may be involved in the delivery process. For example, in an oncology department, when the authorized user (AU) prescribes a teletherapy treatment, the delivery process may involve a team of medical professionals such as an authorized medical physicist (AMP), a dosimetrist, and a radiation therapist. Treatment planning may involve a number of measurements, calculations, computer-generated treatment plans, patient simulations, portal film verifications, and beam-modifying devices to deliver the prescribed dose. Therefore, instructions must be clearly communicated to the professional team members with constant attention devoted to detail during the treatment process. Complicated processes of this nature require good planning and clear, understandable procedures. To help ensure that all personnel involved in the treatment fully understand instructions in the WD or treatment plan, the licensee should instruct all workers to seek guidance if they do not understand how to carry out the WD. Specifically, workers should ask if they have any questions about what to do or how it should be done before administration, rather than continuing a procedure when there is any doubt.

Licensees should also consider verification of WDs or treatment plans by at least one qualified person (e.g., an oncology physician, AMP, nuclear medicine technologist, or radiation therapist), preferably other than the individual who prepared the dose, the dosage, or the treatment plan.

The administration of radioactive materials, including the administration of accelerator-produced radioactive materials and discrete sources of radium-226, can involve a number of treatment modalities (e.g., radiopharmaceutical therapy, teletherapy, brachytherapy, gamma stereotactic radiosurgery (GSR), and future emerging technologies). For each such modality for which 10 CFR 35.40 requires, or would require, a WD (as defined in 10 CFR 35.2), the licensee should develop, implement, and maintain written procedures to meet the requirements and/or objectives of 10 CFR 35.40, 35.41, and 35.63, outlined below:

- Have an AU date and sign a WD, prior to the administration, that includes the information in 10 CFR 35.40(b), including the name of the patient or human research subject;

- Verify the identity of the patient or human research subject prior to each administration;

- Verify that the administration is in accordance with the treatment plan, if applicable, and the WD;

- Check both manual and computer-generated dose calculations;

- Verify that any computer-generated dose calculations are correctly transferred into the consoles of therapeutic medical devices; and

- Determine and record the activity of the radiopharmaceutical dosage or radiation dose before medical use.

The following procedures are provided as assistance in meeting the above objectives.

Procedures for Any Therapeutic Dose or Dosage of a Radionuclide, Including Doses or Dosages of Accelerator-Produced Radioactive Materials and Discrete Sources of Radium-226, or Any Dosage of Quantities Greater than 30 Microcuries of I-131 Sodium Iodide

Develop, implement, and maintain the following procedures to meet the objectives of 10 CFR 35.40 and 10 CFR 35.41:

- An AU must date and sign a WD prior to the administration of any dose or dosage. Written directives may be maintained in patients' charts.

- Prior to administering a dose or dosage, the identity of a patient or human research subject will be positively verified as the individual named in the WD. Examples of positive patient identity verification include examining the patient's ID bracelet, hospital ID card, driver's license, or Social Security card. Asking or calling the patient's name does not constitute positive patient identity verification.

- The specific details of the administration will be verified, including the dose or dosage, in accordance with the WD or treatment plan. All components of the WD (radionuclide, total dose or dosage, etc.) will be confirmed by the person administering the dose or dosage to verify agreement with the WD. Appropriate verification methods include: measuring the activity in the dose calibrator, checking the serial number of the sealed sources behind an appropriate shield, using color-coded sealed sources, or using clearly marked storage locations.

Additional Procedures for Sealed Therapeutic Sources and Devices Containing Sealed Therapeutic Sources (this now includes sources containing accelerator-produced radioactive materials or discrete sources of radium-226)

Licensees are required under 10 CFR 35.40 and 10 CFR 35.41 to have WDs for certain administrations of doses and to have procedures for administrations for which a WD is required. Model procedures for meeting these requirements appear below.

A.   To ensure that the dose is delivered in accordance with the WD, the AU (and the neurosurgeon for GSR therapy) must date and sign (indicating approval of) the treatment plan that provides sufficient information and direction to meet the objectives of the WD.

B.   For sealed sources inserted into the patient's body, radiographs or other comparable images (e.g., computerized tomography) will be used as the basis for verifying the position of the nonradioactive dummy sources and calculating the administered dose before administration. However, for some brachytherapy procedures, the use of various fixed geometry applicators (e.g., appliances or templates) may be required to establish the location of the temporary sources and to calculate the exposure time (or, equivalently, the total dose) required to administer the prescribed brachytherapy treatment. In these cases, radiographs or other comparable images may not be necessary, provided the position of the sources is known prior to insertion of the radioactive sources and calculation of the exposure time (or, equivalently, the total dose).

C.   Dose calculations will be checked before administering the prescribed therapy dose. An AU or a qualified person under the supervision of an AU (e.g., an AMP, oncology physician, dosimetrist, or radiation therapist), preferably one who did not make the original calculations, will check the dose calculations. Methods for checking the calculations include the following:

   1.   For computer-generated dose calculations, examining the computer printout to verify that correct input data for the patient was used in the calculations (e.g., source strength and positions).

   2.   For computer-generated dose calculations entered into the therapy console, verifying correct transfer of data from the computer (e.g., channel numbers, source positions, and treatment times).

   3.   For manually-generated dose calculations, verifying:

      a.   No arithmetical errors;

      b.   Appropriate transfer of data from the WD, treatment plan, tables, and graphs;

      c.   Appropriate use of nomograms (when applicable); and

      d.   Appropriate use of all pertinent data in the calculations.

The therapy dose will be manually calculated to a single key point and the results compared to the computer-generated dose calculations. If the manual dose calculations are performed using computer-generated outputs (or vice versa), verify the correct output from one type of calculation (e.g., computer) to be used as an input in another type of calculation (e.g., manual). Parameters such as the transmission factors for wedges and applicators and the source strength of the sealed source used in the dose calculations will be checked.

D. After implantation but before completion of the procedure: record in the WD the radionuclide, treatment site, number of sources, and total source strength and exposure time (or the total dose) as required by 10 CFR 35.40(b)(6). For example, after insertion of permanent implant brachytherapy sources, an AU should promptly record the actual number of radioactive sources implanted and the total source strength. The WD may be maintained in the patient's chart.

E. Acceptance testing will be performed by a qualified person (e.g., an AMP) on each treatment planning or dose calculating computer program that could be used for dose calculations. Acceptance testing will be performed before the first use of a treatment planning or dose calculating computer program for therapy dose calculations. Each treatment planning or dose calculating computer program will be assessed based on specific needs and applications. A check of the acceptance testing will also be performed after each source replacement or when spot check measurements indicate that the source output differs by more than 5% from the output obtained at the last full calibration corrected mathematically for radioactive decay.

F. Independent checks on full calibration measurements will be performed. The independent check will include an output measurement for a single specified set of exposure conditions and will be performed within 30 days following the full calibration measurements. The independent check will be performed by either:

    1. An individual who did not perform the full calibration (the individual will meet the requirements specified in 10 CFR 35.51) using a dosimetry system other than the one that was used during the full calibration (the dosimetry system will meet the requirements specified in 10 CFR 35.630); or

    2. An AMP (or an oncology physician, dosimetrist, or radiation therapist who has been properly instructed) using a thermoluminescence dosimetry service available by mail that is designed for confirming therapy doses and that is accurate within 5%.

G. For GSR, particular emphasis will be directed on verifying that the stereoscopic frame coordinates on the patient's skull match those of the treatment plan.

H. A physical measurement of the teletherapy output will be made under applicable conditions prior to administration of the first teletherapy fractional dose, if the patient's treatment plan includes: (1) field sizes or treatment distances that fall outside the range of those measured in the most recent full calibration; or (2) transmission factors for beam-modifying devices (except nonrecastable and recastable blocks, bolus and compensator materials, and split-beam blocking devices) not measured in the most recent full calibration measurement.

I.   A weekly chart check will be performed by a qualified person under the supervision of an AU (e.g., an AMP, dosimetrist, oncology physician, or radiation therapist) to detect mistakes (e.g., arithmetical errors, miscalculations, or incorrect transfer of data) that may have occurred in the daily and cumulative dose administrations from all treatment fields or in connection with any changes in the WD or treatment plan.

J.   Treatment planning computer systems using removable media to store each patient's treatment parameters for direct transfer to the treatment system will have each card labeled with the corresponding patient's name and identification number. Such media may be reused (and must be relabeled) in accordance with the manufacturer's instructions.

Review of Administrations Requiring a Written Directive (this now includes administrations of accelerator-produced radioactive materials or discrete sources of radium-226)

Conduct periodic reviews of each applicable program area (e.g., radiopharmaceutical therapy, high-dose-rate brachytherapy, implant brachytherapy, teletherapy, gamma stereotactic radiosurgery, and emerging technologies). The number of patient cases to be sampled should be based on the principles of statistical acceptance sampling and be representative of each treatment modality performed in the institution (e.g., radiopharmaceutical, teletherapy, brachytherapy and gamma stereotactic radiosurgery).

If feasible, the persons conducting the review should not review their own work. If this is not possible, two people should work together as a team to conduct the review of that work. Regularly review the findings of the periodic reviews to ensure that the procedures for administrations requiring a WD are effective.

As required by 10 CFR 35.41, a determination will be made as to whether the administered radiopharmaceutical dosage or radiation dose was in accordance with the WD or treatment plan, as applicable. When deviations from the WD are found, the cause of each deviation and the action required to prevent recurrence should be identified.

Reports of Medical Events (this now includes reports of events involving accelerator-produced radioactive materials or discrete sources of radium-226)

Notify by telephone the NRC Operations Center[1] no later than the next calendar day after discovery of a medical event and submit a written report to the appropriate NRC Regional Office listed in 10 CFR 30.6 within 15 days after the discovery of the medical event, as required by 10 CFR 35.3045. Also notify the referring physician and the patient as required by 10 CFR 35.3045.

---

[1]The commercial telephone number of the NRC Operations Center is (301) 816-5100. The Center will accept collect calls.

APPENDIX T

Model Procedures for Safe Use of Unsealed Licensed Material

This Appendix was originally developed for medical uses only. With the implementation of the EPAct and the addition of NARM materials and nonmedical uses, such as authorizations under 10 CFR 30.32(j) to medical use licenses, the procedures in this Appendix may have to be supplemented to address the new materials and nonmedical uses.

Model Procedures for Safe Use of Unsealed Licensed Material

With the implementation of the EPAct, the NRC now has regulatory authority over accelerator-produced radioactive materials and discrete sources of radium-226. Therefore, the procedures for the safe use of unsealed licensed material also apply to the medical use of accelerator-produced radioactive materials and discrete sources of radium-226 after NRC's waiver of August 31, 2005, is terminated for medical use facilities. The NRC waiver that applied to Government agencies, Federally recognized Indian tribes, Delaware, the District of Columbia, Puerto Rico, the U.S. Virgin Islands, Indiana, Wyoming, and Montana was terminated on November 30, 2007. The NRC Regional Offices should be contacted to confirm the waiver termination date for other medical use facilities.

This model provides acceptable procedures for safe use of unsealed licensed material used for medical uses. This model addresses some of the procedures for the safe use of unsealed licensed material associated with the production of PET radioactive drugs and their transfer or with other nonmedical uses.

Applicants may either adopt this model procedure or develop their own procedure. If applicants will produce PET radioactive drugs for transfer under 10 CFR 30.32(j) or are authorized for other nonmedical uses, they may need to supplement this model procedure that was developed for medical use for those activities. (Some of the health physics practices listed below may also apply to sealed sources.)

- Wear laboratory coats or other protective clothing at all times in areas where radioactive materials are used.

- Wear disposable gloves at all times while handling radioactive materials.

- Either after each procedure or before leaving the area, monitor hands for contamination in a low-background area using an appropriate survey instrument.

- Use syringe shields for reconstitution of radiopharmaceutical kits and administration of radiopharmaceuticals to patients, except when their use is contraindicated (e.g., recessed veins, infants). In these and other exceptional cases, use other protective methods, such as remote delivery of the dose (e.g., use a butterfly needle).

- Do not eat, store food, drink, smoke, or apply cosmetics in any area where licensed material is stored or used.

- Wear personnel monitoring devices, if required, at all times while in areas where radioactive materials are used or stored. These devices shall be worn as prescribed by the RSO. When not being worn to monitor occupational exposures, personnel monitoring devices shall be stored in the work place in a designated low-background area.

- Wear extremity dosimeters, if required, when handling radioactive material.

- Dispose of radioactive waste only in designated, labeled, and properly shielded receptacles.

- Never pipette by mouth.

- Wipe-test unsealed byproduct material storage, preparation, and administration areas weekly for contamination. If necessary, decontaminate the area.

- Survey with a radiation detection survey meter all areas of licensed material use (which now includes use of accelerator-produced radioactive materials or discrete sources of radium-226), including the generator storage, kit preparation, and injection areas, daily for contamination. If necessary, decontaminate the area. Areas used to prepare and administer therapy quantities of radiopharmaceuticals must be surveyed daily in accordance with 10 CFR 35.70 (except when administering therapy dosages in patients' rooms when patients are confined).

- Store radioactive solutions in shielded containers that are clearly labeled.

- Radiopharmaceutical multi-dose diagnostic and therapy vials must be labeled in accordance with 10 CFR 35.69 and 10 CFR 20.1904.

- Syringes and unit dosages must be labeled in accordance with 10 CFR 35.69 and 10 CFR 20.1904. Mark the label with the radionuclide, the activity, the date for which the activity is estimated, and the kind of materials (i.e., radiopharmaceutical). If the container is holding less than the quantities listed in Appendix C to Part 20, the syringe or vial need only be labeled to identify the radioactive drug (10 CFR 35.69). To avoid mistaking patient dosages, label the syringe with the type of study and the patient's name.

- For prepared dosages, assay each patient dosage in the dose calibrator (or instrument) before administering it (10 CFR 35.63).

- Do not use a dosage if it does not fall within the prescribed dosage range or if it varies more than ±20% from the prescribed dosage, except as approved by an AU.

- When measuring the dosage, licensees need not consider the radioactivity that adheres to the syringe wall or remains in the needle.

- Check the patient's name and identification number and the prescribed radionuclide, chemical form, and dosage before administering. If the prescribed dosage requires a WD, the patient's identity must be verified and the administration must be in accordance with the WD (10 CFR 35.41).

- Always keep flood sources, syringes, waste, and other radioactive material in shielded containers.

- Secure all licensed material, including accelerator-produced radioactive materials and discrete sources of radium-226, when not under the constant surveillance and immediate control of an individual authorized under the NRC license (or such individual's designee).

APPENDIX U

Model Procedure for Release of Patients or Human Research Subjects Administered Radioactive Materials

Model Procedure for Release of Patients or Human Research Subjects Administered Radioactive Materials

With the implementation of the EPAct, the NRC now has regulatory authority over accelerator-produced radioactive materials and discrete sources of radium-226. Therefore, the procedures for releasing patients administered radioactive materials also apply to the medical administration of accelerator-produced radioactive materials and discrete sources of radium-226 after NRC's waiver of August 31, 2005, is terminated for medical use facilities. The NRC waiver that applied to Government agencies, Federally recognized Indian tribes, Delaware, the District of Columbia, Puerto Rico, the U.S. Virgin Islands, Indiana, Wyoming, and Montana was terminated on November 30, 2007. The NRC Regional Offices should be contacted to confirm the waiver termination date for other medical use facilities.

Section 35.75, "Release of Individuals Containing Unsealed Byproduct Material or Implants Containing Byproduct Material," of 10 CFR Part 35, "Medical Use of Byproduct Material," permits a licensee to "authorize the release from its control any individual who has been administered unsealed byproduct material or implants containing byproduct material if the total effective dose equivalent to any other individual from exposure to the released individual is not likely to exceed 5 millisieverts (0.5 rem)." *Note:* As a result of the EPAct, byproduct material now includes accelerator-produced radioactive materials and discrete sources of radium-226.

In this Appendix, the individual or human research subject to whom the radioactive material has been administered is called the "patient."

Release Equation

The activity at which patients could be released was calculated by using, as a starting point, the method discussed in the National Council on Radiation Protection and Measurements (NCRP) Report No. 37, "Precautions in the Management of Patients Who Have Received Therapeutic Amounts of Radionuclides." This report uses the following equation to calculate the exposure until time *t* at a distance *r* from the patient:

Equation U.1:

$$D(t) = \frac{34.6 \, \Gamma \, Q_0 \, T_P \, (1 - e^{-0.693t/T_P})}{r^2}$$

where:
$D(t)$ = Accumulated exposure at time t, in roentgens
34.6 = Conversion factor of 24 hrs/day times the total integration of decay (1.44)
Γ = Specific gamma ray constant for a point source, R/mCi-hr at 1 cm
Q_0 = Initial activity of the point source in millicuries, at the time of the release
T_p = Physical half-life in days
r = Distance from the point source to the point of interest, in centimeters
t = Exposure time in days.

This Appendix uses the NCRP equation (Equation U.1) in the following manner to calculate the activities at which patients may be released.

- The dose to an individual likely to receive the highest dose from exposure to the patient is taken to be the dose to total decay. Therefore, $(1-e^{-0.693t/Tp})$ is set equal to 1.

- It is assumed that 1 roentgen is equal to 10 millisieverts (1 rem).

- The exposure-rate constants and physical half-lives for radionuclides typically used in nuclear medicine and brachytherapy procedures are given in Supplement A of Table U.5 in this Appendix.

- Default activities at which patients may be released are calculated using the physical half-lives of the radionuclides and do not account for the biological half-lives of the radionuclides.

- When release is based on biological elimination (i.e., the effective half-life) rather than just the physical half-life of the radionuclide, Equation U.1 is modified to account for the uptake and retention of the radionuclide by the patient, as discussed in Supplement B.2.

- For radionuclides with a physical half-life greater than 1 day and no consideration of biological elimination, it is assumed that the individual likely to receive the highest dose from exposure to the patient would receive a dose of 25% of the dose to total decay (0.25 in Equation U.2), at a distance of 1 meter. Selection of 25% of the dose to total decay at 1 meter for estimating the dose is based on measurements discussed in the supporting regulatory analysis that indicate the dose calculated using an occupancy factor, E, of 25% at 1 meter is conservative in most normal situations.

- For radionuclides with a physical half-life less than or equal to 1 day, it is difficult to justify an occupancy factor of 0.25, because relatively long-term averaging of behavior cannot be assumed. Under this situation, occupancy factors from 0.75 to 1.0 may be more appropriate.

Thus, for radionuclides with a physical half-life greater than 1 day:

Equation U.2:

$$D(\infty) = \frac{34.6\ \Gamma\ Q_0\ T_p\ (0.25)}{(100\ cm)^2}$$

For radionuclides with a physical half-life less than or equal to 1 day, and if an occupancy factor of 1.0 is used:

Equation U.3:

$$D(\infty) = \frac{34.6\ \Gamma\ Q_0\ T_p\ (1)}{(100\ cm)^2}$$

Equations U.2 and U.3 calculate the dose from external exposure to gamma radiation. These equations do not include the dose from internal intake by household members and members of the public, because the dose from intake by other individuals is expected to be small for most radiopharmaceuticals (less than a few percent), relative to the external gamma dose (see "Internal Dose," of Supplement B). Further, the equations above do not apply to the dose to breast-feeding infants or children who continue to breast-feed. Patients who are breast-feeding an infant or child must be considered separately, as discussed in Item U.1.1, "Release of Patients Based on Administered Activity."

U.1 Release Criteria

Licensees should use one of the following options to release a patient to whom unsealed byproduct material or implants containing byproduct material have been administered in accordance with regulatory requirements. As a result of the EPAct, the unsealed byproduct material or implants now include accelerator-produced radioactive materials or discrete sources of radium-226.

U.1.1 Release of Patients Based on Administered Activity

In compliance with the dose limit in 10 CFR 35.75(a), licensees may release patients from licensee control if the activity administered is no greater than the activity in Column 1 of Table U.1. The activities in Table U.1 are based on a total effective dose equivalent of 5 millisieverts (0.5 rem) to an individual using the following conservative assumptions:

- Administered activity;

- Physical half-life;

- Occupancy factor of 0.25 at 1 meter for physical half-lives greater than 1 day and, to be conservative, an occupancy factor of 1 at 1 meter for physical half-lives less than or equal to 1 day; and

- No shielding by tissue.

The total effective dose equivalent is approximately equal to the external dose because the internal dose is a small fraction of the external dose (see Section B.3, "Internal Dose," of Supplement B). In this case, no record of the release of the patient is required unless the patient is breast-feeding an infant or child, as discussed in Item U.3.2, "Records of Instructions for Breast-Feeding Patients." The licensee may demonstrate compliance by using the records of activity that are already required by 10 CFR 35.40 and 35.63.

If the activity administered exceeds the activity in Column 1 of Table U.1, the licensee may release the patient when the activity has decayed to the activity in Column 1 of Table U.1. In this case, 10 CFR 35.75(c) requires a record because the patient's release is based on the retained activity rather than the administered activity. The activities in Column 1 of Table U.1 were calculated using either Equation U.2 or U.3, depending on the physical half-life of the radionuclide.

If a radionuclide that is not listed in Table U.1 is administered, the licensee can demonstrate compliance with the regulation by maintaining, for NRC inspection, a calculation of the release activity that corresponds to the dose limit of 5 millisievert (0.5 rem). Equation U.2 or U.3 may be used, as appropriate, to calculate the activity Q corresponding to 5 millisieverts (0.5 rem).

The release activities in Column 1 of Table U.1 do not include consideration of the dose to a breast-feeding infant or child from ingestion of radiopharmaceuticals contained in the patient's breast milk. When the patient is breast-feeding an infant or child, the activities in Column 1 of Table U.1 are not applicable to the infant or child. In this case, it may be necessary to give instructions as described in Items U.2.2 and U.2.3 as a condition for release. If failure to interrupt or discontinue could result in a dose to the breast-feeding infant or child in excess of 5 millisieverts (0.5 rem), a record that instructions were provided is required by 10 CFR 35.75(d).

U.1.2 Release of Patients Based on Measured Dose Rate

Licensees may release patients to whom radionuclides have been administered in amounts greater than the activities listed in Column 1 of Table U.1, provided the measured dose rate at 1 meter (from the surface of the patient) is no greater than the value in Column 2 of Table U.1 for that radionuclide. In this case, however, 10 CFR 35.75(c) requires a record because the release is based on considering shielding by tissue.

If a radionuclide not listed in Table U.1 is administered and the licensee chooses to release a patient based on the measured dose rate, the licensee should first calculate a dose rate that corresponds to the 5 millisieverts (0.5 rem) dose limit. If the measured dose rate at 1 meter is no greater than the calculated dose rate, the patient may be released. A record of the release is required by 10 CFR 35.75(c). The dose rate at 1 meter may be calculated from Equation U.2 or U.3, as appropriate, because the dose rate at 1 meter is equal to $\Gamma\, Q\, /\, 10{,}000\ cm^2$.

U.1.3 Release of Patients Based on Patient-Specific Dose Calculations

Licensees may release patients based on dose calculations using patient-specific parameters. With this method, based on 10 CFR 35.75(a), the licensee must calculate the maximum likely dose to an individual exposed to the patient on a case-by-case basis. If the calculated maximum likely dose to an individual is no greater than 5 millisievert (0.5 rem), the patient may be released. Using this method, licensees may be able to release patients with activities greater than those listed in Column 1 of Table U.1 by taking into account the effective half-life of the radioactive material and other factors that may be relevant to the particular case. In this case, a record of the release is required by 10 CFR 35.75(c). If the dose calculation considered retained activity, an occupancy factor less than 0.25 at 1 meter, effective half-life, or shielding by tissue, a record of the basis for the release is required by 10 CFR 35.75(c).

Supplement B contains procedures for performing patient-specific dose calculations, and it describes how various factors may be considered in the calculations.

Table U.1 Activities and Dose Rates for Authorizing Patient Release[†]				
Radionuclide	COLUMN 1 Activity At or Below Which Patients May Be Released		COLUMN 2 Dose Rate at 1 Meter, At or Below Which Patients May Be Released*	
	(GBq)	(mCi)	(mSv/hr)	(mrem/hr)
Ag-111	19	520	0.08	8
Au-198	3.5	93	0.21	21
Cr-51	4.8	130	0.02	2
Cu-64	8.4	230	0.27	27
Cu-67	14	390	0.22	22
Ga-67	8.7	240	0.18	18
I-123	6	160	0.26	26
I-125	0.25	7	0.01	1
I-125 implant	0.33	9	0.01	1
I-131	1.2	33	0.07	7
In-111	2.4	64	0.2	20
Ir-192 implant	0.074	2	0.008	0.8
P-32	**	**	**	**
Pd-103 implant	1.5	40	0.03	3
Re-186	28	770	0.15	15
Re-188	29	790	0.2	20
Sc-47	11	310	0.17	17
Se-75	0.089	2	0.005	0.5
Sm-153	26	700	0.3	30
Sn-117m	1.1	29	0.04	4
Sr-89	**	**	**	**
Tc-99m	28	760	0.58	58
Tl-201	16	430	0.19	19
Y-90	**	**	**	**
Yb-169	0.37	10	0.02	2

Footnotes for Table U-1

[†] The activity values were computed based on 5 millisieverts (0.5 rem) total effective dose equivalent.

* If the release is based on the dose rate at 1 meter in Column 2, the licensee must maintain a record as required by 10 CFR 35.75(c), because the measurement includes shielding by tissue. See Item U.3.1, "Records of Release," for information on records.

** Activity and dose rate limits are not applicable in this case because of the minimal exposures to members of the public resulting from activities normally administered for diagnostic or therapeutic purposes.

Notes: The millicurie values were calculated using Equations U.2 or U.3 and the physical half-life. The gigabecquerel values were calculated using the millicurie values and the conversion factor from millicurie to gigabecquerels. The dose rate values are calculated using the millicurie values and the exposure rate constants.

In general, the values are rounded to two significant figures; however, values less than 0.37 gigabecquerel (10 millicuries) or 0.1 millisievert (10 millirems) per hour are rounded to one significant figure. Details of the calculations are provided in NUREG-1492.

Although non-byproduct materials are not regulated by NRC, information on non-byproduct material is included for the convenience of the licensee.

Agreement State regulations may vary. Agreement State licensees should check with their State regulations before using these values.

U.2   Instructions

This Section provides acceptable instructions for release of patients administered radioactive materials. Licensees may either adopt these model instructions or develop their own instructions to meet the requirements of 10 CFR 35.75.

U.2.1   Activities and Dose Rates Requiring Instructions

Based on 10 CFR 35.75(b), for some administrations the released patients must be given instructions, including written instructions, on how to maintain doses to other individuals ALARA after the patients are released.[1] Column 1 of Table U.2 provides the activity above which instructions must be given to patients. Column 2 provides corresponding dose rates at 1 meter, based on the activities in Column 1. The activities or dose rates in Table U.2 may be used for determining when instructions must be given. If the patient is breast-feeding an infant or child, additional instructions may be necessary (see Item U.2.2, "Additional Instructions for Release of Patients Who Could Be Breast-Feeding After Release").

When patient-specific calculations (as described in Supplement B) are used, instructions must be provided if the calculation indicates a dose greater than 1 millisievert (0.1 rem).

If a radionuclide not listed in Table U.2 is administered, the licensee may calculate the activity or dose rate that corresponds to 1 millisievert (0.1 rem). Equation U.2 or U.3, as appropriate, may be used.

U.2.2   Additional Instructions for Release of Patients Who Could Be Breast-Feeding After Release

The requirement in 10 CFR 35.75(b) that a licensee provide instructions on the discontinuation or the interruption period of breast-feeding, and the consequences of failing to follow the recommendation, presumes the licensee will inquire, as appropriate, regarding the breast-feeding status of the patient.[1] The purpose of the instructions (e.g., on interruption or discontinuation) is

---

[1] NRC does not intend to enforce patient compliance with the instructions nor is it the licensee's responsibility to do so.

to permit licensees to release a patient who could be breast-feeding an infant or child when the dose to the infant or child could exceed 5 millisieverts (0.5 rem) if there is no interruption of breast-feeding.

If the patient could be breast-feeding an infant or child after release, and if a radiopharmaceutical with an activity above the value stated in Column 1 of Table U.3 was administered to the patient, the licensee must give the patient instructions on the discontinuation or interruption period for breast-feeding and the consequences of failing to follow the recommendation. The patient should also be informed if there would be no consequences to the breast-feeding infant or child. Table U.3 also provides recommendations for interrupting or discontinuing breast-feeding to minimize the dose to below 1 millisievert (0.1 rem) if the patient has received certain radiopharmaceutical doses. The radiopharmaceuticals listed in Table U.3 are commonly used in medical diagnosis and treatment.

If a radiopharmaceutical not listed in Table U.3 is administered to a patient who could be breast-feeding, the licensee should evaluate whether instructions or records (or both) are required. If information on the excretion of the radiopharmaceutical is not available, an acceptable method is to assume that 50% of the administered activity is excreted in the breast milk. The dose to the infant or child can be calculated by using the dose conversion factors given for a newborn infant by Stabin (see Reference).

U.2.3 Content of Instructions

The instructions should be specific to the type of treatment given, such as permanent implants or radioiodine for hyperthyroidism or thyroid carcinoma, and they may include additional information for individual situations; however, the instructions should not interfere with or contradict the best medical judgment of physicians. The instructions may include the name of a knowledgeable contact person and that person's telephone number, in case the patient has any questions. Additional instructions appropriate for each modality, as shown in examples below, may be provided (refer to U.2.3.1 and U.2.3.2).

Table U.2	Activities and Dose Rates Above Which Instructions Should Be Given When Authorizing Patient Release*			
Radionuclide	COLUMN 1 Activity Above Which Instructions Are Required		COLUMN 2 Dose Rate at 1 Meter Above Which Instructions Are Required	
	(GBq)	(mCi)	(mSv/hr)	(mrem/hr)
Ag-111	3.8	100	0.02	2
Au-198	0.69	19	0.04	4
Cr-51	0.96	26	0.004	0.4
Cu-64	1.7	45	0.05	5
Cu-67	2.9	77	0.04	4
Ga-67	1.7	47	0.04	4
I-123	1.2	33	0.05	5

Table U.2	Activities and Dose Rates Above Which Instructions Should Be Given When Authorizing Patient Release* *(continued)*			
Radionuclide	COLUMN 1 Activity Above Which Instructions Are Required		COLUMN 2 Dose Rate at 1 Meter Above Which Instructions Are Required	
	(GBq)	(mCi)	(mSv/hr)	(mrem/hr)
I-125	0.05	1	0.002	0.2
I-125 implant	0.074	2	0.002	0.2
I-131	0.24	7	0.02	2
In-111	0.47	13	0.04	4
Ir-192 implant	0.011	0.3	0.002	0.2
P-32	**	**	**	**
Pd-103 implant	0.3	8	0.007	0.7
Re-186	5.7	150	0.03	3
Re-188	5.8	160	0.04	4
Sc-47	2.3	62	0.03	3
Se-75	0.018	0.5	0.001	0.1
Sm-153	5.2	140	0.06	6
Sn-117m	0.21	6	0.009	0.9
Sr-89	**	**	**	**
Tc-99m	5.6	150	0.12	12
Tl-201	3.1	85	0.04	4
Y-90	**	**	**	**
Yb-169	0.073	2	0.004	0.4

Footnotes for Table U.2

* The activity values were computed based on 1 millisievert (0.1 rem) total effective dose equivalent.

** Activity and dose rate limits are not applicable in this case because of the minimal exposures to members of the public resulting from activities normally administered for diagnostic or therapeutic purposes.

Notes: The values for activity were calculated using Equations U.2 or U.3 and the physical half-life. The values given in SI units (gigabecquerel values) were using conversion factors.

In general, values are rounded to two significant figures; however, values less than 0.37 gigabecquerel (10 millicuries) or 0.1 millisievert (10 millirems) per hour are rounded to one significant figure. Details of the calculations are provided in NUREG-1492.

Agreement State regulations may vary. Agreement State licensees should check their State regulations before using these values.

Table U.3	Activities of Radiopharmaceuticals That Require Instructions and Records When Administered to Patients Who Are Breast-Feeding an Infant or Child				
Radionuclide	**COLUMN 1** **Activity Above Which Instructions Are Required**		**COLUMN 2** **Activity Above Which a Record is Required**		**COLUMN 3** **Examples of Recommended Duration of Interruption of Breast-Feeding**
	(MBq)	**(mCi)**	**(MBq)**	**(mCi)**	
I-131 NaI	0.01	0.0004	0.07	0.002	Complete cessation (for this infant or child)
I-123 NaI	20	0.5	100	3	
I-123 OIH	100	4	700	20	
I-123 MIBG	70	2	400	10	24 hours for 370 MBq (10 mCi) 12 hours for 150 MBq (4 mCi)
I-125 OIH	3	0.08	10	0.4	
I-131 OIH	10	0.3	60	1.5	
Tc-99m DTPA	1000	30	6000	150	
Tc-99m MAA	50	1.3	200	6.5	12.6 hours for 150 MBq (4 mCi)
Tc-99m Pertechnetate	100	3	600	15	24 hours for 1,100 MBq (30 mCi) 12 hours for 440 MBq (12 mCi)
Tc-99m DISIDA	1000	30	6000	150	
Tc-99m Glucoheptonate	1000	30	6000	170	
Tc-99m MIBI	1000	30	6000	150	
Tc-99m MDP	1000	30	6000	150	
Tc-99m PYP	900	25	4000	120	
Tc-99m Red Blood Cell *In Vivo* Labeling	400	10	2000	50	6 hours for 740 MBq (20 mCi)
Tc-99m Red Blood Cell *In Vitro* Labeling	1000	30	6000	150	

Table U.3	Activities of Radiopharmaceuticals That Require Instructions and Records When Administered to Patients Who Are Breast-Feeding an Infant or Child *(continued)*				
Radionuclide	**COLUMN 1** Activity Above Which Instructions Are Required		**COLUMN 2** Activity Above Which a Record is Required		**COLUMN 3** Examples of Recommended Duration of Interruption of Breast-Feeding
	(MBq)	**(mCi)**	**(MBq)**	**(mCi)**	
Tc-99m Sulfur Colloid	300	7	1000	35	6 hours for 440 MBq (12 mCi)
Tc-99m DTPA Aerosol	1000	30	6000	150	
Tc-99m MAG3	1000	30	6000	150	
Tc-99m White Blood Cells	100	4	600	15	24 hours for 1,100 MBq (30 mCi) 12 hours for 440 MBq (12 mCi)
Ga-67 Citrate	1	0.04	7	0.2	1 month for 150 MBq (4 mCi) 2 weeks for 50 MBq (1.3 mCi) 1 week for 7 MBq (0.2 mCi)
Cr-51 EDTA	60	1.6	300	8	
In-111 White Blood Cells	10	0.2	40	1	1 week for 20 MBq (0.5 mCi)
Tl-201 Chloride	40	1	200	5	2 weeks for 110 MBq (3 mCi)

Footnotes for Table U.3

* The duration of interruption of breast-feeding is selected to reduce the maximum dose to a newborn infant to less than 1 millisievert (0.1 rem), although the regulatory limit is 5 millisieverts (0.5 rem). The actual doses that would be received by most infants would be far below 1 millisievert (0.1 rem). Of course, the physician may use discretion in the recommendation, increasing or decreasing the duration of interruption.

Notes: Activities are rounded to one significant figure, except when it was considered appropriate to use two significant figures. Details of the calculations are shown in NUREG-1492, "Regulatory Analysis on Criteria for the Release of Patients Administered Radioactive Material."

If there is no recommendation in Column 3 of this table, the maximum activity normally administered is below the activities that require instructions on interruption or discontinuation of breast-feeding.

Agreement State regulations may vary. Agreement State licensees should check their State regulations before using these values.

U.2.3.1 Instructions Regarding Radiopharmaceutical Administrations

For procedures involving radiopharmaceuticals, additional instructions may include the following:

- Maintaining distance from other persons, including separate sleeping arrangements.
- Minimizing time in public places (e.g., public transportation, grocery stores, shopping centers, theaters, restaurants, sporting events).
- Precautions to reduce the spread of radioactive contamination.
- The length of time each of the precautions should be in effect.

The Society of Nuclear Medicine published a pamphlet in 1987 that provides information for patients receiving treatment with radioiodine. This pamphlet was prepared jointly by the Society of Nuclear Medicine and the NRC. The pamphlet contains blanks for the physician to fill in the length of time that each instruction should be followed. Although this pamphlet was written for the release of patients to whom less than 1,110 megabecquerels (30 millicuries) of iodine-131 had been administered, the NRC still considers the instructions in this pamphlet to be an acceptable method for meeting the requirements of 10 CFR 35.75(b), provided the times filled in the blanks are appropriate for the activity and the medical condition.

If additional instructions are required because the patient is breast-feeding, the instructions should include appropriate recommendations on whether to interrupt breast-feeding, the length of time to interrupt breast-feeding, or, if necessary, the discontinuation of breast-feeding. The instructions should include information on the consequences of failure to follow the recommendation to interrupt or discontinue breast-feeding. The consequences should be explained so that the patient will understand that, in some cases, breast-feeding after an administration of certain radionuclides should be avoided. For example, a consequence of procedures involving iodine-131 is that continued breast-feeding could harm the infant's or child's thyroid. Most diagnostic procedures involve radionuclides other than radioiodine and there would be no consequences; guidance should simply address avoiding any unnecessary radiation exposure to the infant or child from breast-feeding. If the Society of Nuclear Medicine's pamphlet is given at release to a patient who is breast-feeding an infant or child, the pamphlet should be supplemented with information specified in 10 CFR 35.75(b)(1) and (2).

The requirement of 10 CFR 35.75(b) regarding written instructions to patients who could be breast-feeding an infant or child is not in any way intended to interfere with the discretion and judgment of the physician in providing detailed instructions and recommendations.

U.2.3.2 Instructions Regarding Implants

For patients who have received implants, additional instructions may include the following:

A small radioactive source has been placed (implanted) inside your body. The source is actually many small metallic pellets or seeds, which are each about 1/3 to 1/4 of an inch long, similar in size and shape to a grain of rice. To minimize exposure to radiation to others from the source inside your body, you should do the following for_____days.

- Stay at a distance of _____ feet from _____.
- Maintain separate sleeping arrangements.
- Minimize time with children and pregnant women.
- Do not hold or cuddle children.
- Avoid public transportation.
- Examine any bandages or linens that come into contact with the implant site for any pellets or seeds that may have come out of the implant site.
- If you find a seed or pellet that falls out:

  — Do not handle it with your fingers. Use something like a spoon or tweezers to place it in a jar or other container that you can close with a lid.

  — Place the container with the seed or pellet in a location away from people.

  — Notify _____ at telephone number _____ .

U.3   Records

U.3.1   Records of Release

There is no requirement for recordkeeping on the release of patients who were released in accordance with Column 1 of Table U.1; however, if the release of the patient is based on a dose calculation that considered retained activity, an occupancy factor less than 0.25 at 1 meter, effective half-life, or shielding by tissue, a record of the basis for the release is required by 10 CFR 35.75(c). This record should include the patient identifier (in a way that ensures that confidential patient information is not traceable or attributable to a specific patient), the radioactive material administered, the administered activity, and the date of the administration. In addition, depending on the basis for release, records should include the following information:

- **For Immediate Release of a Patient Based on a Patient-Specific Calculation:** The equation used, including the patient-specific factors and their bases that were used in calculating the dose to the person exposed to the patient, and the calculated dose. The patient-specific factors (see Supplement B of this Appendix) include the effective half-life and uptake fraction for each component of the biokinetic model, the time that the physical half-life was assumed to apply to retention, and the occupancy factor. The basis for selecting each of these values should be included in the record.

- **For Immediate Release of a Patient Based on Measured Dose Rate:** The results of the measurement, the specific survey instrument used, and the name of the individual performing the survey.

- **For Delayed Release of a Patient Based on Radioactive Decay Calculation:** The time of the administration, the date and time of release, and the results of the decay calculation.

- **For Delayed Release of a Patient Based on Measured Dose Rate:** The results of the survey meter measurement, the specific survey instrument used, and the name of the individual performing the survey.

In some situations, a calculation may be case-specific for a class of patients who all have the same patient-specific factors. In this case, the record for a particular patient's release may reference the calculation for the class of patients.

Records, as required by 10 CFR 35.75(c), should be kept in a manner that ensures the patient's confidentiality; that is, the records should not contain the patient's name or any other information that could lead to identification of the patient. These recordkeeping requirements may also be used to verify that licensees have proper procedures in place for assessing potential third-party exposure associated with and arising from exposure to patients who were administered radioactive material.

U.3.2 Records of Instructions for Breast-Feeding Patients

If failure to interrupt or discontinue breast-feeding could result in a dose to the infant or child in excess of 5 millisieverts (0.5 rem), a record that instructions were provided is required by 10 CFR 35.75(d). Column 2 of Table U.3 states, for the radiopharmaceuticals commonly used in medical diagnosis and treatment, the activities that would require such records when administered to patients who are breast-feeding.

The record should include the patient's identifier (in a way that ensures that confidential patient information is not traceable or attributable to a specific patient), the radiopharmaceutical administered, the administered activity, the date of the administration, and whether instructions were provided to the patient who could be breast-feeding an infant or child.

U.4 Summary Table

Table U.4 summarizes the criteria for releasing patients and the requirements for providing instructions and maintaining records.

Table U.4   Summary of Release Criteria, Required Instructions to Patients, and Records to Be Maintained

Patient Group	Basis for Release	Criteria for Release	Instructions Needed?	Release Records Required?
All patients, including patients who are breast-feeding an infant or child	Administered activity	Administered activity ≤ Column 1 of Table U.1	Yes – if administered activity > Column 1 of Table U.2	No
	Retained activity	Retained activity ≤ Column 1 of Table U.1	Yes – if retained activity > Column 1 of Table U.2	Yes
	Measured dose rate	Measured dose rate ≤ Column 2 of Table U.1	Yes – if dose rate > Column 2 of Table U.2	Yes
	Patient-specific calculations	Calculated dose ≤ 5 mSv (0.5 rem)	Yes – if calculated dose > 1 mSv (0.1 rem)	Yes
Patients who are breast-feeding an infant or child	All the above bases for release		Additional instructions required if: Administered activity > Column 1 of Table U.3 or Licensee calculated dose from breast-feeding > 1 mSv (0.1 rem) to the infant or child	Records that instructions were provided are required if: Administered activity > Column 2 of Table U.3 or Licensee calculated dose from continued breast-feeding > 5 mSv (0.5 rem) to the infant or child

Implementation

The purpose of this section is to provide information to licensees and applicants regarding NRC staff's plans for using this Appendix. Except in those cases in which a licensee proposes an acceptable alternative method for complying with 10 CFR 35.75, the methods described in this Appendix will be used in the evaluation of a licensee's compliance with 10 CFR 35.75.

Supplement A

Table U.5 Half-Lives and Exposure Rate Constants of Radionuclides Used in Medicine		
Radionuclide	Physical Half-Life (days)[1]	Exposure Rate Constant[2] (R/mCi-h at 1 cm)
Ag-111	7.45	0.15
Au-198	2.696	2.3
Cr-51	27.704	0.16
Cu-64	0.529	1.2
Cu-67	2.578	0.58
Ga-67	3.261	0.753
I-123	0.55	1.61
I-125	60.14	1.42
I-125 implant[3]	60.14	1.114
I-131	8.04	2.2
In-111	2.83	3.21
Ir-192 implant[3]	74.02	4.594
P-32	14.29	N/A[5]
Pd-103 implant[4]	16.96	0.865
Re-186	3.777	0.2
Re-188	0.708	0.26
Sc-47	3.351	0.56
Se-75	119.8	2
Sn-117m	13.61	1.48
Sr-89	50.5	N/A[5]
Tc-99m	0.251	0.756
Tl-201	3.044	0.447
Yb-169	32.01	1.83
Y-90	2.67	N/A[5]
Yb-169	32.01	1.83

Footnotes for Table U.5

[1] K. F. Eckerman, A. B. Wolbarst, and A. C. B. Richardson, "Federal Guidance Report No. 11, Limiting Values of Radionuclide Intake and Air Concentration and Dose Conversion Factors for Inhalation, Submersion, and Ingestion," Report No. EPA-520/1-88-020, Office of Radiation Programs, U.S. Environmental Protection Agency, Washington, DC, 1988.

2   Values for the exposure rate constant for Au-198, Cr-51, Cu-64, I-131, Sc-47, and Se-75 were taken from the *Radiological Health Handbook*, U.S. Department of Health, Education, and Welfare, p. 135, 1970. For Cu-67, I-123, In-111, Re-186, and Re-188, the values for the exposure rate constant were taken from D. E. Barber, J. W. Baum, and C. B. Meinhold, "Radiation Safety Issues Related to Radiolabeled Antibodies," NUREG/CR-4444, U.S. NRC, Washington, DC, 1991. For Ag-111, Ga-67, I-125, Sm-153, Sn-117m, Tc-99m, Tl-201, and Yb-169, the exposure rate constants were calculated because the published values for these radionuclides were an approximation, presented as a range, or varied from one reference to another. Details of the calculation of the exposure rate constants are shown in Table A.2 of Appendix A to NUREG-1492, "Regulatory Analysis on Criteria for the Release of Patients Administered Radioactive Material," U.S. NRC, February 1997.

3   R. Nath, A. S. Meigooni, and J. A. Meli, "Dosimetry on Transverse Axes of ^{125}I and ^{192}Ir Interstitial Brachytherapy Sources," Medical Physics, Volume 17, Number 6, November/December 1990. The exposure rate constant given is a measured value averaged for several source models and takes into account the attenuation of gamma rays within the implant capsule itself.

4   A. S. Meigooni, S. Sabnis, R. Nath, "Dosimetry of Palladium-103 Brachytherapy Sources for Permanent Implants," *Endocurietherapy Hyperthermia Oncology*, Volume 6, April 1990. The exposure rate constant given is an "apparent" value (i.e., with respect to an apparent source activity) and takes into account the attenuation of gamma rays within the implant capsule itself.

5   Not applicable (N/A) because the release activity is not based on beta emissions.

References

National Council on Radiation Protection and Measurements (NCRP), "Precautions in the Management of Patients Who Have Received Therapeutic Amounts of Radionuclides," NCRP Report No. 37, October 1, 1970. (Available for sale from the NCRP, 7910 Woodmont Avenue, Suite 400, Bethesda, MD 20814-3095.)

S. Schneider and S. A. McGuire, "Regulatory Analysis on Criteria for the Release of Patients Administered Radioactive Material," NUREG-1492 (Final Report), NRC, February 1997.

M. Stabin, "Internal Dosimetry in Pediatric Nuclear Medicine," in *Pediatric Nuclear Medicine*, edited by S. Treves, Springer Verlag, New York, 1995.

"Guidelines for Patients Receiving Radioiodine Treatment," *Society of Nuclear Medicine*, 1987. This pamphlet may be obtained from the Society of Nuclear Medicine, 1850 Samuel Morse Drive, Reston, VA 20190-5316.

Supplement B

Procedures for Calculating Doses Based on Patient-Specific Factors

A licensee may release a patient to whom an activity with a value higher than the values listed in Column 1 of Table U.1 of this supplement has been administered if dose calculations using patient-specific parameters, which are less conservative than the conservative assumptions, show that the potential total effective dose equivalent to any individual would be no greater than 5 millisieverts (0.5 rem).

If the release of a patient is based on a patient-specific calculation that considered retained activity, an occupancy factor less than 0.25 at 1 meter, biological or effective half-life, or shielding by tissue, a record of the basis of the release is required by 10 CFR 35.75(c). The following equation can be used to calculate doses:

Equation B-1:

$$D(t) = \frac{34.6 \; \Gamma \; Q_0 \; TE \; (1 - e^{-0.693t/T_p})}{r^2}$$

where:
$D(t)$ = Accumulated dose to time t, in rem;
34.6 = Conversion factor of 24 hrs/day times the total integration of decay (1.44);
Γ = Exposure rate constant for a point source, R/mCi x hr at 1 cm;
Q_0 = Initial activity at the start of the time interval;
T_p = Physical half-life, in days;
E = Occupancy factor that accounts for different occupancy times and distances when an individual is around a patient;
r = Distance in centimeters. This value is typically 100 cm; and
t = Exposure time in days.

B.1   Occupancy Factor

B.1.1   Rationale for Occupancy Factors Used to Derive Table U.1

In Table U.1 in this Appendix, the activities at which patients could be released were calculated using the physical half-life of the radionuclide and an occupancy factor at 1 meter of either 0.25 (if the radionuclide has a half-life longer than 1 day) or 1.0 (if the radionuclide has a half-life less than or equal to 1 day). The basis for the occupancy factor of 0.25 at 1 meter is that measurements of doses to family members, as well as considerations of normal human behavior (as discussed in the supporting regulatory analysis (Ref. B-1)), suggest that an occupancy factor of 0.25 at 1 meter, when used in combination with the physical half-life, will produce a generally conservative estimate of the dose to family members when instructions on minimizing doses to others are given.

An occupancy factor of 0.25 at 1 meter may not be appropriate when the physical half-life is less than or equal to 1 day, and hence, the dose is delivered over a short time. Specifically, the assumptions regarding patient behavior that led to an occupancy factor of 0.25 at 1 meter include the assumption that the patient will not be in close proximity to other individuals for several days; however, when the dose is from a short-lived radionuclide, the time that individuals spend in close proximity to the patient immediately following release will be most significant because the dose to other individuals could be a large fraction of the total dose from the short-lived radionuclide. Thus, to be conservative when providing generally applicable release quantities that may be used with little consideration of the specific details of a particular patient's release, the values calculated in Table U.1 were based on an occupancy factor of 1 at 1 meter when the half-life is less than or equal to 1 day. If information about a particular patient implies the assumptions were too conservative, licensees may consider case-specific conditions. Conversely, if young children are present in the household of the patient who is to be discharged, conservative assumptions about occupancy may be appropriate.

B.1.2 Occupancy Factors to Consider for Patient-Specific Calculations

The selection of an occupancy factor for patient-specific calculations will depend on whether the physical or effective half-life of the radionuclide is used and whether instructions are provided to the patient before release. The following occupancy factors, E, at 1 meter, may be useful for patient-specific calculations:

- $E = 0.75$ when a physical half-life, an effective half-life, or a specific time period under consideration (e.g., bladder holding time) is less than or equal to 1 day.

- $E = 0.25$ when an effective half-life is greater than 1 day, if the patient has been given instructions, such as:

    – Maintain a prudent distance from others for at least the first 2 days;
    – Sleep alone in a room for at least the first night;
    – Do not travel by airplane or mass transportation for at least the first day;
    – Do not travel on a prolonged automobile trip with others for at least the first 2 days;
    – Have sole use of a bathroom for at least the first 2 days; and
    – Drink plenty of fluids for at least the first 2 days.

- $E = 0.125$ when an effective half-life is greater than 1 day if the patient has been given instructions, such as:

    – Follow the instructions for $E = 0.25$ above;
    – Live alone for at least the first 2 days; and
    – Have few visits by family or friends for at least the first 2 days.

- In a two-component model (e.g., uptake of iodine-131 using thyroidal and extrathyroidal components), if the effective half-life associated with one component is less than or equal to 1 day but is greater than 1 day for the other component, it is more justifiable to use the occupancy factor associated with the dominant component for both components.

Example 1: Calculate the maximum likely dose to an individual exposed to a patient who has received 2,220 megabecquerels (60 millicuries) of iodine-131. The patient received instructions to maintain a prudent distance from others for at least 2 days, lives alone, drives home alone, and stays at home for several days without visitors.

Solution: The dose to total decay (t = ∞) is calculated based on the physical half-life using Equation B-1. (This calculation illustrates the use of physical half-life. To account for biological elimination, calculations described in the next section should be used.)

$$D(\infty) = \frac{34.6 \ \Gamma \ Q_0 \ T_P \ E}{r^2}$$

Because the patient has received instructions for reducing exposure as recommended for an occupancy factor of E = 0.125, the occupancy factor of 0.125 at 1 meter may be used.

$$D(\infty) = \frac{34.6 \ (2.2 \ R \cdot \ cm^2/mCi \cdot hr)(60mCi)(8.04 \ d)(0.125)}{(100 \ cm)^2}$$

$$D \ (\infty) = 4.59 \ \text{millisieverts} \ (0.459 \ \text{rem})$$

Since the dose is less than 5 millisievert (0.5 rem), the patient may be released, but 10 CFR 35.75(b) requires that instructions be given to the patient on maintaining doses to others as low as is reasonably achievable. A record of the calculation must be maintained, pursuant to 10 CFR 35.75(c), because an occupancy factor of less than 0.25 at 1 meter was used.

B.2   Effective Half-Life

A licensee may take into account the effective half-life of the radioactive material to demonstrate compliance with the dose limits for individuals exposed to the patient that are stated in 10 CFR 35.75. The effective half-life is defined as:

Equation B-2:

$$T_{eff} = \frac{T_b \times T_p}{T_b + T_p}$$

where: T_b = Biological half-life of the radionuclide and
T_p = Physical half-life of the radionuclide.

The behavior of iodine-131 can be modeled using two components: extrathyroidal iodide (i.e., existing outside of the thyroid) and thyroidal iodide following uptake by the thyroid. The effective half-lives for the extrathyroidal and thyroidal fractions (i.e., F_1 and F_2, respectively) can be calculated with the following equations.

Equation B-3:

$$T_{1eff} = \frac{T_{b1} \times T_P}{T_{b1} + T_P}$$

Equation B-4:

$$T_{2eff} = \frac{T_{b2} \times T_P}{T_{b2} + T_P}$$

where: T_{b1} = Biological half-life for extrathyroidal iodide;
T_{b2} = Biological half-life of iodide following uptake by the thyroid; and
T_p = Physical half-life of iodine-131.

However, simple exponential excretion models do not account for: (a) the time for the iodine-131 to be absorbed from the stomach to the blood; and (b) the holdup of iodine in the urine while in the bladder. Failure to account for these factors could result in an underestimate of the dose to another individual. Therefore, this supplement makes a conservative approximation to account for these factors by assuming that, during the first 8 hours after the administration, about 80% of the iodine administered is removed from the body at a rate determined only by the physical half-life of iodine-131.

Thus, an equation to calculate the dose from a patient administered iodine-131 may have three components. First is the dose for the first 8 hours (0.33 day) after administration. This component comes directly from Equation B-1, using the physical half-life and a factor of 80%. Second is the dose from the extrathyroidal component from 8 hours to total decay. In this component, the first exponential factor represents the activity at t = 8 hours based on the physical half-life of iodine-131. The second exponential factor represents the activity from t = 8 hours to total decay based on the effective half-life of the extrathyroidal component. The third component, the dose from the thyroidal component for 8 hours to total decay, is calculated in the same manner as the second component. The full equation is shown as Equation B-5.

Equation B-5:

$$D(\infty) = \frac{34.6\, \Gamma\, Q_0}{(100\,cm)^2} \left\{ E_1\, T_p\, (0.8)(1 - e^{-0.693(0.33)/T_p}) \right.$$
$$\left. + e^{-0.693(0.33)/T_p} E_2\, F_1\, T_{1eff} + e^{-0.693(0.33)/T_p} E_2\, F_2\, T_{2eff} \right\}$$

where: F_1 = Extrathyroidal uptake fraction;
F_2 = Thyroidal uptake fraction;
E_1 = Occupancy factor for the first 8 hours; and
E_2 = Occupancy factor from 8 hours to total decay.

All the other parameters are as defined in Equations B-1, B-3, and B-4. Acceptable values for F_1, T_{1eff}, F_2, and T_{2eff} are shown in Table U.6 for thyroid ablation and treatment of thyroid remnants after surgical removal of the thyroid for thyroid cancer. If these values have been measured for a specific individual, the measured values may be used.

The record of the patient's release required by 10 CFR 35.75(c) is described in Item U.3.1 of this Appendix.

Example 2, Thyroid Cancer: Calculate the maximum likely dose to an individual exposed to a patient to whom 5550 megabecquerels (150 millicuries) of iodine-131 have been administered for the treatment of thyroid remnants and metastasis.

Solution: In this example, the dose will be calculated by using Equation B-5 to account for the elimination of iodine-131 from the body, based on the effective half-lives appropriate for thyroid cancer. The physical half-life and the exposure rate constant are from Table U.5. The uptake fractions and effective half-lives are from Table U.6. An occupancy factor, E, of 0.75 at 1 meter, will be used for the first component because the time period under consideration is less than 1 day; however, for the second and third components, an occupancy factor of 0.25 will be used, because: (1) the effective half-life associated with the dominant component is greater than 1 day; and (2) patient-specific questions were provided to the patient to justify the occupancy factor (see Section B.1.2, "Occupancy Factors to Consider for Patient-Specific Calculations," of this Supplement).

Table U.6	Uptake Fractions and Effective Half-Lives for Iodine-131 Treatments			
	Extrathyroidal Component		**Thyroidal Component**	
Medical Condition	**Uptake Fraction F_1**	**Effective Half-Life T_{1eff} (day)**	**Uptake Fraction F_2**	**Effective Half-Life T_{2eff} (day)**
Hyperthyroidism	0.20[1]	0.32[2]	0.80[1]	5.2[1]
Post-Thyroidectomy for Thyroid Cancer	0.95[3]	0.32[2]	0.05[3]	7.3[2]

Footnotes for Table U.6

[1] M. G. Stabin et al., "Radiation Dosimetry for the Adult Female and Fetus from Iodine-131 Administration in Hyperthyroidism," *Journal of Nuclear Medicine*, Volume 32, Number 5, May 1991. The thyroid uptake fraction of 0.80 was selected as one that is seldom exceeded by the data shown in Figure 1 in this referenced document. The effective half-life of 5.2 days for the thyroidal component was derived from a biological half-life of 15 days, which was obtained from a straight-line fit that accounts for about 75% of the data points shown in Figure 1 of the *Journal of Nuclear Medicine* document.

[2] International Commission on Radiological Protection (ICRP), "Radiation Dose to Patients from Radiopharmaceuticals," ICRP Publication No. 53, March 1987. (Available for sale from Pergamon Press, Inc., Elmsford, NY 10523.) The data in that document suggest that the extrathyroidal component effective half-life in normal subjects is about 0.32 days. Lacking other data, this value is applied to hyperthyroid and thyroid cancer

patients. For thyroid cancer, the thyroidal component effective half-life of 7.3 days is based on a biological half-life of 80 days (adult thyroid), as suggested in the ICRP document.

[3] The thyroidal uptake fraction of 0.05 was recommended by Dr. M. Pollycove, M.D., NRC Medical Visiting Fellow, as an upper-limit post-thyroidectomy for thyroid cancer.

Substituting the appropriate values into Equation B-5, the dose to total decay is:

$$D(\infty) = \frac{(34.6)\ (2.2)\ (150)}{(100\ cm)^2}\ \{(0.75)\ (8.04)\ (0.8)\ (1 - e^{-0.693\ (0.33)\ /\ 8.04})$$

$$+ e^{-0.693\ (0.33)\ /\ 8.04}\ (0.25)\ (0.95)\ (0.32)$$

$$+ e^{-0.693\ (0.33)\ /\ 8.04}\ (0.25)\ (0.05)\ (7.3)\}$$

$$D(\infty) = 3.40\ mSv\ (0.340\ rem)$$

Therefore, thyroid cancer patients to whom 5550 megabecquerels (150 millicuries) of iodine-131 or less have been administered would not have to remain under licensee control and could be released under 10 CFR 35.75, assuming that the foregoing assumptions can be justified for the individual patient's case and that the patient is given instructions. Patients administered somewhat larger activities could also be released immediately if the dose is not greater than 5 millisieverts (0.5 rem).

In the example above, the thyroidal fraction, $F_2 = 0.05$, is a conservative assumption for persons who have had surgery to remove thyroidal tissue. If F_2 has been measured for a specific patient, the measured value may be used.

Example 3, Hyperthyroidism: Calculate the maximum likely dose to an individual exposed to a patient to whom 2035 megabecquerels (55 millicuries) of iodine-131 have been administered for the treatment of hyperthyroidism (i.e., thyroid ablation).

Solution: In this example, the dose will again be calculated using Equation B-5, Table U.5, and Table U.6, to account for the elimination of iodine-131 from the body by using the effective half-lives appropriate for hyperthyroidism. An occupancy factor, E, of 0.25 at 1 meter will be used for the second and third components of the equation because patient-specific instructions were provided to justify the occupancy factor (see Section B.1.2, "Occupancy Factors to Consider for Patient-Specific Calculations").

Substituting the appropriate values into Equation B-5, the dose to total decay is:

$$D(\infty) = \frac{(34.6)\ (2.2)\ (55)}{(100\ cm)^2}\ \{(0.75)\ (8.04)\ (0.8)\ (1 - e^{-0.693\ (0.33)\ /\ 8.04})$$

$$+ e^{-0.693\ (0.33)\ /\ 8.04}\ (0.25)\ (0.20)\ (0.32)$$

$$+ e^{-0.693\ (0.33)\ /\ 8.04}\ (0.25)\ (0.80)\ (5.2)\}$$

$$D(\infty) = 4.86\ mSv\ (0.486\ rem)$$

Therefore, hyperthyroid patients to whom 2035 megabecquerels (55 millicuries) of iodine-131 have been administered would not have to remain under licensee control and could be released under 10 CFR 35.75 when the occupancy factor of 0.25 in the second and third components of the equation is justified.

In the example above, the thyroidal fraction $F_2 = 0.8$ is a conservative assumption for persons who have this treatment for hyperthyroidism. If F_2 has been measured for a specific patient, the measured value may be used.

B.3   Internal Dose

For some radionuclides, such as iodine-131, there may be concerns that the internal dose of an individual from exposure to a released patient could be significant. A rough estimate of the maximum likely committed effective dose equivalent from internal exposure can be calculated from Equation B-6.

Equation B-6:

$$D_i = Q\,(10^{-5})(DCF)$$

where:

D_i = Maximum likely internal committed effective dose equivalent to the individual exposed to the patient in rem;

Q = Activity administered to the patient in millicuries;

10^{-5} = Assumed fractional intake; and

DCF = Dose conversion factor to convert an intake in millicuries to an internal committed effective dose equivalent (such as tabulated in Reference B-2).

Equation B-6 uses a value of 10^{-5} as the fraction of the activity administered to the patient that would be taken in by the individual exposed to the patient. A common rule of thumb is to assume that no more than 1 millionth of the activity being handled will become an intake to an individual working with the material. This rule of thumb was developed in Reference B-3 for cases of worker intakes during normal workplace operations, worker intakes from accidental exposures, and public intakes from accidental airborne releases from a facility, but it does not specifically apply to cases of intake by an individual exposed to a patient. However, two studies (Refs. B-4 and B-5) regarding the intakes of individuals exposed to patients administered iodine-131 indicated that intakes were generally of the order of 1 millionth of the activity administered to the patient and that internal doses were far below external doses. To account for the most highly exposed individual and to add a degree of conservatism to the calculations, a fractional transfer of 10^{-5} has been assumed.

Example 4, Internal Dose: Using the ingestion pathway, calculate the maximum internal dose to a person exposed to a patient to whom 1221 megabecquerels (33 millicuries) of iodine-131 have been administered. The ingestion pathway was selected because it is likely that most of the intake would be through the mouth or through the skin, which is most closely approximated by the ingestion pathway.

Solution: This is an example of the use of Equation B-6. The dose conversion factor DCF for the ingestion pathway is 53 rem/millicurie from Table 2.2 of Reference B-2.

Substituting the appropriate values into Equation B-6, the maximum internal dose to the person is:

$$D_i = (33 \text{ mCi})(10^{-5})(53 \text{ rem/mCi})$$

$$D_i = 0.17 \text{ mSv } (0.017 \text{ rem})$$

Using Equation B-1 and assuming the patient has received instructions for reducing exposure as recommended for an occupancy factor of 0.25, the external dose is approximately 5 mSv (0.5 rem). Thus, the internal dose is about 3% of the external dose due to gamma rays. Internal doses may be ignored in calculations of total dose if they are likely to be less than 10% of the external dose because the internal dose due to this source is small in comparison to the magnitude of uncertainty in the external dose.

The conclusion that internal contamination is relatively unimportant in the case of patient release was also reached by the NCRP. The NCRP addressed the risk of intake of radionuclides from patients' secretions and excreta in NCRP Commentary No. 11, "Dose Limits for Individuals Who Receive Exposure from Radionuclide Therapy Patients" (Ref. B-6). The NCRP concluded, "Thus, a contamination incident that could lead to a significant intake of radioactive material is very unlikely." For additional discussion on the subject, see Reference B-1.

Example 5, Internal Dose: Calculate the maximum internal dose to a person exposed to a patient to whom 5550 megabecquerels (150 millicuries) of iodine-131 have been administered for the treatment of thyroid remnants and metastasis.

Solution: In this example, the dose is again calculated using Equation B-6 and selecting the ingestion pathway. Substituting the appropriate values into Equation B-6, the maximum internal dose to the person is:

$$D_i = (150 \text{ mCi})(10^{-5})(53 \text{ rem/mCi})$$

$$D_i = 0.80 \text{ mSv } (0.08 \text{ rem})$$

In this case, the external dose to the other person from Example 2, Thyroid Cancer, was approximately 3.4 millisieverts (0.34 rem), while the internal dose would be about 0.80 millisievert (0.08 rem). Thus, the internal dose is about 24% of the external gamma dose. Therefore, the internal and external doses must be summed to determine the total dose; 4.2 millisieverts (0.42 rem).

References for Supplement B

B-1.   S. Schneider and S. A. McGuire, "Regulatory Analysis on Criteria for the Release of Patients Administered Radioactive Material," U.S. NRC, NUREG-1492, February 1997.

B-2.   K. F. Eckerman, A. B. Wolbarst, and A. C. B. Richardson, "Limiting Values of Radionuclide Intake and Air Concentration and Dose Conversion Factors for Inhalation, Submersion, and Ingestion," Federal Guidance Report No.11, U. S. Environmental Protection Agency, Washington, DC, 1988.

B-3.   A. Brodsky, "Resuspension Factors and Probabilities of Intake of Material in Process (or 'Is 10^{-6} a Magic Number in Health Physics?')," *Health Physics*, Volume 39, Number 6, 1980.

B-4.   R. C. T. Buchanan and J. M. Brindle, "Radioiodine Therapy to Out-patients – The Contamination Hazard," *British Journal of Radiology*, Volume 43, 1970.

B-5.   A. P. Jacobson, P. A. Plato, and D. Toeroek, "Contamination of the Home Environment by Patients Treated with Iodine-131," *American Journal of Public Health*, Volume 68, Number 3, 1978.

B-6.   National Council on Radiation Protection and Measurements, "Dose Limits for Individuals Who Receive Exposure from Radionuclide Therapy Patients," Commentary No. 11, February 28, 1995.

Regulatory Analysis

"Regulatory Analysis on Criteria for the Release of Patients Administered Radioactive Material" (NUREG-1492, February 1997) provides the regulatory basis and examines the costs and benefits. A copy of NUREG-1492 is available for inspection and copying for a fee at NRC's Public Document Room, 2120 L Street NW, Washington, DC. Copies may be purchased at current rates from the U.S. Government Printing Office, P.O. Box 37082, Washington, DC 20402-9328 (telephone (202) 512-2249), or from the National Technical Information Service by writing NTIS at 5285 Port Royal Road, Springfield, VA 22161.

APPENDIX V

Guidance for Mobile Medical Services

Guidance for Mobile Medical Services

With the implementation of the EPAct, the NRC now has regulatory authority over accelerator-produced radioactive materials and discrete sources of radium-226. Therefore, all the requirements for mobile medical services also apply to the mobile medical use of accelerator-produced radioactive materials and discrete sources of radium-226 after NRC's waiver of August 31, 2005, is terminated for medical use facilities. The NRC waiver that applied to Government agencies, Federally recognized Indian tribes, Delaware, the District of Columbia, Puerto Rico, the U.S. Virgin Islands, Indiana, Wyoming, and Montana was terminated on November 30, 2007. The NRC Regional Offices should be contacted to confirm the waiver termination date for other medical use facilities.

Before submitting information to the NRC, review Section 5.2 of this document for guidance on identifying and protecting sensitive information. All security-related information in the application should be identified and properly marked.

Mobile medical service providers must comply with all applicable sections of 10 CFR Part 35 as well as DOT regulations with regard to approved source holders, placement of sources in approved containers prior to their transport, and hazardous materials training. For example, mobile medical service providers offering remote afterloaders must comply with Subpart H of 10 CFR Part 35.

Type and Location of Use

In general, there are two types of mobile medical service. One type is transportation and use of byproduct material within a transport vehicle (e.g., in-van use). A second type is transportation of byproduct material to a client's facility for use within a client's facility by the mobile medical service's employees (i.e., transport and use). As a result of the EPAct, byproduct material now includes accelerator-produced radioactive materials and discrete sources of radium-226.

Whether a PET mobile medical service provider that uses a "quiet room" in the client's facility is authorized for "in-van use" or " transport and use" depends on whether the PET patients meet the criteria for release in 10 CFR 35.75 while they are in the "quiet room." If they do not, then the "quiet room" is an area of use for the mobile service licensee.

For the first and second types, which include use by the service provider, the service provider should apply for full service authorization. Service providers who only transport and store a therapy device need only apply for authorization for possession and transport of the byproduct material. In this case, when the service provider is only transporting the therapy device for use, the client must possess a license for medical use of the byproduct material. Additionally, in this case, the client is authorized to provide the patient treatments and is responsible for all aspects of the byproduct material use and patient treatments upon transfer of the byproduct material to the client's possession.

For all types, licensed activities must be conducted in accordance with the regulations for compliance with 10 CFR 35.80(a), which states that the licensee will obtain a letter signed by the management of each of its clients for which services are rendered. The letter will permit the use of byproduct material at the client's address and will clearly delineate the authority and responsibility of each entity. This agreement must be applicable for the entire period of time over which the service is to be provided. The letter will be retained for 3 years after the last provision of service, as required by 10 CFR 35.80(c) and 10 CFR 35.2080. Additionally, as required by 10 CFR 35.80(a)(4), the licensee must survey to ensure compliance with the requirements in 10 CFR Part 20 (e.g., ensure that all byproduct material, including radiopharmaceuticals, sealed sources, and all associated wastes, have been removed) before leaving a client's address.

The locations of use for mobile medical services are of two basic types. One type of location is the base location where licensed material is received, stored, and sometimes used. The other type of location is the temporary job site at client facilities. The following two sections describe the type of information necessary for base locations and temporary job sites.

Base Location

The base location (e.g., central radiopharmaceutical laboratory or storage location for the remote afterloader) for the mobile medical service must be specified. The base facility may be located in a medical institution, noninstitutional medical practice, commercial facility, or mobile van. Applicants should specify in what type of facility the proposed base facility is located. A mobile licensee cannot provide a service to a private practice (nonlicensee) located within a licensed medical institution (e.g., hospital). As required by 10 CFR 30.33 and 10 CFR 35.12, applicants must submit a description and diagram(s) of the proposed base facility and associated equipment in accordance with Items 8.14 through 8.19 of this report. The description and diagram of the proposed facility should demonstrate that the building (or van) is of adequate construction and design to protect its contents from the elements (e.g., high winds, rain), ensures security of licensed material to prevent unauthorized access (e.g., control of keys), and ensures that radiation levels in unrestricted areas are in compliance with 10 CFR 20.1301. Include a diagram showing the location of the licensed material (which now also includes accelerator-produced radioactive materials and discrete sources of radium-226), receipt, and use areas, and identify all areas adjacent to restricted areas, including areas above and below the restricted areas. For storage locations within a van, the description of the van should address radiation levels in the van driver's compartment to demonstrate compliance with 10 CFR 20.1201, "Occupational dose limits for adults."

- Applicants may request multiple base locations. Radioactive material must be delivered only to a facility licensed to receive the type of radioactive material ordered.

- Base locations can include the use of a mobile van. When the base facility is in the van, and there is no permanent structure for the byproduct material storage, provide for the following:

  — Secured off-street parking under licensee control. Public rights-of-way are not considered part of the address of the client;

  — Secured storage facilities available for storage of byproduct material and radioactive waste if the van is disabled; and

— Byproduct material (which now also includes accelerator-produced radioactive materials and discrete sources of radium-226) delivered (if necessary) directly to the van only if the van is occupied by licensee personnel at the time of delivery.

- If a base facility is located in a residential area, provide the following information:

  — Justification of the need for a private residence location rather than for a commercial location.

  — Documentation of the agreement between the residence owner and the licensee. It is essential that the mobile medical service have access to the facility in the event of contamination. Provisions for decontamination of the mobile medical service van, etc., on the client property (if necessary) will be included. Documentation from both parties will illustrate the agreement between the client and the mobile medical service.

  — A description of the program demonstrating compliance with 10 CFR 20.1301, "Dose limits for individual members of the public."

  — Verification that restricted areas do not contain residential quarters.

- Perform surveys necessary to show that exposure rates do not exceed 2 mrem in any 1 hour nor 100 mrem per year.

Client Site

This section applies only to therapeutic uses of byproduct material (which now also includes accelerator-produced radioactive materials and discrete sources of radium-226). For all types of therapy uses, the medical institutions, hospitals, or clinics and their addresses that comprise the client sites for mobile medical services must be listed.

For self-contained byproduct material services (e.g., in-van), the following additional facility information should be provided:

- For therapy treatments with byproduct material (e.g., high dose-rate remote afterloader), a separate drawing for each client site showing the location of the treatment device/vehicle in relation to all nearby roads, sidewalks, structures, and any other locations accessible by members of the public;

- A signed agreement, as delineated in the letter required by 10 CFR 35.80(a), that location of the device/vehicle will be on client-owned or controlled property;

- The protection from vehicular traffic that could adversely affect patient treatment(s), that could be accomplished either by locating the facility away from all vehicular traffic or by using barriers. Any protective measures must be shown on the facility/site drawings provided.

- A description of the emergency lighting system that automatically activates on detection of the loss of primary power during patient remote afterloader treatments. The system must provide sufficient light to perform any possible emergency procedures, including the removal of a detached or stuck source that remains within the patient.

If transportable services will be provided to the client's site for use within the client's facility by the mobile medical service's employees, the following client facility information and commitment should be provided:

- A detailed description and diagram(s) of the proposed use facility (e.g., client site) and associated equipment in accordance with Items 8.14 through 8.19 of this report. The description and diagram of the proposed use facility must demonstrate that the facility is of adequate construction and design to protect its contents from the elements (e.g., high winds, rain), ensure security of licensed material to prevent unauthorized access, and ensure that radiation levels in unrestricted areas are in compliance with 10 CFR 20.1301. Include a diagram showing the location of the equipment, receipt, and use areas, and identify all areas adjacent to restricted areas.

- A commitment, as delineated in the letter required by 10 CFR 35.80(a), that the mobile medical service licensee has full control of the treatment room during byproduct material use for each client.

- The initial installation records and function checks of a remote afterloader device for each site of use, as required by 10 CFR 35.633, 10 CFR 35.643, and 10 CFR 35.647.

For a transport-only mobile medical service for therapy devices that are transported to the client's facility, used by the client's staff (under their own license), and removed by the service provider, ensure the following:

- Each client is properly licensed for medical use of byproduct material (which now also includes accelerator-produced radioactive materials and discrete sources of radium-226). If applicable, licensees should ensure that each client has received the necessary initial and, if appropriate, recurrent training for the specific make and model of the remote afterloader device being provided. If the above applicable conditions are not met, the mobile medical service licensee must not transfer the remote afterloader device to the client.

- No signed agreement with a client may state or imply any assumption of responsibility on the part of the mobile medical service for the use of byproduct material for patient treatments. This includes such activities as dosage measurements, source calibrations, and remote afterloader device operational checks. Although these and other services may be provided to the client by the mobile medical service if the mobile medical service is specifically licensed to provide such services, the client (licensee) retains all of the responsibilities related to the use of the byproduct material for patient treatments. The responsibilities for supervising individuals who use the byproduct material, set forth in 10 CFR 35.27, transfer to the client's AUs upon transfer of the device to the client by the mobile medical service provider.

- The initial installation of a remote afterloader device at the client site may be performed by either the mobile medical service provider or the client, but all device function checks are the responsibility of the client (i.e., the licensee authorized to provide patient treatments at the client site).

- As required by 10 CFR 30.51, a formal record of the transfer of control of the byproduct material from the mobile medical service provider to the client, and from the client back to the mobile medical service provider, must be made for each transfer of byproduct material. A signed receipt of each transfer must be made and retained for inspection for 3 years.

Supervision

In addition to the requirements in 10 CFR 19.12, 10 CFR 35.27 requires that instructions be given to supervised individuals in written radiation protection procedures, written directive procedures, regulations, and license conditions with respect to the use of byproduct material (which now also includes accelerator-produced radioactive materials and discrete sources of radium-226). Additionally, 10 CFR 35.27 requires the supervised individual to:

- Follow the instructions of the supervising AU for medical uses of byproduct material;

- Follow the instructions of the supervising ANP or supervising AU for preparation of byproduct material for medical uses;

- Follow the written radiation protection procedures and written directive procedures established by the licensee; and

- Comply with the provisions of 10 CFR Part 35 (e.g., 10 CFR 35.80 and 10 CFR 35.647 (if applicable)), and the license conditions with respect to the mobile medical use of byproduct material.

Training for Individuals Working in or Frequenting Restricted Areas

Drivers and technologists (or therapists) will be properly trained in applicable transportation regulations and emergency procedures in addition to the training requirements of 10 CFR 19.12, 10 CFR 35.27, 10 CFR 35.310, 10 CFR 35.410, and 10 CFR 35.610 (as applicable). The training for these individuals will include, at a minimum, DOT regulations, shielding, ALARA, and basic radiation protection.

Survey Instrument and Dose Measurement Instrument Checks

As required by 10 CFR 35.80, instruments should be checked for proper operation before use at each address of use. Dosage measurement instruments should be checked before medical use at each address of use or on each day of use, whichever is more frequent. Additionally, all other transported equipment (e.g., cameras) should be checked for proper function before medical use at each address of use.

Order and Receipt of Byproduct Material (which now also includes accelerator-produced radioactive materials and discrete sources of radium-226)

Byproduct material will be delivered by a supplier to the base location or to the client's address if the client is licensed to receive the type of byproduct material ordered. Delivery of byproduct material to a van that is not occupied by the mobile medical service personnel will not be permitted.

Alternatively, licensees may pick up the byproduct material (e.g., radiopharmaceuticals) from the supplier (e.g., nuclear pharmacy) en route to client facilities.

Emergency Procedures

Develop, implement, and maintain emergency procedures, in accordance with the Radiation Protection Program required by 10 CFR 20.1101. Indicate typical response times of the RSO and AU in the event of an incident and develop and implement procedures that include emergency response regarding an accident scenario. An accident is defined as a vehicle collision or other event, such as wind, water, or fire, that results in damage to exterior or interior portions of the vehicle or the byproduct material used in the mobile medical service. The transportation emergency response plan should cover both the actions to be taken by the mobile medical service provider's headquarters emergency response personnel and the "on-scene" hazardous-material (HAZMAT)-trained personnel, and it will be readily available to both transport vehicle personnel and headquarters emergency-response contacts. The plan should include the following:

- A 24-hour emergency contact telephone number for the mobile medical service provider's emergency response personnel;

- The emergency contact numbers for NRC's Operation Center and all appropriate State radiological protection agencies;

- Procedures for restricting access to the transport vehicle until surveys have been made to determine if any radiological hazards exist;

- Procedures for retrieving and securing any byproduct material, including a sealed source that may become detached and/or dislodged to the extent that a radiological hazard is created, which may require one or more emergency shielded source containers;

- Predetermined (calculated) exposure rates for an unshielded therapy source (if applicable) as a function of distance for use in controlling the exposures of emergency response personnel to the maximum extent possible under various emergency response scenarios;

- Preplanned decontamination procedures, including ready access to all necessary materials;

- A calibrated, operational survey meter maintained in the cab of the transporting vehicle, which may be used at an accident scene for conducting surveys;

- Security of the transport vehicle against unauthorized access, including the driver's compartment; and

- Procedures to ensure that following any accident, no patient treatments with remote afterloaders will occur until all systems pertaining to radiation safety have been tested and confirmed to be operational by the RSO or AMP. If any problem is found, including remote afterloader device interlocks and operation, the remote afterloader device or facility will be repaired and re-certified by the device vendor prior to return to service. In addition, a copy of the report, generated in accordance with 10 CFR 30.50, will be provided to clients following any accident in which there is actual or possible damage to the client's facility or the device.

Note: The type of response should be consistent with the level of the incident. The response may range from telephone contact for minor spills to prompt onsite response (less than 3 hours) to events such as a medical event or lost radioactive material.

Transportation

Develop, document, and implement procedures to assure that the following takes place:

- Radioactive material is transported in accordance with 49 CFR Parts 170–189. Procedures will include:

    — Use of approved packages,
    — Use of approved labeling,
    — Conduct of proper surveys,
    — Complete and accurate shipping papers,
    — Bracing of packages,
    — Security provisions, and
    — Written emergency instructions.

- Management (or management's designee) will perform audits, at least annually, of transportation documentation (e.g., shipping papers and survey reports) and activities at client facilities.

- Licensed material (which now also includes accelerator-produced radioactive materials and discrete sources of radium-226) is secured during transport and use at the client's facilities.

- Radioactive waste is handled properly during transport. Describe the method of storage and final disposal.

- The transport vehicle, including the driver's compartment, if separate, will be secured at all times from any unauthorized access when the vehicle is unattended.

Note: The necessary DOT Type 7A package certification for remote afterloader devices is established by prior approval of the appropriate sealed source and device sheets; however, if the remote afterloader device is damaged in any way during use or transport, then the integrity of the DOT Type 7A packaging may be compromised, and the device must not be used or transported until checked by the vendor and certified as retaining its integrity as a Type 7A package.

Radioactive Waste Management (this now includes waste containing accelerator-produced radioactive materials and discrete sources of radium-226)

If waste will be stored in vans, the vans must be properly secured and posted as byproduct material storage locations. Ensure that the van will be secured against unauthorized access and that the waste storage location will be posted as a byproduct material storage area.

Develop, document, and implement final waste disposal procedures in accordance with Section 8.28 of this report.

Excreta from individuals undergoing medical diagnosis or therapy with radioactive material may be disposed of without regard to radioactivity if it is discharged into the sanitary sewer system, in accordance with 10 CFR 20.2003. However, collecting excreta from patients in a van restroom with a holding tank is not considered direct disposal into the sanitary sewer system. If restroom facilities are provided in the van for patient use, submit the following information for NRC review:

- A description of the structure of the tank holding facility and the location of the tank in relation to members of the public, workers in the van, and the driver of the van; a description of procedures to assess the tank for possible leakage; and a description of any restroom ventilation if any I-131 will be held in the tank.

- A description of procedures to ensure doses to occupational workers and members of the public will not exceed the exposure limits in 10 CFR 20.1201 and 20.1301, that the external surfaces of the van do not exceed 2 mrem/hour, and that doses to members of the public and workers are maintained ALARA, including considerations of external dose rates in the restroom caused by the proximity of the holding tank to the toilet.

- A description of procedures for emptying and disposing of the contents of the holding tank, including the frequency of disposal, who empties the tank into the sanitary sewer system, and the location of disposal into the sanitary sewer, including precautions taken to minimize contamination in this process.

Mobile Medical Services With Remote Afterloader Devices

Because the movement of the remote afterloader device from one location to another increases the risk of electro-mechanical component failures or misalignments, it is important that the proper operation of the device be fully checked after each such relocation. Therefore, develop, document, and implement the following procedures to determine if a device is operating properly before the commencement of patient treatments:

- Conduct safety checks on a remote afterloader device and facility. The procedure will include the periodic spot checks required by 10 CFR 35.643 and the additional spot checks required by 10 CFR 35.647 before use at each address of use. Additionally, the procedure should include provisions for prompt repair of any system not operating properly.

- The pretreatment operational function checks after each device move should include a review of any device alarm or error message and, if necessary, a resolution of problems indicated by such messages.

- Such tests should be performed in accordance with written procedures.

- As required by 10 CFR 35.2647 and 10 CFR 35.2643, records showing the results of the above safety checks must be maintained for NRC inspection and review for a period of 3 years.

- Perform surveys of the source housing and areas adjacent to the treatment room following relocation of a high dose-rate unit. These surveys should include the source housing with the source in the shielded position and all areas adjacent to the treatment room with the source in the treatment position.

APPENDIX W

Model Procedure for Waste Disposal by Decay-In-Storage, Generator Return, and Licensed Material Return

Model Procedure for Waste Disposal by Decay-In-Storage, Generator Return, and Licensed Material Return

With the implementation of the EPAct, the NRC now has regulatory authority over accelerator-produced radioactive materials and discrete sources of radium-226. Therefore, the procedures for waste disposal by decay-in-storage, generator return, and licensed material return also apply to the medical accelerator-produced radioactive materials and discrete sources of radium-226 after NRC's waiver of August 31, 2005, is terminated for medical use facilities. The NRC waiver that applied to Government agencies, Federally recognized Indian tribes, Delaware, the District of Columbia, Puerto Rico, the U.S. Virgin Islands, Indiana, Wyoming, and Montana was terminated on November 30, 2007. The NRC Regional Offices should be contacted to confirm the waiver termination date for other medical use facilities.

This model provides acceptable procedures for waste disposal. Note that some short half-life radionuclide products (e.g., Tc-99m/Mo-99 generator columns and some Y-90 microspheres) contain long half-life contaminants that may preclude disposal by decay-in-storage. Applicants may either adopt these model procedures or develop alterative procedures to meet the requirements of Subpart K to 10 CFR Part 20, 10 CFR 20.1101, and 10 CFR 35.92.

Model Procedure for Decay-In-Storage (this now includes decay-in-storage of accelerator-produced radioactive materials)

Section 10 CFR 35.92 describes the requirements for decay-in-storage. Storage should be designed to allow for segregation of wastes with different half-lives (e.g., multiple shielded containers). Containers should have shielded covers to maintain occupational exposure at ALARA levels. Storage areas must be in a secure location.

- If possible, use separate containers for different types of waste (e.g., needles and syringes in one container, other injection paraphernalia such as swabs and gauze in another, and unused dosages in a third container). Because the waste will be surveyed with all shielding removed, the containers in which the waste will be placed must not provide any radiation shielding for the material.

- When the container is full, seal it and attach an identification tag that includes the date sealed and the longest-lived radionuclide in the container. The container may then be transferred to the decay-in-storage area.

- Prior to disposal as in-house waste, monitor and record the results of monitoring of each container as follows:

  – Use a survey instrument that is appropriate for the type and energy of the radiation being measured.

  – Check the radiation detection survey meter for proper operation and current calibration status.

  – Monitor in a low-level radiation (<0.05 millirem per hour) area away from all sources of radioactive material, if possible.

  – Remove any shielding from around the container or generator column.

− Monitor, at contact, all surfaces of each individual container.

− Remove or deface any radioactive material labels (unless the containers will be managed as biomedical waste after they have been released from the licensee as described in 10 CFR 35.92).

− Discard as in-house waste only those containers that cannot be distinguished from background radiation. Containers may include trash bags full of waste, generator columns, and biohazard (needle) boxes. Record the disposal date, the survey instrument used, the background dose rate, the dose rate measured at the surface of each waste container, and the name of the individual who performed the disposal.

Containers that can be distinguished from background radiation levels must be returned to the storage area for further decay or transferred to an authorized byproduct material recipient.

Model Procedure for Returning Generators to the Manufacturer (this now includes generators containing accelerator-produced radioactive materials)

Used Mo/Tc-99m or Sr-82/Ru-82 generators may be returned to the manufacturer. This permission does not relieve licensees from the requirement to comply with 10 CFR Part 71 and DOT regulations. Perform the following actions when returning generators:

- Retain the records needed to demonstrate that the package qualifies as a DOT Specification 7A container,

- Assemble the package in accordance with the manufacturer's instructions,

- Perform the dose-rate and removable-contamination measurements,

- Label the package and complete the shipping papers in accordance with the manufacturer's instructions, and

- Retain records of receipts and transfers in accordance with 10 CFR 30.51.

Model Procedure for Return of Licensed Material to Authorized Recipients (this now includes return of accelerator-produced radioactive materials or discrete sources of radium-226)

Note: Licensees authorized to produce and noncommercially transfer PET radioactive drugs to consortium members are not authorized to receive unused dosages or empty syringes or vials from consortium members.

Perform the following steps when returning licensed material (which now also includes accelerator-produced radioactive materials and discrete sources of radium-226) to authorized recipients:

- In accordance with 10 CFR 30.41(a)(5), confirm that persons are authorized to receive byproduct material prior to transfer (e.g., obtain a copy of the transferee's NRC license or Agreement State license that authorizes the byproduct material).

- Retain the records needed to demonstrate that the package qualifies as a DOT Specification 7A container.

- Assemble the package in accordance with the manufacturer's instructions.

- Perform the dose-rate and removable-contamination measurements.

- Label the package and complete the shipping papers in accordance with the manufacturer's instructions.

- Retain records of receipts and transfers in accordance with 10 CFR 30.51.

APPENDICES X-Z

RECORDKEEPING AND REPORTING REQUIREMENTS AND DOT RULES FOR SHIPPING

APPENDIX X

Recordkeeping Requirements

Recordkeeping Requirements

With the implementation of the EPAct, the NRC now has regulatory authority over accelerator-produced radioactive materials and discrete sources of radium-226. Therefore, the recordkeeping requirements below also apply to the medical uses of accelerator-produced radioactive materials and discrete sources of radium-226 after NRC's waiver of August 31, 2005, is terminated for medical use facilities. The NRC waiver that applied to Government agencies, Federally recognized Indian tribes, Delaware, the District of Columbia, Puerto Rico, the U.S. Virgin Islands, Indiana, Wyoming, and Montana was terminated on November 30, 2007. The NRC Regional Offices should be contacted to confirm the waiver termination date for other medical use facilities.

Table X.1 Typical Records and Retention Times			
Record	Survey Requirement	Recordkeeping Requirement	Retention Period
Results of surveys and calibrations	20.1501; 20.1906(b)	20.2103(a)	3 years
Results of surveys to determine dose from external sources		20.2103(b)(1)	duration of license
Results of measurements and calculations used to determine individual intakes		20.2103(b)(2)	duration of license
Results of air samplings, surveys, and bioassays	20.1703(c)(1); 20.1703(c)(2)	20.2103(b)(3)	duration of license
Results of measurements and calculations used to evaluate the release of radioactive effluents to the environment		20.2103(b)(4)	duration of license
Determination of prior occupational dose		20.2104	duration of license
Planned special exposure	20.1206	20.2105	duration of license
Individual monitoring results	20.1502	20.2106	duration of license
Dose to individual members of the public	20.1301	20.2107	duration of license
Waste disposal	20.2002; 20.2003; 20.2004; 20.2005	20.2108	duration of license
Records of receipt of byproduct material		30.51(a)(1)	duration of possession and 3 years after transfer
Records of transfer of byproduct material		30.51(a)(2)	3 years after transfer
Records of disposal of byproduct material		30.51(a)(3)	duration of license

Table X.1 Typical Records and Retention Times (continued)			
Record	**Survey Requirement**	**Recordkeeping Requirement**	**Retention Period**
Authority and responsibilities of Radiation Protection Program	35.24(a)	35.2024	5 years
Radiation Protection Program changes	35.26(a)	35.2026	5 years
Written directives	35.40	35.2040	3 years
Procedures for administrations requiring a written directive	35.41(a)	35.2041	duration of license
Calibrations of instruments used to measure activity of unsealed byproduct material	35.60	35.2060	3 years
Radiation survey instrument calibrations	35.61	35.2061	3 years
Dosages of unsealed byproduct material for medical use	35.63	35.2063	3 years
Leak tests and inventory of sealed sources and brachytherapy sources	35.67(b)	35.2067	3 years
Surveys for ambient radiation exposure rate	35.70	35.2070	3 years
Release of individuals containing unsealed byproduct material or implants containing byproduct material	35.75	35.2075	3 years
Mobile medical services	35.80(a)(1)	35.2080	3 years
Decay-in-storage	35.92	35.2092	3 years
Molybdenum-99 or strontium-82 or strontium-85 concentrations	35.204(b)	35.2204	3 years
Safety instruction	35.310; 35.410; 35.610	35.2310	3 years
Surveys after source implant and removal	35.404; 35.604	35.2404	3 years
Brachytherapy source accountability	35.406	35.2406	3 years
Calibration measurements of brachytherapy sources	35.432	35.2432	3 years
Decay of strontium-90 sources for ophthalmic treatments	35.433	35.2433	life of source
Installation, maintenance, adjustment, and repair of remote afterloader units, teletherapy units, and gamma stereotactic radiosurgery units	35.604	35.2605	3 years

Table X.1	Typical Records and Retention Times (continued)		
Record	Survey Requirement	Recordkeeping Requirement	Retention Period
Safety procedures	35.610(a)(4); 35.610(d)(2)	35.2610	duration of possession of specified equipment
Dosimetry equipment used with remote afterloader units, teletherapy units, and gamma stereotactic radiosurgery units	35.630	35.2630	duration of license
Teletherapy, remote afterloader, and gamma stereotactic radiosurgery full calibrations	35.632; 35.633; 35.635	35.2632	3 years
Periodic spot-checks of teletherapy units	35.642	35.2642	3 years
Periodic spot-checks of remote afterloader units	35.643	35.6243	3 years
Periodic spot-checks of gamma stereotactic radiosurgery units	35.645	35.6245	3 years
Additional technical requirements for mobile remote afterloader units	35.647	35.6247	3 years
Surveys of therapeutic treatment units	35.652	35.2652	duration of use of unit
5-year inspection for teletherapy and gamma stereotactic radiosurgery units	35.655	35.2655	duration of use of unit

APPENDIX Y

Reporting Requirements

Reporting Requirements

With the implementation of the EPAct, the NRC now has regulatory authority over accelerator-produced radioactive materials and discrete sources of radium-226. Therefore, the reporting requirements below also apply to the medical uses of accelerator-produced radioactive materials and discrete sources of radium-226 after NRC's waiver of August 31, 2005, is terminated for medical use facilities. The NRC waiver that applied to Government agencies, Federally recognized Indian tribes, Delaware, the District of Columbia, Puerto Rico, the U.S. Virgin Islands, Indiana, Wyoming, and Montana was terminated on November 30, 2007. The NRC Regional Offices should be contacted to confirm the waiver termination date for other medical use facilities.

Table Y.1	Typical NRC Notifications and/or Reports		
Event	**Telephone Notification**	**Written Report**	**Regulatory Requirement**
Reports to individual workers	none	annually	10 CFR 19.13(b)
Reports to former individual workers	none	upon request	10 CFR 19.13(c)
Notification of special circumstances to individuals	none	30 days	10 CFR 19.13(d)
Reports to worker terminating employment	none	upon request	10 CFR 19.13(e)
Theft or loss of material	immediate	30 days	10 CFR 20.2201(a)(1)(i)
Whole body dose greater than 0.25 Sv (25 rems)	immediate	30 days	10 CFR 20.2202(a)(1)(i), 10 CFR 20.2203 (a)
Extremity dose greater than 2.5 Sv (250 rems)	immediate	30 days	10 CFR 20.2202(a)(1)(iii), 10 CFR 20.2203 (a)
Whole body dose greater than 0.05 Sv (5 rems) in 24 hours	24 hours	30 days	10 CFR 20.2202(b)(1)(i), 10 CFR 20.2203 (a)
Extremity dose greater than 0.5 Sv (50 rems) in 24 hours	24 hours	30 days	10 CFR 20.2202(b)(1)(iii), 10 CFR 20.2203(a)
Doses in excess of specified criteria	none	30 days	10 CFR 20.2203(a)(2)
Levels of radiation or concentrations of radioactive material in excess of specified criteria	none	30 days	10 CFR 20.2203(a)(3)
Planned special exposures	none	30 days	10 CFR 20.2204
Report to individuals of exceeding dose limits	none	30 days	10 CFR 20.2205
Report of individual monitoring	none	annually	10 CFR 20.2206

Table Y.1 Typical NRC Notifications and/or Reports

Event	Telephone Notification	Written Report	Regulatory Requirement
Defect in equipment that could create a substantial safety hazard	2 days	30 days	10 CFR 21.21(d)(3)(i)
Event that prevents immediate protective actions necessary to avoid exposure to radioactive materials that could exceed regulatory limits	immediate	30 days	10 CFR 30.50(a)
Equipment is disabled or fails to function as designed when required to prevent radiation exposure in excess of regulatory limits	24 hours	30 days	10 CFR 30.50(b)(2)
Unplanned fire or explosion that affects the integrity of any licensed material or device, container, or equipment with licensed material	24 hours	30 days	10 CFR 30.50(b)(4)
Licensee permits individual to work as AU, ANP, or AMP	none	30 days	10 CFR 35.14(a)
AU, ANP, or AMP discontinues performance of duties under license or has a name change	none	30 days	10 CFR 35.14(b)(1)
Licensee's mailing address changes	none	30 days	10 CFR 35.14(b)(2)
Licensee's name changes without constituting a transfer of control	none	30 days	10 CFR 35.14(b)(3)
Licensee adds or changes areas of 10 CFR 35.100 or 35.200 use of byproduct material identified in application or license if the change or addition did not involve movement of a PET radionuclide production facility or transfer line from a PET radionuclide production facility	none	30 days	10 CFR 35.14(b)(4)
Medical event	1 day	15 days	10 CFR 35.3045
Dose to embryo or nursing child	1 day	15 days	10 CFR 35.3047
Leaking source	none	5 days	10 CFR 35.3067

Note: Telephone notifications shall be made to the NRC Operations Center at 301-951-0550, except as noted.

APPENDIX Z

Summary of DOT Requirements for Transportation of Type A or Type B Quantities of Licensed Material

Summary of DOT Requirements for Transportation of Type A or Type B Quantities of Licensed Material

With the implementation of the EPAct, the NRC now has regulatory authority over accelerator-produced radioactive materials and discrete sources of radium-226. Therefore, the requirements for transportation of licensed material also apply to transportation of accelerator-produced radioactive materials and discrete sources of radium-226 after NRC's waiver of August 31, 2005, is terminated for medical use facilities. The NRC waiver that applied to Government agencies, Federally recognized Indian tribes, Delaware, the District of Columbia, Puerto Rico, the U.S. Virgin Islands, Indiana, Wyoming, and Montana was terminated on November 30, 2007. The NRC Regional Offices should be contacted to confirm the waiver termination date for other medical use facilities.

Licensed material must be transported in accordance with the Department of Transportation (DOT) regulations. The major areas in the DOT regulations that are most relevant for transportation of Type A or Type B quantities of licensed material are:

- Table of Hazardous Materials and Special Provisions, 49 CFR 172.101: Purpose and use of hazardous materials table;

- Shipping Papers, 49 CFR 172.200-204: Applicability, general entries, description of hazardous material on shipping papers, additional description requirements, shipper's certification;

- Package Marking, 49 CFR 172.300, 49 CFR 172.301, 49 CFR 172.303, 49 CFR 172.304, 49 CFR 172.310, 49 CFR 172.324: Applicability, general marking requirements for nonbulk packaging, prohibited marking, marking requirements, radioactive material, hazardous substances in nonbulk packaging;

- Package Labeling, 49 CFR 172.400, 49 CFR 172.401, 49 CFR 172.403, 49 CFR 172.406, 49 CFR 172.407, 49 CFR 172.436, 49 CFR 172.438, 49 CFR 172.440: General labeling requirements, prohibited labeling, Class 7 (radioactive) material, placement of labels, label specifications, radioactive white-I label, radioactive yellow-II label, radioactive yellow-III label;

- Placarding of Vehicles, 49 CFR 172.500, 49 CFR 172.502, 49 CFR 172.504, 49 CFR 172.506, 49 CFR 172.516, 49 CFR 172.519, 49 CFR 172.556: Applicability of placarding requirements, prohibited and permissive placarding, general placarding requirements, providing and affixing placards: highway, visibility and display of placards, general specifications for placards, "RADIOACTIVE" placard;

- Emergency Response Information, 49 CFR 172.600, 49 CFR 172.602, 49 CFR 172.604: Applicability and general requirements, emergency response information, emergency response telephone number;

- Training, 49 CFR 172.702, 49 CFR 172.704: Applicability and responsibility for training and testing, training requirements;

- Shippers – General Requirements for Shipments and Packaging, 49 CFR 173.403, 49 CFR 173.410, 49 CFR 173.411, 49 CFR 173.412, 49 CFR 173.413, 49 CFR 173.415,

49 CFR 173.416, 49 CFR 173.433, 49 CFR 173.435, 49 CFR 173.441, 49 CFR 173.471, 49 CFR 173.475, 49 CFR 173.476: Definitions, general design requirements, industrial packages, additional design requirements for Type A packages, requirements for Type B packages, authorized Type A packages, authorized Type B packages, requirements for determining A1 and A2 values for radionuclides and for the listing of radionuclides on shipping papers and labels, table of A1 and A2 values for radionuclides, radiation level limitations, requirements for NRC-approved packages, quality control requirements prior to each shipment of Class 7 (radioactive) materials, approval of special form Class 7 (radioactive) materials; and

- Carriage by Public Highway, 49 CFR 177.816, 49 CFR 177.817, 49 CFR 177.834(a), 49 CFR 177.842: Driver training, shipping papers, general requirements (packages secured in a vehicle), Class 7 (radioactive) material.

For additional transportation information, licensees may consult DOT's "A Review of the Department of Transportation Regulations for Transportation of Radioactive Materials," or at the DOT Web site http://www.dot.gov.

APPENDIX AA

Production and Noncommercial Distribution by the Medical Facility of PET Radioactive Drugs to Consortium Members under Authorization of 10 CFR 30.32(j)

PURPOSE OF APPENDIX

The purpose of this Appendix is to provide guidance to the medical use applicant, in a "consortium" as defined in 10 CFR 30.4, that is requesting authority under 10 CFR 30.32(j) for the production and noncommercial distribution of PET radioactive drugs to other medical use licensees within the consortium. The information required by the regulations and addressed in this Appendix is specific to this authorization and supplements information required for other uses of byproduct material provided in the applicant's medical use byproduct materials license application.

Section 10 CFR 30.4 states: "Consortium means an association of medical use licensees and a PET radionuclide production facility in the same geographical area that jointly own or share in the operation and maintenance cost of the PET radionuclide production facility that produces PET radionuclides for use in producing radioactive drugs within the consortium for noncommercial distributions among its associated members for medical use. The PET radionuclide production facility within the consortium must be located at an educational institution or a Federal facility or a medical facility."

The regulatory requirements for what an application from educational institutions, Federal facilities, and medical facilities must include for authorization to produce PET radioactive drugs for noncommercial distribution to licensees in a consortium are found in 10 CFR 30.32(j). Regulatory requirements for licensees with this specific authorization are found in 10 CFR 30.34(j). The noncommercial distribution of PET radioactive drugs can be requested as an additional authorization on a licensee's current byproduct material possession license (e.g., educational institution or Federal facility broad-scope or limited specific license). The information associated with the Radiation Safety Program specifically needed for producing PET radioactive drugs can be found in this volume and the current version of NUREG-1556, Volume 13, "Consolidated Guidance About Materials Licenses: Program-Specific Guidance About Commercial Radiopharmacy Licenses." To avoid duplication, sections in this Appendix refer the applicant to the appropriate sections in this volume or Volume 13.

It should be noted that, as stated in 10 CFR 30.34(j)(1), the authorization under 10 CFR 30.32(j) to produce PET radioactive drugs for noncommercial distribution to medical use licensees in a consortium does not relieve the applicant or licensee from complying with applicable FDA, other Federal, and State requirements governing radioactive drugs.

CONSORTIUM CRITERIA

This Appendix addresses only the authorization in 10 CFR 30.32(j) for medical facilities to noncommercially transfer (distribute) PET radioactive drugs to medical use licensees in the medical facility's consortium. Therefore, the staff must have sufficient information to make the necessary determination that the licensee is a member of a consortium that meets the definition in 10 CFR 30.4, and that the applicant will distribute the PET radioactive drugs only to medical use licensees in its consortium. To assist the staff in making this determination, the applicant should describe this consortium. This description should focus on the regulatory requirements. This includes a description of the geographical area in which the members are located. Even if the names of the individual members of the consortium are provided, the applicant should provide documentation of the terms of the association demonstrating the joint ownership or sharing of the operation and maintenance cost of the PET radionuclide production facility. This documentation may include, but may not be limited to, copies of signed agreements or contracts indicating roles and responsibilities of all of the individuals/entities involved.

The applicant for authorization under 10 CFR 30.32(j) for the production of PET radioactive drugs is required to be a consortium member but is not required to be the consortium member that has the PET radionuclide production facility. The applicant is required by 10 CFR 30.32(j)(1) to either request authorization for the production of PET radionuclides, if the applicant has the PET radionuclide production facility and does not have a license for it, or provide evidence of an existing license issued under 10 CFR Part 30 or the Agreement State requirements for a PET radionuclide production facility within its consortium from which it receives PET radionuclides.

Response from the Applicant:

- Identify the medical use members of the consortium or provide a description of the criteria for consortium membership.

- Describe the geographical area in which the members are located.

- Provide documentation of the terms of the association, demonstrating the joint ownership or sharing of the operation and the maintenance cost of the PET radionuclide production facility.

- Request authorization for the production of PET radionuclides, if the applicant has the PET radionuclide production facility but does not have a license for it.

- Provide evidence of an existing license issued under 10 CFR Part 30 or Agreement State requirements for a PET radionuclide production facility within its consortium from which it receives PET radionuclides.

QUALIFICATION TO PRODUCE PET RADIOACTIVE DRUGS

Section 10 CFR 30.32(j)(2) requires that the applicant be qualified to produce PET radioactive drugs for medical use by providing evidence that meets one of the following criteria:

- Registered with the FDA as the owner or operator of a drug establishment that engages in the manufacture, preparation, propagation, compounding, or processing of a drug under 21 CFR 207.20(a);

- Registered or licensed with a State agency as a drug manufacturer;

- Licensed as a pharmacy by a State Board of Pharmacy;

- Operating as a nuclear pharmacy within a Federal medical institution; or

- A PET drug production facility registered with a State agency.

Response from the Applicant: Follow the guidance in Section 5.2 of this document to determine if the response includes security-related sensitive information and needs to be marked accordingly.

- Provide documentation of registration with the U.S. Food and Drug Administration as the owner or operator of a drug establishment that engages in the manufacture, preparation, propagation, compounding, or processing of a drug under 21 CFR 207.20(a); or

- Provide a copy of a State agency registration or license as a drug manufacturer; or

- Provide a copy of the State Board of Pharmacy pharmacy license; or

- Provide evidence of operation as a nuclear pharmacy within a Federal medical institution; or

- Provide a copy of a State agency registration as a PET drug production facility.

RADIOACTIVE MATERIALS AND USES

10 CFR 30.32(j)(4) requires the applicant to identify the PET radioactive drugs authorized under 10 CFR 30.32(j) for production and noncommercial distribution and requires the applicant to submit information on the radionuclide in the PET radioactive drug, including its chemical and physical form. Because applicants are only authorized for the noncommercial distribution of these PET radioactive drugs, the applicant should request authorization to receive potentially contaminated "empty" radiation transport shields back from consortium members. It is the responsibility of the other medical use consortium licensees under 10 CFR 20.2001 to properly dispose of licensed materials such as unused dosages and residual radioactivity remaining in syringes and vials that were received from the licensee authorized to produce and transfer PET radioactive drugs to consortium members.

Response from the Applicant:

- Identify the radionuclide, including the chemical and physical form, for each PET radioactive drug produced under this authorization.

- Request authorization to receive potentially contaminated "empty" radiation transport shields back from consortium members.

INDIVIDUALS RESPONSIBLE FOR RADIOACTIVE SAFETY PROGRAM AND THEIR TRAINING AND EXPERIENCE

Individuals responsible for the Radiation Safety Program for the production of PET radioactive drugs and their transfer are the applicant's (or licensee's) Radiation Safety Officer (RSO) and the authorized individual(s) responsible during the processing of the PET radionuclides into PET radioactive drugs. The applicant's RSO and authorized individuals must be qualified to use the material as required by 10 CFR 30.33(a)(3). If these individuals are already identified on an NRC or Agreement State license or permit issued by an NRC or Agreement State licensee for similar materials and uses, they may already be qualified to use the quantities, materials, and uses by experience with radiation safety practices similar to those associated with the process of producing PET radioactive drugs. In order to demonstrate that these individuals are qualified by their training and experience to use these materials for the purpose requested, as required by 10 CFR 30.33(a)(3), the applicant should describe the additional training and experience these individuals have for quantities, materials, and radiation safety considerations that differ substantially from the authorization(s) on current licenses or permits. If the applicant is producing the PET radioactive drugs in a pharmacy, the applicant, under 10 CFR 30.32(j)(3), must have an Authorized Nuclear Pharmacist (ANP). The applicant should refer to Section 8.13 of this document for guidance on the minimum training and experience requirements for an ANP. The applicant may chose to use NRC Form 313A (ANP) to document the individual's training and experience. See Appendix B for a copy of this form and Appendix D for instructions in completing it.

A licensee that produces PET radioactive drugs under a 10 CFR 30.32(j) authorization in a pharmacy is permitted under 10 CFR 30.34(j)(3) to allow an individual that meets the board certification requirements in 10 CFR 35.55(a), is listed on a license as an ANP, or is recognized in an appropriate licensee or master materials licensee permit as an ANP, to begin work as an ANP, provided NRC is notified within 30 days of the individual beginning work and specified information is provided to the NRC.

Response from the Applicant:

- Identify the individuals responsible for the Radiation Safety Program and describe their training and experience using similar quantities, materials, and uses of radioactive materials.

- Describe the RSO's additional training and experience if the quantities, materials, and radiation safety considerations differ substantially from existing authorizations.

- Describe the authorized individual's additional training and experience if the quantities, materials, and radiation safety considerations differ substantially from existing authorizations.

- If producing the PET radioactive drugs in a pharmacy, identify at least one individual who meets the requirements of an ANP and document that his/her training and experience meet the requirements in 10 CFR 35.55 for a new ANP or in 10 CFR 35.57 for an experienced ANP. The NRC Form 313A (ANP) may be used to document this information for a new ANP.

TRAINING FOR INDIVIDUALS WORKING IN OR FREQUENTING RESTRICTED AREAS

Individuals working with licensed material must receive radiation safety training commensurate with their assigned duties and specific to the licensee's Radiation Safety Program. In addition, those individuals who, in the course of employment, are likely to receive in a year a dose in excess of 100 mrem (1 mSv) must be instructed according to 10 CFR 19.12. A commitment to provide this training for individuals working in or frequenting restricted areas should already be included as part of the medical use possession license application.

Additionally, to meet the requirements in 49 CFR 172.704, applicants must commit to providing training for individuals that will be involved in the preparation and transport of hazardous materials packages. The training must include:

- General awareness and familiarization training designed to provide familiarity with DOT requirements and to enable the employee to recognize and identify hazardous materials;

- Function-specific training concerning the DOT requirements that are specifically applicable to the functions the employee performs (e.g., if the employee's duties require affixing DOT radioactive labels to packages, the employee must receive training in DOT's regulations governing package labeling); and

- Safety training concerning emergency response information, measures to protect the employee and other employees from the hazards associated with the hazardous materials to which they may be exposed in the workplace, and methods of avoiding accidents, such as the proper procedures for handling packages containing hazardous materials.

The hazardous materials training must be provided initially, and every 3 years thereafter. Records of training must be maintained. *Note:* When the licensee uses a common carrier to transport radioactive materials packages, the common carrier, not the licensee, is responsible for ensuring its employees have DOT-required HAZMAT training (49 CFR 172.702).

If the PET radioactive drugs will be produced under the supervision of an ANP or the applicant's qualified medical use AU, the supervised individuals who will prepare PET radioactive drugs under 10 CFR 30.32(j)(3)(ii) must be instructed in the preparation of byproduct material for medical use, as appropriate to that individual's involvement with byproduct material, and are required to follow the instructions of the supervising AU or ANP regarding the preparation of byproduct material for medical use, written radiation protection procedures established by the licensee, the relevant regulations in 10 CFR Part 35, and license conditions.

Response from the Applicant:

For personnel involved in the preparation and transport of hazardous materials, the applicant should submit the following statement:

"We have developed and will implement and maintain written procedures for training personnel involved in hazardous materials package preparation and transport that meet the requirements in 49 CFR 172.704."

For supervised individuals preparing radioactive drugs, the applicant does not need to provide a response. Supervision will be reviewed during inspection.

FACILITIES AND EQUIPMENT

Applicants should have already provided information under Section 9 of this document regarding the facilities and equipment used for the medical use facility. As part of this information, to demonstrate that the facilities and equipment are adequate to protect the public health and safety, as required by 10 CFR 30.33(a)(2), the applicant must provide a description of the facilities and equipment used for the production of PET radioactive drugs and the noncommercial distribution to consortium members. Therefore, applicants should describe the equipment and/or method used to physically transfer (e.g., transfer lines) PET radiochemicals to the chemical synthesis equipment for radioactive drug production and then to the dispensing area. The description should include shielding used for the transfer of radioactive materials and the shielding equipment (e.g., hot cells) and remote handling equipment used for chemical synthesis and/or preparing the PET radioactive drugs for noncommercial transfer. Due to the short half-lives of positron-emitting radionuclides, PET radioactive drug production facilities generally produce high amounts of activity (curies), which could lead to fairly high activities (millicuries) of effluents released in the air if the proper engineering controls are not used. Examples of some engineering controls that should be used would include exhaust filtration (e.g., HEPA and carbon filters) and/or containment for decay of effluents. It is also recommended that a continuous "real-time" effluent (stack) monitor be installed at the facility. Appendix R of the current version of NUREG-1556, Volume 13, "Consolidated Guidance About Materials Licenses: Program-Specific Guidance About Commercial Radiopharmacy Licenses," provides more information on effluent monitoring. Note that the majority of the radioactive effluents at a PET radioactive drug production facility are produced during the chemical synthesis process of the PET radioactive drug.

Response from the Applicant: As part of the description of facilities and equipment for the PET radioactive drug production and noncommercial transfer areas and the facilities diagram for this area, include the following:

- Descriptions of the area(s) assigned for the production or receipt, storage, preparation, measurement, and distribution of produced PET radioactive drugs and the location(s) for radioactive waste storage;

- Sufficient detail in the diagram to indicate locations of shielding and/or shielding equipment (e.g., hot cells for positron-emitting radionuclides), the proximity of radiation sources to unrestricted areas, and other items related to radiation safety, such as remote handling equipment and area monitors;

- A general description of the ventilation system, including representative equipment such as glove boxes or fume hoods. Pertinent airflow rates, differential pressures, filtration equipment, and monitoring systems should be described in terms of the minimum performance to be achieved. Confirm that such systems will be employed for the production, use, or storage of radioactive drugs; and

- Verification that ventilation systems ensure that effluents are ALARA, are within the dose limits of 10 CFR 20.1301, and are within the ALARA constraints for air emissions established under 10 CFR 20.1101(d).

RADIATION SAFETY PROGRAM

To receive authorization under 10 CFR 30.32(j) for the PET radioactive drug production and noncommercial transfer operations, applicants must provide sufficient information to demonstrate that they meet the criteria in 10 CFR 30.33(a)(2) for these activities. Applicants should refer to the Radiation Safety Program described in Sections 8.21 through 8.28 and 8.32 through 8.48 of this document.

Dosage Measurement System

Section 10 CFR 30.34(j)(2)(ii) requires a licensee to possess and use instrumentation to measure the radioactivity of the PET radioactive drugs intended for noncommercial distribution to members of its consortium and have procedures for use of the instrumentation. Under 10 CFR 30.34(j)(2)(ii), licensees are also required to measure, by direct measurement or by a combination of measurements and calculations, the amount of radioactivity in dosages of alpha-, beta-, or photon-emitting radioactive drugs prior to transfer for commercial distribution. Also the licensee must perform tests before initial use, periodically, and following repair, on each instrument for accuracy, linearity, and geometry dependence, as appropriate for the use of the instrument; make adjustments when necessary; and check each instrument for constancy and proper operation at the beginning of each day of use. Generally, PET radionuclides can be measured using direct measurement only and do not require calculations to be performed, which is often required for beta-emitting radionuclides.

In addition to checking all systems each day of use for constancy to ensure continued proper operation of the system, other appropriate tests may include accuracy (for the range of energies to be measured), linearity (for the range of activities to be measured), and geometry dependence (for the range of volumes and product containers). Licensees should assay patient dosages in the same type of vial or syringe and geometry as used to determine the correct dose calibrator settings. The use of vials or syringes other than those used for geometry dependence may result in measurement errors. Also, the applicant should ensure that it possesses a sufficient number of such instruments to allow for periods when instruments are out of service for repair and calibration.

Response from the Applicant:

- Describe instrumentation to measure the radioactivity of the PET radioactive drugs intended for noncommercial distribution to members of its consortium.

- Describe the types of systems (measurement or combination of measurement and calculation) intended for the measurement of PET radioactive drugs.

- For each dose measurement system used to measure the amount of radioactivity in PET radioactive drugs, state: "We have developed, and will implement and maintain a written procedure for the performance of dose measurement system checks and tests that meets the requirements in 10 CFR 30.34(j)(2)(ii)."

Radioactive Drug Labeling for Distribution

Section 30.34(j)(2)(i) of 10 CFR Part 30 requires the licensee for the noncommercial transfer of PET radioactive drugs to label each transport radiation shield to include the radiation symbol and include the words "CAUTION, RADIOACTIVE MATERIAL" OR "DANGER, RADIOACTIVE MATERIAL," the name of the radioactive drug or its abbreviation, and the quantity of radioactivity at a specified date and time. The term "transport radiation shield" refers to the primary shield for the radioactive drug, which may include the syringe, vial, or syringe or vial shield. In order to demonstrate that the shielding is appropriate for the safe handling and storage of radioactive drugs as required to comply with 10 CFR 30.32(j)(4), the transport radiation shield should be constructed of material appropriate for the isotope to be transferred for noncommercial distribution. The "transport radiation shield" does not refer to the outer suitcase, packaging, or other carrying device used as the transportation packaging for DOT purposes, even though the transportation packaging may provide some radiation shielding.

The licensee must also label each syringe, vial, or other container (e.g., generator) used to hold PET radioactive drugs for noncommercial transfer to consortium members. The label must include the words "CAUTION, RADIOACTIVE MATERIAL" OR "DANGER, RADIOACTIVE MATERIAL." The label must also include an identifier that ensures the syringe, vial, or other container can be correlated with the information on the transport radiation shield label. Identifiers may include the prescription number, the name of the radioactive drug or its abbreviation, the name of the patient, or the clinical procedure.

Response from the Applicant:

- Describe all labels, indicating the colors to be used, that will accompany the products and describe where each label is placed (e.g., on the transport radiation shield or the container used to hold the radioactive drug).

- Confirm that the required labels will be affixed to all transport radiation shields and each container used to hold the radioactive drugs.

Radioactive Drug Shielding for Noncommercial Transfer

Under 10 CFR 30.32(j)(4), the applicant must provide information to demonstrate that shielding provided for each radioactive drug to be noncommercially distributed is appropriate for safe handling and storage by the consortium members. The applicant must provide appropriate transport radiation shields for the primary container of each PET radioactive drug that it intends to distribute. The shielding must be adequate for the types and quantities of radioactive materials that the applicant intends to distribute. Typically, transport radiation shields used to carry radioactive drugs include two-piece, shielded syringe and vial containers (or "pigs"). Facilities have used lead and tungsten shields for gamma/photon-emitting materials. The applicant should select appropriate shielding materials and dimensions to ensure not only that occupational doses are ALARA, but also that the transport radiation shield can be easily handled.

Response from the Applicant: For each PET radioactive drug to be noncommercially distributed,

- Indicate the radionuclide and the maximum activity for each type of container (e.g., vial, syringe),

- Describe the type and thickness of the "transport radiation shield" provided for each type of container, and

- Indicate the maximum radiation level to be expected at the surface of each transport radiation shield when the radioactive drug container is filled with the maximum activity.

Note: With respect to the transport radiation shield, it is not acceptable to state that the applicant will comply with DOT regulations. The dose rate limits that DOT imposes apply to the surface of the package, not the surface of the "transport radiation shield."

Transportation

For the transportation of PET radioactive drugs to consortium members, the required transportation information should be consistent with the information provided for the production and distribution of accelerator-produced radionuclides.

The types and quantities of PET radionuclides in PET radioactive drugs shipped by noncommercial transfer to other medical use licensees in the consortium will usually meet the criteria for shipment in a "Type A" package, as defined by the DOT. The requirements for these packages include the provisions for shipping papers, packaging design standards, package marking and labeling, and radiation and contamination level limits. For applicants who transport their own packages, the packages must be secured to prevent shifting (e.g., blocked and braced), and shipping papers must be used and located properly in the driver's compartment (49 CFR 173.448 and 49 CFR 177.817).

Packaging used for the noncommercial transfer of PET radioactive drugs should be similar to those used by commercial radiopharmacies. These packages will normally meet the criteria for "Type A" quantities, which must meet specified performance standards to demonstrate that they will maintain the integrity of containment and shielding under normal conditions of transport. Such packages will normally withstand minor accident situations and rough handling conditions. The testing criteria for Type A packages are listed in 49 CFR 173.465. Before offering a Type A package for shipment, the shipper is responsible for ensuring that the package has been tested to meet the criteria for the contents and the configuration to be shipped and for maintaining a certificate of testing (49 CFR 173.415). Shippers are not required to personally test the packages but must ensure that the testing was performed before use and maintain a record of the testing.

An outline of DOT and NRC requirements is included in Appendix Z.

Response from Applicant: No response is required. The licensee's program for the transportation of radioactive materials will be reviewed during inspection.

WASTE MANAGEMENT

Radioactive waste generated as part of the production of PET radioactive drugs for noncommercial distribution to consortium members must be disposed of in accordance with regulatory requirements and license conditions. In order to comply with the regulations in 10 CFR Part 20 and 10 CFR 30.51, appropriate records of waste disposal must be maintained. Section 8.29 (Item 11: Waste Management) of this document provides guidance on the information required for handling waste.

Return Waste

It is the responsibility of the other medical use consortium licensees to dispose of unused dosages, empty syringes, and vials received from the licensee authorized to produce and transfer PET radioactive drugs to its consortium members. Under 10 CFR 20.2001, these consortium members can only send radioactive waste to individuals authorized to receive it, and the licensee authorized to produce and transfer PET radioactive drugs to consortium members will not be authorized to receive returned, used or unused, radioactive drugs from consortium members. Therefore, only "empty" radiation transport shield packages can be returned to the production facility.

APPENDIX BB

January 2003 Summary of Public Comments on the August 1998 and March 2002 Drafts and NRC Responses

Summary of Public Comments on the August 1998 and March 2002 Drafts and NRC Responses

The initial draft of NUREG-1556, Volume 9, was published for public comment in August 1998. A revised draft was published in March 2002. Appendix Z of the March 2002 draft included a summary of comments on the 1998 draft and NRC responses. The NRC held two public workshops, on April 25 and April 30, 2002, to receive stakeholder comments on the March 2002 draft. The NRC also received written public comments during a 60-day comment period (April 5 to June 4, 2002). A summary and analysis of the set of comments for the August 1998 draft, as well as the set of comments for the March 2002 draft, were published as a separate document, Appendix BB to NUREG-1556, Volume 9 (January 2003), and this document is available on NRC's Web site http://www.nrc.gov in the Electronic Reading Room. Interested parties may also check NRC's Web site on the Medical Use of Byproduct Material http://www.nrc.gov/miau/med-use-toolkit.html.

APPENDIX CC

List of Documents Considered in the Development of this NUREG

List of Documents Considered in the Development of this NUREG

This report incorporates and updates the guidance previously found in the Regulatory Guides (RG), Policy and Guidance Directives (P&GD), and Information Notices (IN) listed in the table below. When this report is issued in final form, the documents in the table will be considered superseded and should not be used. Other references were also used in this report and are listed in "References."

Some sections of the guidance include references to other documents that may be useful to the applicant. Appendix CC provides a complete list of documents referenced in the guidance. While specific availability information is included for some reference documents, the documents also may be accessed at the NRC Public Document Room, which is located at NRC Headquarters in Rockville, Maryland, or the NRC Electronic Reading Room at www.nrc.gov. See the Notice of Availability on the inside front cover of this report for more information.

Document Identification	Title	Date
RG 10.8, Revision 2	Guide for the Preparation of Applications for Medical Use Programs.	8/87
Appendix X to RG 10.8, Revision 2	Guidance on Complying With New Part 20 Requirements.	6/92
Draft RG DG-0009	Supplement to Regulatory Guide 10.8, Revision 2, Guide for the Preparation of Applications for Medical Use Programs.	3/97
Draft RG FC 414-4	Guide for the Preparation of Applications for Licenses for Medical Teletherapy Programs.	12/85
P&GD FC 87-2	Standard Review Plan (SRP) for License Applications for the Medical Use of Byproduct Material.	12/87
Supplement 1 to P&GD FC 86-4; Revision 1	Mobile Remote Afterloading Brachytherapy Licensing Module.	5/97
P&GD FC 86-4, Revision 1	Information Required for Licensing Remote Afterloading Devices.	9/93
Addendum to Revision 1 to P&GD FC 86-4	Information Required for Licensing Remote Afterloading Devices – Increased Source Possession Limits.	7/95
P&GD 3-15	Standard Review Plan for Review of Quality Management Programs.	6/95
RG 8.39	Release of Patients Administered Radioactive Materials.	4/97
RG 8.33	Quality Management Program.	10/91
P&GD 3-17 (previously 16)	Review of Training and Experience Documentation Submitted by Proposed Physician User Applicants.	
RG 8.23	Radiation Safety Surveys at Medical Institutions, Revision 1.	1/81

The additional references listed below were used.

References

Title 10, Code of Federal Regulations

1. Part 2 – Rules of Practice for Domestic Licensing Proceedings and Issuance of Orders.

2. Part 19 – Notices, Instructions, and Reports to Workers; Inspections and Investigations.

3. Part 20 – Standards for Protection Against Radiation.

4. Part 21 – Reporting of Defects and Noncompliance.

5. Part 30 – Rules of General Applicability to Domestic Licensing of Byproduct Material.

6. Part 31 – General Domestic Licenses for Byproduct Material.

7. Part 32 – Specific Domestic Licenses to Manufacture or Transfer Certain Items Containing Byproduct Material.

8. Part 33 – Specific Domestic Licenses of Broad Scope for Byproduct Material.

9. Part 35 – Medical Use of Byproduct Material.

10. Part 40 – Domestic Licensing of Source Material.

11. Part 70 – Domestic Licensing of Special Nuclear Material.

12. Part 71 – Packaging and Transportation of Radioactive Material.

13. Part 150 – Exemptions and Continued Regulatory Authority in Agreement States and in Offshore Waters Under Section 274.

14. Part 170 – Fees for Facilities, Materials, Import and Export Licenses, and Other Regulatory Services Under the Atomic Energy Act of 1954, as Amended.

15. Part 171 – Annual Fees for Reactor Licenses and Fuel Cycle Licenses and Materials Licenses, Including Holders of Certificates of Compliance, Registrations, and Quality Assurance Program Approvals and Government Agencies Licensed by the NRC.

Title 49, Code of Federal Regulations

1. Part 172 – Hazardous Materials Table, Special Provisions, Hazardous Materials Communications, Emergency Response Information, and Training Requirements.

2. Part 173 – Shippers – General Requirements for Shipments and Packaging.

3. Part 177 – Carriage by Public Highway.

4. Part 178 – Specifications for Packaging.

NRC Regulatory Guides (RG)

1. RG 1.86 – Termination of Operating Licenses for Nuclear Reactors, June 1974.

2. RG 3.66 – Standard Format and Content of Financial Assurance Mechanisms Required for Decommissioning Under 10 CFR Parts 30, 40, 70, and 72, June 1990.

3. RG 7.10 – Revision 1 – Establishing Quality Assurance Programs for Packaging Used in the Transport of Radioactive Material, June 1986. (Superseded by NUREG 1556, Volume 2, "Consolidated Guidance About Materials Licenses: Program-Specific Guidance About Radiography Licenses," August 1998.)

4. RG 8.4 – Direct-Reading and Indirect-Reading Pocket Dosimeters, February 1973.

5. RG 8.7 – Instructions for Recording and Reporting Occupational Radiation Exposure Data, Revision 1, June 1992.

6. RG 8.9 – Acceptable Concepts, Models, Equations, and Assumptions for a Bioassay Program, Revision 1, June 1993.

7. RG 8.10 – Operating Philosophy for Maintaining Occupational Radiation Exposures As Low As Is Reasonably Achievable, Revision 1-R, September 1975.

8. RG 8.13 (Final) – Instruction Concerning Prenatal Radiation Exposure, June 1999.

9. RG 8.18 – Information Relevant to Ensuring that Occupational Radiation Exposures at Medical Institutions Will Be As Low As Reasonably Achievable, Revision 1, October 1982. (Superseded by NUREG 1736, "Consolidated Guidance: 10 CFR Part 20 - Standards for Protection Against Radiation," October 2001.)

10. RG 8.21 – Health Physics Surveys for Byproduct Material at NRC-Licensed Processing and Manufacturing Plants.

11. RG 8.25 – Air Sampling in the Workplace, Revision 1, June 1992.

12. RG 8.29 – Instruction Concerning Risks from Occupational Radiation Exposure, Revision 1, February 1996.

13. RG 8.34 – Monitoring Criteria and Methods to Calculate Occupational Radiation Doses, July 1992.

14. RG 8.36 – Radiation Dose to the Embryo/Fetus, July 1992.

15. RG 10.2 – Guidance to Academic Institutions Applying for Specific Byproduct Material Licenses of Limited Scope, Revision 1, December 1976. (Superseded by NUREG 1556, Volume 7, "Consolidated Guidance About Materials Licenses: Program-Specific Guidance About Academic, Research and Development, and Other Licenses of Limited Scope," December 1999.)

16. RG 10.5 (Draft) – Applications for Type A Licenses of Broad Scope, October 1994.

17. RG 10.8 – Revision (Draft NUREG-1569 - never published), Program-Specific Guidance for Medical Use Licensees, 1997.

18. RG FC 412-4 (Draft) – Guide for the Preparation of Applications for the Use of Radioactive Materials in Leak-Testing Services, June 1985.

19. RG FC 413-4 (Draft) – Guide for the Preparation of Applications for Licenses for the Use of Radioactive Materials in Calibrating Radiation Survey and Monitoring Instruments, June 1985.

NRC Information Notices (IN) and Regulatory Issue Summaries (RIS)

1. IN 89-25, Revision 1 – Unauthorized Transfer of Ownership or Control of Licensed Activities.

2. IN 94-09 – Release of Patients with Residual Radioactivity.

3. IN 94-70 – Issues Associated with Use of Strontium-89 and Other Beta-Emitting Radiopharmaceuticals.

4. IN 96-28 – Suggested Guidance Relating to Development and Implementation of Corrective Action.

5. IN 97-30 – Control of Licensed Material During Reorganizations, Employee-Management Disagreements, and Financial Crises.

6. IN 99-24 – Broad-Scope Licensees' Responsibilities for Reviewing and Approving Unregistered Sealed Sources and Devices.

7. IN 99-33 – Management of Wastes Contaminated with Radioactive Materials.

8. RIS 2002-06 – Evaluating Occupational Dose for Individuals Exposed to NRC-Licensed Material and Medical X-Rays, April 2002.

9. RIS 2002-10 – Revision of the Skin Dose Limit in 10 CFR Part 20, July 2002.

10. RIS 2005-31 – Control of Security-Related Sensitive Unclassified Non-Safeguards Information Handled by Individuals, Firms, and Entities Subject to NRC Regulation of the Use of Source, Byproduct, and Special Nuclear Material.

NRC Fuel Cycle (FC) and Policy and Guidance Directives (P&GD)

1. FC 412-4 (Draft) – Guide for the Preparation of Applications for the Use of Radioactive Materials in Leak-Testing Services, June 1985. (Superseded by NUREG 1556, Volume 18, "Consolidated Guidance About Materials Licenses: Program-Specific Guidance About Service Provider Licenses," November 2000.)

2. FC 413-4 (Draft) – Guide for the Preparation of Applications of Licenses for the Use of Radioactive Materials in Calibrating Radiation Survey and Monitoring Instruments, June 1985. (Superseded by NUREG 1556, Volume 18, "Consolidated Guidance About Materials Licenses: Program-Specific Guidance About Service Provider Licenses," November 2000.)

3. P&GD 1-23 – Guidance for Multi-Site Licenses, April 1996. (Superseded by NUREG 1556, Volume 20, Consolidated Guidance About Materials Licenses: Program-Specific Guidance About Administrative Licensing Procedures," December 2000.)

4. P&GD FC 90-2, Revision 1 – Standard Review Plan for Evaluating Compliance with Decommissioning Requirements, April 1991. (Superseded by NUREG 1757, Volume 3, "Consolidated NMSS Decommissioning Guidance: Financial Assurance, Recordkeeping, and Timeliness," September 2003.

5. P&GD 8-11 – NMSS Procedures for Reviewing Declarations of Bankruptcy, August 1996. (Superseded by NUREG 1556, Volume 15, "Consolidated Guidance About Materials Licenses: Program-Specific Guidance About Changes of Control and About Bankruptcy Involving Byproduct, Source, or Special Nuclear Material Licenses," November 2000.)

NRC NUREGs

1. NUREG-0267, Revision 1 – Principles and Practices for Keeping Occupational Radiation Exposures at Medical Institutions As Low As Reasonably Achievable, October 1982.

2. NUREG-1134 – Radiation Protection Training for Personnel Employed in Medical Facilities, May 1985.

3. NUREG-1400 – Air Sampling in the Workplace, September, 1993.

4. NUREG-1492 – Regulatory Analysis on Criteria for the Release of Patients Administered Radioactive Material, February 1997.

5. NUREG-1539 – Methodology and Findings of the NRC's Materials Licensing Process Redesign, April 1996.

6. NUREG-1541 (Draft) – Process and Design for Consolidating and Updating Materials Licensing Guidance, April 1996.

7. NUREG-1556, Volume 3, Rev. 1 – Consolidated Guidance About Materials Licenses: Applications for Sealed Source and Device Evaluation and Registration, April 2004.

8. NUREG-1556, Volume 11 – Consolidated Guidance About Materials Licenses: Program-Specific Guidance About Licenses of Broad Scope.

9. NUREG-1556, Volume 15 – Consolidated Guidance About Materials Licenses: Program-Specific Guidance About Changes of Control and About Bankruptcy Involving Byproduct, Source, or Special Nuclear Material Licenses, November, 2000.

10. NUREG-1556, Volume 18 – Consolidated Guidance About Materials Licenses: Program- Specific Guidance About Service Provider Licenses, November 2000.

11. NUREG-1556, Volume 20 – Consolidated Guidance About Materials Licenses: Guidance About Administrative Licensing Procedures, December 2000.

12. NUREG-1556, Volume 21 – Consolidated Guidance About Materials Licenses: Program- Specific Guidance About Possession Licenses for Production of Radioactive Sources using an Accelerator.

13. NUREG-1600 – General Statement of Policy and Procedures for NRC Enforcement Actions, June 1995, and Compilation of NRC Enforcement Policy as of September 10, 1997.

14. NUREG/CR-4444 – Radiation Safety Issues Related to Radiolabeled Antibodies, 1991.

15. NUREG/CR-4884 – Interpretation of Bioassay Measurement, July 1987.

16. NUREG-CR-5631, PNL-7445, Rev. 2 – Contribution of Maternal Radionuclide Burdens to Prenatal Radiation Doses, 1996.

17. NUREG-CR-6276 – Quality Management in Remote Afterloading Brachytherapy, October 1994.

18. NUREG/CR-6323 – Relative Risk Analysis in Regulating the Use of Radiation-Emitting Medical Devices: A Preliminary Application, September 1995.

19. NUREG/CR-6324 – Quality Assurance for Gamma Knives, September 1995.

20. NUREG-1736 – Consolidated Guidance: 10 CFR Part 20 – Standards for Protection Against Radiation, 2001.

21. NUREG 1757 – Consolidated NMSS Decommissioning Guidance, September 2003.

National Council on Radiation Protection and Measurements (NCRP) Publications
(Available from NCRP, 7910 Woodmont Avenue, Suite 400, Bethesda, MD 20814-3095, or ordered electronically from http://www.ncrp.com)

1. NCRP Report No. 30 – Safe Handling of Radioactive Materials, 1964.

2. NCRP Report No. 37 – Precautions in the Management of Patients Who Have Received Therapeutic Amounts of Radionuclides, 1970.

3. NCRP Report No. 40 – Protection Against Radiation from Brachytherapy Sources, 1972.

4. NCRP Report No. 49 – Structural Shielding Design and Evaluation for Medical Use of X-Rays and Gamma Rays of Energies up to 10 MeV, 1976.

5. NCRP Report No. 57 – Instrumentation and Monitoring Methods for Radiation Protection, 1978.

6. NCRP Report No. 58 – A Handbook of Radioactivity Measurement Procedures, Second Edition, 1985.

7. NCRP Report No. 69 – Dosimetry of X-Ray and Gamma-Ray Beams for Radiation Therapy in the Energy Range 10 keV to 50 MeV, 1981.

8. NCRP Report No. 71 – Operational Radiation Safety – Training, 1983.

9. NCRP Report No. 87 – Use of Bioassay Procedures for Assessment of Internal Radionuclide Deposition, February 1987.

10. NCRP Report No. 88 - Radiation Alarms and Access Control Systems, 1986.

11. NCRP Report No. 102 – Medical X-Ray, Electron Beam and Gamma Ray Protection for Energies up to 50 MeV (Equipment Design, Performance and Use), 1989.

12. NCRP Report No. 105 – Radiation Protection for Medical and Allied Health Personnel, 1989.

13. NCRP Report No. 107 – Implementation of the Principle of As Low As Reasonably Achievable (ALARA) for Medical and Dental Personnel, 1990.

14. NCRP Report No. 111 – Developing Radiation Emergency Plans for Academic, Medical, and Industrial Facilities, 1991.

15. NCRP Report No. 112 – Calibration of Survey Instruments Used in Radiation Protection for the Assessment of Ionizing Radiation Fields and Radioactive Surface Contamination, 1991.

16.   NCRP Commentary No. 11 – Dose Limits for Individuals Who Receive Exposure from Radionuclide Therapy Patients, February 1995.

International Commission on Radiological Protection (ICRP) Publications
(Published by Elsevier Science: www.elsevierhealth.com/journals/icrp.)

1.   ICRP Report No. 26 – Recommendations of the International Commission on Radiological Protection, 1977.

2.   ICRP Report No. 30 – Limits for Intakes of Radionuclides by Workers, 1978.

3.   ICRP Report No. 35 – General Principles of Monitoring for Radiation Protection of Workers, 1982.

4.   ICRP Publication No. 53 – Radiation Dose to Patients from Radiopharmaceuticals, 1987.

5.   ICRP Publication 54 – Individual Monitoring for Intake of Radionuclides by Workers: Design and Interpretation, 1987.

ANSI Standards (Available from the American National Standards Institute, 25 West 43rd Street, 4th Floor, New York, NY 10036 or ordered electronically from http://www.ansi.org)

1.   ANSI N13.4-1971 (R1983) – Specification of Portable X- or Gamma-Radiation Survey Instruments.

2.   ANSI N13.5-1972 (R1989) – Performance and Specifications for Direct Reading and Indirect Reading Pocket Dosimeters for X- and Gamma-Radiation.

3.   ANSI N13.6-1966 (R1999) – Practice for Occupational Radiation Exposure Records Systems.

4.   ANSI N14.5-1987 – Leakage Tests on Packages for Shipment of Radioactive Materials.

5.   ANSI N42.12-1994 – Calibration and Usage of Thallium-Activated Sodium Iodide Detector Systems for Assay of Radionuclides.

6.   ANSI N42.13-1986 (R1993) – Calibration and Usage of Dose Calibrator Ionization Chambers for the Assay of Radionuclides.

7.   ANSI N42.15-1990 – Performance Verification of Liquid Scintillation Counting Systems.

8.   ANSI N42.17A-1989 – Performance Specifications for Health Physics Instrumentation-Portable Instrumentation for Use in Normal Environmental Conditions.

9.   ANSI N322 – Inspection and Test Specifications for Direct and Indirect Reading Quartz Fiber Pocket Dosimeters.

10.   ANSI N323A-1997 – Radiation Protection Instrumentation Test and Calibration, Portable Survey Instruments.

11.   ANSI N449.1-1978 (R1984) – Procedures for Periodic Inspection of Cobalt-60 and Cesium-137 Teletherapy Equipment.

American Association of Physicists in Medicine (AAPM) Reports (Available from Medical Physics Publishing (MPP), 4513 Vernon Boulevard, Madison, WI 53705-4964 or ordered electronically from http://www.medicalphysics.org. Readers may wish to contact AAPM to determine if more recent documents or reports on these topics have been issued by AAPM. Such documents should be reviewed by applicants for compliance with 10 CFR Part 35 prior to use.)

1.   AAPM Task Group No. 21 – A Protocol for the Determination of Absorbed Dose from High-Energy Photon and Electron Beams, 1984.

2.   AAPM Report No. 41 – Remote Afterloading Technology (Remote Afterloading Technology Task Group No. 41), 1993.

3.   AAPM Report No. 46 – Comprehensive QA for Radiation Oncology, (Radiation Therapy Committee Task Group No. 40), 1994.

4.   AAPM Report No. 51 – Dosimetry of Interstitial Brachytherapy Sources, (Radiation Therapy Committee Task Group No. 43), 1995.

5.   AAPM Report No. 54 – Stereotactic Radiosurgery, (Radiation Therapy Committee Task Group No. 42), 1995.

6.   AAPM Report No. 59 – Code of Practice for Brachytherapy Physics, (Radiation Therapy Committee Task Group No. 56), 1997.

7.   AAPM Report No. 61 – High Dose Rate Brachytherapy Treatment Delivery, (Radiation Therapy Committee Task Group No. 59), 1998.

8.   AAPM Report No. 67 – Protocol for Clinical Reference Dosimetry of High-Energy Photon and Electron Beams, Medical Physics 26(9), pp. 1847-1870, (Radiation Therapy Committee Task Group No. 51) September, 1999.

Other Technical Publications

1.   International Commission on Radiation Units and Measurements (ICRU), "Certification of Standardized Radioactive Sources," Report No. 12, 1968.

2.   U.S. Department of Health, Education, and Welfare, "Radiological Health Handbook," 1970.

3.   R. C. T. Buchanan and J. M. Brindle, "Radioiodine Therapy to Out-patients – The Contamination Hazard," *British Journal of Radiology*, Volume 43, 1970.

4.   International Atomic Energy Agency (IAEA), "Monitoring of Radioactive Contamination on Surfaces," Technical Report Series No. 120, 1970.

5.   IAEA, "Handbook on Calibration of Radiation Protection Monitoring Instruments," Technical Report Series No. 133, 1971.

6.   A. P. Jacobson, P. A. Plato, and D. Toeroek, "Contamination of the Home Environment by Patients Treated with Iodine-131," *American Journal of Public Health*, Volume 68, Number 3, 1978.

7.   A. Brodsky, "Resuspension Factors and Probabilities of Intake of Material in Process (or 'Is 10-6 a Magic Number in Health Physics?')," *Health Physics*, Volume 39, Number 6, 1980.

8.   Bureau of Radiological Health, "Radiation Safety in Nuclear Medicine: A Practical Guide," Department of Health and Human Services (HHS) Publication FDA 82-8180, November 1981.

9.   Center for Devices and Radiological Health, "Recommendations for Quality Assurance Programs in Nuclear Medicine Facilities," HHS Publication FDA 85-8227, October 1984.

10. S. R. Jones, "Derivation and Validation of a Urinary Excretion Function for Plutonium Applicable over Ten Years Post Intake," *Radiation Protection Dosimetry*, Volume 11, No. 1, 1985.

11. "Guidelines for Patients Receiving Radioiodine Treatment," *Society of Nuclear Medicine*, 1987.

12. J. R. Johnson and D. W. Dunford, "GENMOD – A Program for Internal Dosimetry Calculations," AECL-9434, Chalk River Nuclear Laboratories, Chalk River, Ontario, 1987.

13. K. F. Eckerman, A. B. Wolbarst, and A. C. B. Richardson, "Federal Guidance Report No. 11, Limiting Values of Radionuclide Intake and Air Concentration and Dose Conversion Factors for Inhalation, Submersion, and Ingestion," Report No. EPA-520/1-88-020, 1988.

14. K. W. Skrable et al., "Intake Retention Functions and Their Applications to Bioassay and the Estimation of Internal Radiation Doses," *Health Physics Journal*, Volume 55, No. 6, 1988.

15. A. S. Meigooni, S. Sabnis, R. Nath, "Dosimetry of Palladium-103 Brachytherapy Sources for Permanent Implants," *Endocurietherapy Hyperthermia Oncology*, Volume 6, April 1990.

16. R. Nath, A. S. Meigooni, and J. A. Meli, "Dosimetry on Transverse Axes of 125I and 192Ir Interstitial Brachytherapy Sources," *Medical Physics*, Volume 17, Number 6, November/December 1990.

17. M. G. Stabin et al., "Radiation Dosimetry for the Adult Female and Fetus from Iodine-131 Administration in Hyperthyroidism," *Journal of Nuclear Medicine*, Volume 32, Number 5, May 1991.

18. P. Early, D. B. Sodee, "Principles and Practice of Nuclear Medicine," 2nd ed., 1995.

19. M. Stabin, "Internal Dosimetry in Pediatric Nuclear Medicine," *Pediatric Nuclear Medicine*, 1995.

20. "Intravascular Brachytherapy – Guidance for Data to be Submitted to the Food and Drug Administration In Support of Investigational Device Exemption (IDE) Applications," Draft Version 1.3, 1996.

21. R. O. Dunkelberger, II, "Which Probe Should I Use," *Baltimore-Washington Health Physics Society Newsletter*.

APPENDIX DD

Summary of Comments Received on Draft Revision 2 of NUREG-1556, Volume 9

Summary of Comments Received on Draft Revision 2 of NUREG-1556, Volume 9

Note: The page number references associated with each comment under the location heading refers to the page numbers in the July 2007 NUREG-1556 Draft Report for Comment version of Volume 9, Revision 2, "Consolidated Guidance about Materials Licenses: Program-Specific Guidance About Medical Use Licenses." Comments were requested on the specific changes in this NUREG related to the expanded definition of byproduct material, the NARM Rule, and revision of the NRC Form 313A series of forms. Therefore, generally, only comments related to these topics were considered. Comments that were raised related to other issues will be evaluated during any future revision of this NUREG.

Table DD.1 Comments from the State of Wisconsin, Dated August 29, 2007

Location	Subject	Comment
Section 8.5 (Page 8-10, Other Material-table)	Ra-226, unsealed	Delete "Ra-226, unsealed, 1 millicurie" as this would not be included under the new definition of byproduct material. It is also improbable that a medical licensee would request to use unsealed Ra-226.

NRC Staff Response: Unsealed radium-226 is included in the new definition of byproduct material. The NRC now has regulatory authority over discrete sources of Ra-226. The term discrete sources includes both sealed and unsealed material and this point is included in several places in this and other sections of the guidance. The NRC recognizes that it is improbable that medical licensees would use unsealed Ra-226 but it is included in the guidance because the use of discrete sources of Ra-226 for medical use is not prohibited by the regulations.

Location	Subject	Comment
Section 8.9 (Page 8-18, 35.1000 Use)	Ra-226, unsealed	Delete "unsealed Ra-226 or" as this would not be included under the new definition of byproduct material. It is also improbable that a medical licensee would request to use unsealed Ra-226 for medical use as Ra-226 is a known bone-seeker.

NRC Staff Response: See the previous response.

Location	Subject	Comment
Section 8.10 (Page 8-21)	grandfathered individuals and new 10 CFR 35.57 criteria	The last sentence in this item needs clarification. It states that authorized users (generic, i.e. all types) that meet the criteria in 10 CFR 35.57 qualify under NRC's waiver of August 31, 2005 and can be "grandfathered" in as authorized users. What does this mean to the licensee or reviewer? It would be more straight-forward to state that authorized users of "accelerator-produced radioactive material, discrete sources of Ra-226, or both" (per page 8-23 for the RSO) can be "grandfathered" and then explain any limiting conditions, for example, what is the effective period of NRC's waiver of August 7, 2005?

NRC Staff Response: The NRC staff agrees that without references to specific paragraphs under the revised 10 CFR 35.57, the guidance in this section is confusing. References to specific 10 CFR 35.57 paragraphs are added for clarity. These paragraphs include all conditions for grandfathering. The intent of the last sentence is to remind applicants and license reviewers that nuclear pharmacists, medical physicists, physicians, podiatrists, and dentists that used the newly defined byproduct material before November 30, 2007, and under NRC's waiver of August 31, 2005, are considered ANPs, AMPs, and AUs for the uses performed, before or under the waiver.

Location	Subject	Comment
Section 8.11 (Page 8-23)	10 CFR 35.57(a)(3)	This section and its Response from Applicant section references 35.57(a)(3). There is no (a)(3) in the current 10 CFR 35 regulation.

NRC Staff Response: The reference to 35.57(a)(3) is correct. Paragraph (a)(3) was added to 10 CFR 35.57 as part of the amendment to NRC's regulations to include jurisdiction over discrete sources of radium-226, accelerator-produced radioactive materials, and discrete sources of naturally occurring radioactive material, as required by the EPAct of 2005. See 72 FR 55864, Oct. 1, 2007, for the final rule.

Location	Subject	Comment
Section 8.12 (Pages 8-27 and 8-28)	10 CFR 35.57(b)(3).	This section and its Response from Applicant section reference 35.57(b)(3). There is no (b)(3) in the current 10 CFR 35 regulation.

NRC Staff Response: The reference to 35.57(b)(3) is correct; paragraph (b)(3) was added to 10 CFR 35.57 as part of the amendment to NRC's regulations to include jurisdiction over discrete sources of radium-226, accelerator-produced radioactive materials, and discrete sources of naturally occurring radioactive material, as required by the EPAct of 2005.

Location	Subject	Comment
Section 8.13 (Page 8-32)	10 CFR 35.57(a)(3)	This Section and its Response from Applicant section reference 35.57(a)(3). There is no (a)(3) in the current 10 CFR 35 regulation.

NRC Staff Response: As indicated in a response to a prior comment, the reference to 35.57(a)(3) is correct; paragraph (a)(3) was added to 10 CFR 35.57 as part of the amendment to NRC's regulations to include jurisdiction over discrete sources of radium-226, accelerator-produced radioactive materials, and discrete sources of naturally occurring radioactive material, as required by the EPAct of 2005.

Location	Subject	Comment
Section 8.14 (Pages 8-34 and 8-35)	10 CFR 35.57(a)(3)	This section and its Response from Applicant section references 35.57(a)(3). There is no (a)(3) in the current 10 CFR 35 regulation.

NRC Staff Response: See previous response.

Location	Subject	Comment
Section 8.25 (Page 8-58)	Ra-226, unsealed	Delete "and Ra-226." Unsealed Ra-226 would not be considered byproduct material under the new definition. It is improbable that unsealed Ra-226 would be used under a medical license.

NRC Staff Response: As discussed earlier, unsealed Ra-226 is included in the new definition of byproduct material. The term discrete sources includes both sealed and unsealed material. The NRC recognizes that it is improbable that medical licensees would use unsealed Ra-226 but it is included because its medical use is not prohibited by the regulations.

Location	Subject	Comment
Appendix AA (Page AA-4 and AA-5)	Department of Transportation (DOT) Security training	There is no mention of DOT Security training as required in (4).

NRC Staff Response: This comment is not related to the NARM regulations or the revision of NRC Form 313A, and is therefore beyond the scope of this revision. The comment will be evaluated during any future revision of this NUREG.

Table DD.2 Comments from Darrell R. Fisher, Dated August 30, 2007

Location	Subject	Comment
Entire Volume	General Comment	This is a 458-page document. Some of the document seems to be excessively wordy and repetitive. Any effort to provide a shorter, briefer guidance document would probably be appreciated by prospective and current licensees.

NRC Staff Response: This comment is not related to the NARM regulations or the revision of NRC Form 313A, and is therefore beyond the scope of this revision. The comment will be evaluated during any future revision of this NUREG. Note, however, that this guidance document contains information that is repeated in different sections because it is used by a diverse body of medical users that may focus only on sections applicable to them and miss common information if it only appeared once.

Location	Subject	Comment
Entire document	General Comment	Also throughout, this reviewer noted several inconsistent uses of "mCi" and "mci" for millicuries.

NRC Staff Response: The corrections were made.

Location	Subject	Comment
Overview (Page 1-7)	alpha particle quality factor	Add text as underlined: The quality factor <u>used in 10 CFR 20</u> for alpha particles is 10. This will show the reader that the value 10 was taken from 10 CFR 20. Clarification is needed because the RBE for alpha emitters is determined experimentally and may vary widely for given circumstances. Quality factor is an upper limit on the RBE, chosen by committee, and the quality factor recommended by ICRP is 20. Therefore, it would be helpful for the reader to know why the NRC uses a value of 10 for the quality factor in this document, and where it was obtained.

NRC Staff Response: The clarification was made.

Location	Subject	Comment
Section 8.14 (Page 8-34 Authorized Medical Physicist (AMP))	AMP	The text states that "an AMP is directly involved with the calculation and administration of the radiation dose." Instead, the text should read: "an AMP is directly involved with radiation therapy treatment planning." The AMP would not normally be involved in administration of therapy radiation to a patient.

NRC Staff Response: The statement in the guidance is a general statement of the AMP's role in the use of byproduct material in high-dose remote-afterloader, teletherapy, and gamma stereotactic radiosurgery units for medical uses. The AMP's role is not restricted only to direct involvement in radiation therapy treatment planning, but includes other medical physics tasks during the administration and after the completion of the procedure. The text was revised to clarify that the AMP may not administer the dose but is directly involved with medical physics tasks associated with the administration of the radiation dose.

Location	Subject	Comment
Section 8.17 (Page 8-42)	survey instruments	The text states that "Usually, it is not necessary for a licensee to possess a survey meter solely for use during sealed source diagnostic procedures, unless the procedure involves localization of radioactive seeds, since it is not expected that a survey will be performed each time such a procedure is performed." However, the text fails to mention the importance of having a survey meter on hand during and after a brachytherapy seed implant to look for seeds that may have been misplaced, that may have fallen to the floor, or that may remain in equipment after the procedure. Furthermore, these seeds are used for therapeutics, not diagnostic procedures as the text incorrectly suggests.

NRC Staff Response: The paragraph only addresses sealed source diagnostic procedures. The text correctly indicates that some radioactive seeds that would normally be considered for therapeutic brachytherapy use are now being used for diagnostic purposes to mark the tissue and tissue boundaries for surgical removal. The word "procedure" has been replaced with the phrase "diagnostic study" to correct any ambiguity, as warranted. Also, the first paragraph in the discussion in this section addresses the need for survey instruments for traditional brachytherapy seed use.

Location	Subject	Comment
Section 8.18 (Page 8-43, Dose Calibrator)	dosages vs. activity	Throughout this section, the text implies that a dose calibrator measures dosages. More correctly, the dose calibrator measures activity or radioactivity, not dosages. The text describes "instruments (e.g., dose calibrators) used to measure patient dosages." Instead, it would be more correct to state that the dose calibrators are used "to measure the radioactivity present in a syringe, tube, vial, or capsule." Further, the text states that "As described in 10 CFR 35.63, dosage measurement is required for licensees who prepare patient dosages." More correctly, this sentence should read "…measurement is required for licensees who prepare radiopharmaceuticals for administration to patients.

NRC Staff Response: No change was made in response to this comment. The use of the term "dosage" in this section conforms with NRC's use of the term in 10 CFR Part 35; e.g., in the 10 CFR 35.2 definition of prescribed dosage and in 10 CFR 35.63, "Determination of dosages of unsealed byproduct material for medical use." The dose calibrator is used to measure the activity of the patient dosage. While the dose calibrator may be used to measure activity in syringes, tubes, vials, capsules, or any other container, the intent of this section is to describe the requirements for use of the dose calibrator and other equipment to measure activities of unsealed byproduct material that will be administered for medical use; i.e., for patient and human research subject dosages.

Location	Subject	Comment
Section 8.20 (Page 8-47)	typo	Correct spelling should be as follows (underlined): "When patients are treated with I-131 sodium iodide, sources of contamination include . . ."

NRC Staff Response: The correction was made.

Location	Subject	Comment
Section 8.39 (Page 8-78)	beta radiation vs gamma and brehmssthralung	The statement is given that: "The change in emphasis when an operation or autopsy is to be performed is due to the possible exposure of the hands and face to relatively intense beta radiation." I question the accuracy of this statement. The beta dose from tissue surfaces is only a small part of the total from within the body. Most of the beta radiation is locally absorbed, except for small amounts present on tissue surfaces. The major dose to man is still gamma from brehmsstralung, even during an operation or an autopsy. The skin dose to hands is negligible. I have experience doing this kind of work with the assistance from radiation monitoring specialists.

NRC Staff Response: The sentence was revised to clarify that the intent of the statement is not to diminish the importance of the exposure due to gamma radiation and bremsstralung but to heighten awareness of potential beta exposure.

Location	Subject	Comment
Section 8.40 (Page 8-79)	accountability and security	The statement is given that: "Licensed materials must be tracked from 'cradle to grave,' from receipt (from another licensee or from its own radionuclide production facility) to its eventual transfer/disposal in order to ensure accountability; identify when licensed material could be lost, stolen, or misplaced; and ensure that possession limits listed on the license are not exceeded." However, there seems to be a thought disconnect between proper tracking "from cradle to grave" and inadvertent losses of material by theft. Further, it will NOT BE POSSIBLE, in advance, to anticipate how licensed material would be lost, stolen, or misplaced if the licensee is doing everything possible to prevent loss and theft. There could be almost an unlimited number of ways that theft or loss could happen and times when it could happen. Further, there seems to be a disconnect between the concept of theft or loss and the concept of possession-limit tracking. I recommend that you separate the distinctly different concepts of <u>tracking</u> possession limits and <u>safeguarding</u> against theft or loss.

NRC Staff Response: The statement was not intended to mean the licensee can predict when loss or theft of licensed material will happen, but rather that licensees should identify when licensed material is missing and document its last confirmed possession of the licensed material. The purpose of this sentence is to reinforce that a licensee is accountable for the material it possesses from receipt to transfer/disposal. The sentence has been revised to clarify this intent.

Table DD.3 Comments from the Washington University in St. Louis, Dated August 31, 2007

Location	Subject	Comment
Section 8.5 (Page 8-11), Section 8.6 (Page 8-12), and Section 8.7 (Page 8-13)	10 CFR 30.32(g)	The NRC's guidance is inconsistent in this draft on how a licensee should add a Ra-226 or NARM sealed source or device to its NRC license when that source or device does not have an SSDR certificate. The information NRC requests in 10 CFR 30.32(g)(1) and (2) may not be readily available to the applicants if they purchased the source from someone else. If NRC asks for this information under 10 CFR 30.32(g)(3) from every applicant possessing the sealed source, then it appears that NRC will be receiving multiple requests to do a safety evaluation for the same sealed source model. We **recommend** that NRC work directly with the sealed source manufacturers to begin conducting safety evaluations and issuing SSDR certificates. Guidance for applicants who only possess these sealed sources should be to provide NRC with the manufacturer name, source model number, and general physical description; e.g., Ge-68 rod source 1/4" diameter & 8" long.

NRC Staff Response: The NRC staff does not agree that the guidance in this draft is inconsistent. The guidance clarifies that if the applicant has a NARM or Ra-226 sealed source, device, or source and device combination that is not in the SSDR, the applicant must either submit all the information required in 10 CFR 30.32(g)(2), if the information is available, or if not, provide the information required in 10 CFR 30.32(g)(3). The information required under 10 CFR 30.32(g)(1) and (2) applies to all sealed sources, devices, and sealed source-device combinations. As part of the NARM rule, a new paragraph (3) was added to 10 CFR 30.32(g) that allows for the licensing of sealed sources and devices containing NARM that were manufactured before the effective date of the rule and for which all of the information required in 10 CFR 30.32(g)(1) and (2) is not available. Without this provision, an applicant who wanted to use the NARM source or device that was not registered in the SS&D Registry would have been required to submit all of the safety information identified in 10 CFR 32.210(c), because this information had not been submitted already by the manufacturer or distributor as part of registering the source or device. When all the information required by 10 CFR 32.210(c) is not available, 10 CFR 30.32(g)(3) delineates additional information that will be required to license a NARM source or device.

The NRC will be working with sealed source and device manufacturers to include NARM and Ra-226 sealed sources in the SSDR. The NRC recognizes that a number of "legacy" sources containing these materials were produced by manufacturers that are no longer in business or stopped making the devices some time ago. These are the sources for which NRC expects to receive information under the provisions of 10 CFR 30.32(g)(3).

Location	Subject	Comment
Section 8.42 (Page 8-80)	National Source Tracking regulations	"Sealed Source Inventory" section was not updated in this draft guidance, but we **recommend** that NRC update this section in Rev. 2 to reflect the guidance needed by medical use licensees to meet the new National Source Tracking regulations (71 FR 65686, November 8, 2006).

NRC Staff Response: This comment is not related to the NARM regulations or the revision of NRC Form 313A, and is therefore beyond the scope of this revision. The comment will be evaluated during any future revision of this NUREG. Note, however, that NRC postponed the implementation date of the National Source Tracking regulations from the effective date stated in the Federal Register referenced in the comment.

Location	Subject	Comment
Section 8.5 (Page 8-7)	Activation products	How will NRC deal with very short-lived radioactive materials (e.g., half-life less than 2 minutes) that may be activated to activities exceeding the requested limits? Should the license application state that possession limits apply to incidentally activated radioactive materials with half-lives greater than or equal to 2 minutes? This is a repeat question from WU comment letter for Volume 21, dated July 3, 2007.

NRC Staff Response: This comment pertains to radionuclide production and not medical use. As indicated in NUREG-1556, Volume 21, the possession limits in question will be addressed on the radionuclide production license. The radionuclides possessed on the medical use license will be those used for medical use and will not include the short half-life incidentally activated radioactive materials produced during the production of the radionuclides by the accelerator.

Location	Subject	Comment
Section 8.5 (Page 8-7)	Identifying incidental contaminants in radioactive drugs	What guidance does NRC give license applicants for 10 CFR 32.72 distribution of radionuclides that may contain other radionuclide contaminants? Should not guidance on how to describe these potential contaminants be included in this document? Examples of these types of radiopharmaceuticals that are widely used include: Sm-153 Quadramet which can include Eu-154 and Eu-155 Tl-201 Thallous Chloride which can include Tl-200, Tl-202 and Pb-203 In-111 Indium Chloride which can include In-114m and Zn-65 This is a repeat question from WU comment letter for Volume 13, Rev. 1, dated August 1, 2007.

NRC Staff Response: Guidance for applicants seeking a license for distribution of radionuclides under the provisions of 10 CFR 32.72 is not within the scope of this medical use guidance document. This is not an issue for medical use licenses because most radionuclides used for medical uses described in 10 CFR 35.100, 35.200, or 35.300, such as those mentioned in the comment, are requested and listed as "any byproduct material" permitted by 10 CFR 35.100, 35.200, or 35.300, respectively. In other cases, when NRC has to list individual radionuclides, the NRC authorizes the possession and use of the main radionuclide and assumes that the contaminants are part of the main radionuclide's characteristics. For example, the NRC recognizes that certain long-lived radionuclides are present in small amounts in the Mo-99/Tc-99m generators and list only Molybdenum-99 (Mo-99) and Tc-99m. Also, Tc-99m is listed although small amounts of Mo-99m may also be present. In generator elutions, a minimum activity of contaminant is permitted and must be tested for.

Location	Subject	Comment
Sections 8.10 (Page 8-21), Section 8.11 (Page 8-23), Section 8.12 (Page 8-27), Section 8.13 (Page 8-32), and Section 8.14 (Page 8-34)	10 CFR 35.57 Grandfathering of individuals and preceptor statements	As NRC is preparing to "grandfather" individuals who have used accelerator-produced radionuclides to be an ANP (or an AU, AMP or RSO), there is an opportunity to bring the training and experience criteria for ANPs (AUs, AMPs and RSOs) more in line with the preceptor definition. We agree that a preceptor statement from a current ANP is appropriate for those individuals seeking to become an ANP by the alternative pathway. WU strongly **recommends** that the NRC Staff and, in particular, the Nuclear Regulatory Commissioners reconsider the need for an ANP preceptor statement for those individuals who are board-certified by an NRC-recognized specialty board. Each of the specialty boards recognized by the NRC have proven to the NRC that their board-eligible candidates meet the training and experience requirements for the type(s) of medical use for which they are recognized. In order to sit for a board exam, an individual requires the recommendation of a sponsor who verifies the individual has met all of the requirements to become board-certified. While this sponsor may not be an ANP, the sponsor is responsible to the board for recommending only individuals who meet the board's, and therefore the NRC's, requirements. Successful completion of the board exam by the individual gives further verification of the individual's training and experience. WU believes the current regulations imposing the additional requirement of an ANP preceptor statement is an unnecessary redundancy that has greatly complicated the process of approving an individual as an ANP, and has led to the trivialization of long-established radiopharmacy board-certification. This is a repeat recommendation made in WU comment letter for Volume 13, Rev. 1, dated August 1, 2007.

NRC Staff Response: Any revisions to the training and experience requirements would require a revision to NRC's current regulations. Therefore, this comment is beyond the scope of this guidance document revision.

Location	Subject	Comment
Section 8.10 (Page 8-21), Section 8.11 (Page 8-23), Section 8.12 (Page 8-27), Section 8.13 (Page 8-32), and Section 8.14 (Page 8-34)	10 CFR 35.57 Grandfathering of individuals	We appreciate that NRC has taken care to ensure the continuing access of PET imaging techniques by allowing the "grandfathering" of individuals who have used accelerator-produced radionuclides to become ANPs (or AUs, AMPs or RSOs). We believe that NRC also "grandfathering" individuals who have received board-certification prior to NRC's recognition of a specialty board would be in line with the grandfathering for medical use of the new byproduct materials. In certain cases, such as those individuals who have been board certified by the American Board of Health Physics (ABHP) prior to January 1, 2005, and never named as RSO on a NRC or Agreement State license, an individual could not currently be named as an RSO based on their board-certification even though the ABHP made no changes in its certification process to receive NRC-recognition. WU also strongly **recommends** that NRC allow grandfathering of all individuals who were board-certified prior to NRC-recognition for any specialty boards which receive NRC-recognition prior to the required implementation date, August 9, 2009, for the new byproduct definition. This is a repeat recommendation made in WU comment letter for Volume 13, Rev. 1, dated August 1, 2007.

NRC Staff Response: No change was made to the guidance in response to this comment. The comment is beyond the scope of this guidance document revision and would require rulemaking. No changes were made to the training and experience requirements in 10 CFR Part 35, Subparts B, D, E, F, G, and H in the NARM rule. The guidance accurately reflects the rule text changes to 10 CFR 35.57 which address "grandfathering" of individuals that only used NARM and discrete sources of Ra-226 before or under NRC's waiver of August 31, 2005.

Location	Subject	Comment
General Comment	NRC Form 313A series	As stated in this draft guidance, NRC is committed to risk-informed, performance-based regulation, guidance, inspection and enforcement. We believe the latest revision of NRC 313A forms documenting training and experience, plus the preceptor statement, indicate that NRC is moving towards prescriptive "requirements" in the name of "guidance" which has the effect of impeding individuals from being approved as an RSO, an authorized user, an authorized nuclear, pharmacists, or an authorized medical physicist.

NRC Staff Response: The NRC staff does not agree with this comment. NRC is committed to risk-informed performance-based activities. The NRC Form 313A series of forms provide applicants with a means of documenting the training and experience requirements in the regulations. The forms do not require any information that is not required in the regulations, and the preceptor forms include only attestations required by the regulations. Also, the latest revision does not ask for any new information that was not a part of the previous form.

Location	Subject	Comment
General Comment	Regulation by guidance	We also see further indication of NRC's tendency to "regulate via guidance" as it appears in the recent NRC guidance on licensing the Leksell Gamma Knife® Perfexion™ (guidance document not dated, but medical generic communications sent notice of availability on August 8, 2007). This new gamma knife guidance states that this new device must be licensed under 35.1000 rather than 35.600, but does not justify why the existing 35.600 regulations do not adequately cover the radiation safety considerations for the new gamma knife device. We agree that specific training for a new gamma knife device that has expanded treatment capabilities is required, but we do not agree this change in device capability warrants a change in the type of medical use. By telling licensees to consider the use of this new gamma knife device as 35.1000, NRC will be imposing unnecessary training and experience documentation of individuals who are currently approved for another gamma knife, and vice versa.

NRC Staff Response: This comment is beyond the scope of this guidance document revision. Guidance for the 10 CFR 35.1000 uses is posted on the NRC public Web site and updated when necessary to address comments from stakeholders.

Location	Subject	Comment
General Comment	NRC 313A series of Forms and 10 CFR 35.1000 guidance on Leksell Gamma Knife® Perfexion	We ask that NRC evaluate the current NRC 313A forms, and the current guidance on licensing the Leksell Gamma Knife® Perfexion™, with regard to NRC's policy promoting risk-informed, performance-based regulation, guidance, inspection and enforcement. We note that NRC did not ask for public comment on these documents, nor has NRC taken full advantage of the expert review that NRC's Advisory Committee on the Medical Use of Isotopes could provide NRC if given the time to really partner with the NRC Staff in developing these documents. We are concerned that NRC is moving away from these valuable review processes. As medical use of PET and other accelerator-produced radionuclides come under NRC authority, the problems we are experiencing with training and experience documentation, and with NRC issuing minimally reviewed prescriptive guidance, will be compounded.

NRC Staff Response: This comment is beyond the scope of this guidance document revision. However, note that the Advisory Committee on the Medical Use of Isotopes discussed and provided comments on the NRC 313A series of forms and the guidance for completing the forms. Guidance for the 10 CFR 35.1000 Leksell Gamma Knife® Perfexion™ is posted on the NRC public Web site and is updated when necessary to address comments from stakeholders.

Location	Subject	Comment
Section 3 (Page 3-1)	Management	Definition of "Management" should be similar to that found in Volume 11 (Broad Scope). We suggest it should be the same for all NUREG 1556 volumes, and thus be modified to read: "'Management' refers to the processes for conduct and control of a Radiation Safety Program and to the individuals who are responsible for those processes and have *authority to provide necessary resources* to ensure safety and to achieve regulatory compliance."

NRC Staff Response: No change was made in response to this comment. The definition of "management" used in NUREG 1556, Volume 9, as indicated in Section 3 of this document, is the same as the definition of "management" found in 10 CFR 35.2, "Definitions."

Location	Subject	Comment
Section 8.5 (Page 8-8)	Typo	The word, cyclotron, is misspelled in the footnote.

NRC Staff Response: The correction was made.

Location	Subject	Comment
Section 8.16 (Page 8-40)	Updating references	We suggest that the references should be updated: replace NCRP Report 49 with NCRP Report 147; replace NCRP Report 102 with NCRP Report 151; and add NCRP Report 144 "Radiation Protection for Particle Accelerator Facilities" to the list.

NRC Staff Response: This comment is beyond the scope of this revision and will be addressed during a subsequent revision of the guidance.

Location	Subject	Comment
Appendix B	NRC Form 313A series	Will this appendix have the current NRC 313A forms in the final version, or will you just point to the NRC website for the current forms?

NRC Staff Response: The current NRC Form 313A series of forms will be included in the hard copy of NUREG 1556, Vol, 9, Revision 2. In the electronic version of Volume 9, Revision 2, the NRC Form 313As will be hyperlinked to the forms on the NRC Web site within the medical use license toolkit. The most current version of the forms will always be on NRC's public Web site so that they are available to applicants if they are revised before subsequent revisions are made to the NUREG.

Location	Subject	Comment
Appendix AA	Medical licensees noncommercial distribution to PET radioactive drugs to other medical use licensees in its consortium	This appendix appears to be the same kind of guidance as in NUREG 1556 Volume 21 draft Appendix P, but is not word for word the same. Will these two appendices be made identical in the final publications of these two NUREG 1556 volumes?

NRC Staff Response: Although the basic information in both is the same, the two appendices will not be identical because Appendix AA of NUREG-1556, Volume 9, Revision 2, will refer the medical use applicant primarily to the Radiation Safety Program requirements in this volume. While NUREG-1556, Volume 21, Appendix Q, can be used by medical use facilities, as well as the nonmedical use facilities at educational institutions and Federal facilities, it refers the applicant to the radiation safety requirements in other volumes.

Location	Subject	Comment
NARM Rule	Review of implementation	In reviewing the draft of Volume 21 and updates for Volumes 13 and 9 of the NUREG 1556 guidance documents, we noted that only NRC Staff plus one former state regulator were involved in the drafting of these documents. We appreciate that the NRC Staff have been under a tight time schedule to provide these much needed guidance documents in advance of the final rule being published. We have also faced this time pressure in being allowed only 30 days to review and comment on these guidance documents. Because of the limited involvement by people who have safely produced and worked with cyclotron-produced radioactive materials for many years, we hope that the NRC Staff accepts the recommendations made by this community. In the May 14, 2007, NRC memo announcing that the Commission had approved implementation of the final rule, they made this recommendation: "The staff should conduct a review of the effectiveness of this rulemaking after it has gained some experience with implementing the new regulations. This review should occur no sooner than 18 months after the effective date of the rule and include recommendations for studies or rule changes that may be needed to more effectively implement the EPAct." We support and suggest that the NRC more fully include the newly regulated community in this effort.

NRC Staff Response: This comment is beyond the scope of this guidance document revision. The staff will review the effectiveness of the rulemaking through licensing, inspection, and stakeholder interactions. The NRC will work with the Agreement States during this review process.

Table DD.4 Comments from the Department of Environmental Quality, State of Michigan, Dated September 7, 2007

Location	Subject	Comment
Section 8-16 (Page 8-37) and Section 8-33 (Page 8-71)	"quiet rooms"	Patients administered F-18 fluorodeoxyglucose (FDG) are typically kept in a "quiet room" for an hour because the diagnostic procedure requires the patient to remain still to make sure the FDG uptake in normal muscles is as low as possible. The location of quiet rooms and adjacent areas should be specifically mentioned in the text.

NRC Staff Response: The NRC staff is identifying the "quiet room" in this guidance as a potential area of radiation exposure that the applicant should include in the facility diagram and consider under its ALARA program. The text references "quiet rooms" and adjacent areas. Text has also been added to address the appropriate public dose criteria for areas in and around a "quiet room."

Location	Subject	Comment
Section 8.23 (Page 8-55)	Web link	Replace the link to NIST, http://ts.nist.gov/ts/htdocs/210/214/scopes/programs.htm with http://ts.nist.gov/Standards/scopes/programs.htm .

NRC Staff Response: The link was revised.

Location	Subject	Comment
Section 8.37 (Page 8-74) and Appendix V (Page V-1)	"quiet room" shielding	Some mobile positron emission tomography (PET) services inject the patient in the van but have the patient wait in a quiet room in each facility. A shielding evaluation for these quiet rooms needs to be performed to determine if additional shielding is required to meet public dose limits for adjacent occupants.

NRC Staff Response: The reference to a "quiet room" was added to the discussion on public dose limits in Section 8.33. Section 8.37 was revised to indicate when PET mobile service licensees must consider the "quiet room" an area of use based upon the patient release criteria in 10 CFR 35.75. The applicant should refer to Sections 8.1 through 8.31 regarding information that must be provided. These sections provide guidance regarding the "quiet room," including the need to identify this area on the facility diagram, the need to identify its location, the need to consider the design of the "quiet room" under the applicant's ALARA program, and the need to review Section 8.33 to determine the appropriate public dose limits for adjacent occupants to determine if additional shielding is needed. Appendix V was revised to add a discussion of whether the "quiet room" is an area of use under the license.

Location	Subject	Comment
Appendix CC (Pages CC-6, 7, 8)		The commenter noted that there are NCRP, ICRP, and ANSI documents that have been revised and replaced with more recent documents.

NRC Staff Response: The reference to ANSI 13.1 was revised. Revision of other documents is beyond the scope of this revision and will be addressed during subsequent revisions of the document, as may be warranted.

Location	Subject	Comment
Entire document	typos	The commenter noted a number of "typos" throughout the document.

NRC Staff Response: The typos have been corrected.

Table DD.5 NRC Staff Identified Comments

Location	Subject	Comment
Section 8.5 (Page 8-11)	10 CFR 30.32(g) Guidance	The following sentence contained an error. "Applicants requesting authorization for the medical use of a discrete source of Ra-226 (which includes a sealed source of Ra-226) or other NARM sources or devices containing NARM sources that do not have the information described above, or the information required in 10 CFR 30.32(g)(3) (e.g., manufacturer and model number from an SSDR certificate) should consult with the appropriate NRC Regional Office to discuss the contents of their application."

NRC Staff Response: This error has been corrected. The words in the parentheses "manufacturer and model number from an SSDR certificate" should have come after the words "information described above." Applicants would only use the provisions of 10 CFR 30.32(g)(3) if the NARM source or device is not in the SSDR or the applicant does not have enough information to determine if there is an SSDR certificate. Further, while applicants may contact the Regional Office at any time, they do not have to contact the Regional Office if they can use any of the provisions in 10 CFR 30.32(g).

Location	Subject	Comment
Section 8.6 (Page 8-12)	SSDR	The following sentence created confusion regarding sources, devices, or source-device combinations not in the SSDR: "Applicants must provide the manufacturer's name and model number for each requested sealed source and device so that NRC can verify that they have been evaluated in an SSDR certificate or specifically approved on a license."

NRC Staff Response: The sentence was revised to change "that" they have been evaluated in an SSDR certificate to "whether" they have been evaluated in an SSDR certificate.

Location	Subject	Comment
Appendix B (Pages B-18, 22, and 29)	AU Preceptor Attestations	Stakeholders mistakenly assumed that the attestation required in the preceptor attestation refers to the physician's general clinical competency in areas not under NRC regulation.

NRC Staff Response: The NRC staff revised NRC Forms 313A (AUD), 313A (AUS), and 313A (AUT) in Appendix B by adding a note that checking the attestation boxes did not mean that the preceptor was attesting to the individual's "general clinical competency." This clarification was also added to Appendix D under the discussion of the preceptor attestations.

Location	Subject	Comment
Appendix D (Pages D-7 and D-10)	Information on supervising individual for specific training	The language clarifying when the applicant had to provide information on the supervising individual providing the specific training required under 10 CFR 35.50(e) and 35.51(c), was not consistent.

NRC Staff Response: The guidance was revised for consistency.

Location	Subject	Comment
Section 8.10 (Page 8-21), Section 8.11 (Page 8-23), Section 8.12 (Pages 8-27 and 8-28), Section 8.13 (Page 8-32), Section 8.14 (Pages 8-34 and 8-35), Appendix C (Pages C-9, C-12, C-14, and C-16), Appendix D (Page D-1), and Appendix E (Pages E-10, E-11, and E-12)	Experienced individuals using only accelerator-produced or discrete sources on radium-226 before the EPAct of 2005 and termination of NRC's Waiver of August 30, 2005	The guidance regarding individuals eligible to be recognized as an RSO, AU, AMP, or ANP because they only used accelerator produced materials or discreet sources of Radium-226 before the EPAct or termination of NRC's waiver of August 30, 2005, does not include the period before the NRC waiver.

NRC Staff Response: All the references have been changed to reflect that individuals using these materials before and during the waiver are included.

Location	Subject	Comment
Appendix U (Page U-22)	Footnotes	The last digit in the numbers under "Thyrodial Component, Effective Half-Life" should reference numbers for footnotes 1 and 2 respectively.

NRC Staff Response: The correction was made.

Location	Subject	Comment
Appendix H	NRC Form 314	The OMB date has expired

NRC Staff Response: The old form was replaced with the new form, including the new OMB expiration date.

NRC FORM 335
(9-2004)
NRCMD 3.7

U.S. NUCLEAR REGULATORY COMMISSION

1. REPORT NUMBER
(Assigned by NRC, Add Vol., Supp., Rev., and Addendum Numbers, if any.)

NUREG-1556, Volume 9, Rev. 2

BIBLIOGRAPHIC DATA SHEET

(See instructions on the reverse)

2. TITLE AND SUBTITLE	3. DATE REPORT PUBLISHED	
NUREG-1556, Volume 9, Rev. 2, Consolidated Guidance About Materials Licenses, "Program-Specific Guidance About Medical Use Licenses"	MONTH	YEAR
	January	2008
Final Report	4. FIN OR GRANT NUMBER	
	NA	

5. AUTHOR(S)	6. TYPE OF REPORT
Donna-Beth Howe, NRC/FSME Michelle Beardsley, NRC/RI Sarah Bakhsh, NRC/RIII	Final
	7. PERIOD COVERED *(Inclusive Dates)*

8. PERFORMING ORGANIZATION - NAME AND ADDRESS *(If NRC, provide Division, Office or Region, U.S. Nuclear Regulatory Commission, and mailing address; if contractor, provide name and mailing address.)*

Office of Federal and State Materials and Environmental Management Programs
U.S. Nuclear Regulatory Commission
Washington D.C. 20555-0001

9. SPONSORING ORGANIZATION - NAME AND ADDRESS *(If NRC, type "Same as above"; if contractor, provide NRC Division, Office or Region, U.S. Nuclear Regulatory Commission, and mailing address.)*

Same as above

10. SUPPLEMENTARY NOTES

11. ABSTRACT *(200 words or less)*

This document contains information that is intended to assist those preparing applications for licenses for the medical use of byproduct material. Revision 2 of NUREG-1556, Volume 9, provides additional guidance to reflect regulatory changes made by the Naturally Occurring and Accelerator-Produced Material (NARM) Rule, "Requirements for Expanded Definition of Byproduct Material," replaces NRC Form 313A with six new NRC Form 313A forms, makes additional changes to enhance clarification of the training and experience requirements, and removes all references to and information contained in 10 CFR Part 35, Subpart J, because it expired on October 25, 2005. This document also provides information needed to request authorization for preparation of Positron Emission Tomography (PET) radioactive drugs for noncommercial distribution to other members of a consortium.

12. KEY WORDS/DESCRIPTORS *(List words or phrases that will assist researchers in locating the report.)*	13. AVAILABILITY STATEMENT
Medical Noncommercial distribution of PET Drugs PET NARM Training and Experience NRC Form 313 and 313A Part 35 Radium-226	unlimited
	14. SECURITY CLASSIFICATION
	(This Page)
	unclassified
	(This Report)
	unclassified
	15. NUMBER OF PAGES
	16. PRICE

www.ingramcontent.com/pod-product-compliance
Lightning Source LLC
Chambersburg PA
CBHW081427170526
45166CB00008B/2122

* 9 7 8 1 5 0 0 2 0 9 0 4 9 *